T4-BBU-890

# THE EVOLUTION OF SEX AND ITS CONSEQUENCES

Edited by
S. C. Stearns

1987
**Birkhäuser Verlag**
**Basel · Boston**

LIBRARY
College of St. Francis
JOLIET, ILLINOIS

**Editor**

Stephen C. Stearns
Zoologisches Institut der
Universität Basel
Rheinsprung 9
CH-4051 Basel

Illustration of the front cover by S. Bousani, Basel

**Library of Congress Cataloging-in-Publication Data**

The evolution of sex and its consequences
   (Experientia. Supplementum ; vol. 55)
   Includes index
   1. Sex (Biology) 2. Evolution. I. Stearns, S. C. (Stephen C.), 1946– . II. Series: Experientia.
Supplementum ; v. 55. [DNLM: 1. Evolution. 2. Sex. HQ 60 E93]
QH481.E96 1987 574.1'6 87-15847
ISBN 0-8176-1807-4 (U.S.)


**CIP-Kurztitelaufnahme der Deutschen Bibliothek**

**The evolution of sex and its consequences** / ed. by S. C. Stearns. – Basel ; Boston ; Stuttgart :
Birkhäuser, 1987.
   (Experientia : Supplementum ; Vol. 55)
   ISBN 3-7643-1807-4 (Basel . . .)
   NE: Stearns, Stephen C. [Hrsg.]; Experientia / Supplementum
   ISBN 0-8176-1807-4 (Boston . . .)

All rights reserved.
No part of this publication may be reproduced, stored in a retrieval system, or transmitted in
any form or by any means, electronic, mechanical, photocopying, recording or otherwise,
without the prior permission of the copyright owner.

© 1987 Birkhäuser Verlag Basel
ISBN 3-7643-1807-4
ISBN 0-8176-1807-4

1988

University of St. Francis
GEN 574.16 S810

The Evolution of sex and its c

3 0301 00076008 8

**B**

EXS 55:
Experientia Supplementum,
Vol. 55

Birkhäuser Verlag
Basel · Boston

574.16
S810

# Contributors

**Arnold, S. J.** (Department of Biology, University of Chicago, Chicago, Ill 60637, USA)

**Bell, G.** (Biology Department, McGill University, Montréal, P. Québec H3A 1B1, Canada)

**Bierzychudek, P.** (Department of Biology, Pomona College, Claremont, CA 91711, USA)

**Bremermann, H. J.** (Department of Biophysics and Medical Physics, and Department of Mathematics, University of California, Berkeley, CA 94720, USA)

**Bull, J. J.** (Department of Zoology, University of Texas, Austin, TX 78712, USA)

**Charlesworth, D.** (Department of Biology, University of Chicago, Chicago, Ill 60637, USA)

**Couvet, D.** (Centre Emberger, CNRS, Route de Mende, BP 5051, F-34033 Montpellier Cedex, France)

**Fischer, E. A.** (Department of Psychology NI-25, University of Washington, Seattle, Washington 98195, USA)

**Gouyon, P.-H.** (Institut National Agronomique, 16 rue Claude Bernard, F-75231 Paris Cedex 05, France)

**Herbert, P.** (Department of Biological Sciences, University of Windsor, Windsor, Ontario N9B 3PA, Canada)

**Herre, E. A.** (Smithsonian Tropical Research Institute, Apartado 2072, Balboa, Republic of Panama, and Department of Biology, University of Iowa, Iowa City, Iowa 52242, USA)

**Hoekstra, R. F.** (Department of Genetics, University of Groningen, Kerklaan 30, NL-9751 NN Haren, The Netherlands)

**Leigh, E. G. Jr** (Smithsonian Tropical Research Institute, Apartado 2072, Balboa, Republic of Panama)

**Lewis, W. M. Jr** (Center for Limnology, Department of Environmental, Population and Organismic Biology, University of Colorado, Boulder, CO 80309, USA)

**Lloyd, D. G.** (Botany Department, University of Canterbury, Christchurch, New Zealand)

130,764

**Schmeske, D. W.** (Department of Biology, University of Chicago, Chicago, Ill 60637, USA)

**Sork, V. L.** (Department of Biology, University of Chicago, Chicago, Ill 60637, USA)

**Stearns, S. C.** (Zoologisches Institut, Rheinsprung 9, CH-4051 Basel, Switzerland)

# Contents

8

## III. The Major Hypotheses

*G. Bell*

*H. J. Bremermann*

## IV. Experimental Tests

*P. Bierzychudek*

## V. The Comparative Evidence

## VI. The Consequences: 1. Sex-allocation

## VIII.  The Consequences: 3. Selection Arenas

*S. C. Stearns*

# Preface

No area of evolutionary biology offers the curious investigator a more fascinating mixture of strange phenomena and deep intellectual puzzles than the evolution of sex and its consequences. Diversity, behavior, form, color, life cycles, genes, organisms, molecules and geography all play a role. Definitions of fitness and the importance of various units of selection are at stake. The material basis of our own nature enters the discussion at many points, perhaps most clearly when the conversation turns to sexual selection. It therefore comes as no surprise that the evolution of sex has been a central theme in evolutionary biology for more than a decade and shows no signs of fading from the scene. Discussion has never been more vigorous, more imaginative, or more rigorous than it is right now.

This book contributes to that discussion by summarizing the current views of well-qualified authors. In organizing it, I have kept the scope of discussion broad by including, at the beginning, chapters on the evolution of mating types and anisogamy, the costs of sex, and sex-determining mechanisms, and, at the end, on the consequences of sex – sex-allocation theory, sexual selection, and selection arenas. The middle of the book is concerned with leading hypotheses for the evolutionary maintenance of sex and with the evidence for and against these ideas.

This book developed from a multi-author review on the evolution of sex published by Experientia in October, 1985. I thank Hans Mislin for asking me to organize that review, Ric Charnov for help in the early stages, and Cynthia Baer for cheerful and competent editorial assistance. For reviewing some of the new chapters written for the book, I thank Joos van Damme, Dave Queller, Jakob Koella, Sabine Henrich and Arie van Noordwijk. I am most grateful to the authors for their hard work, clear thinking, and punctuality. I corrected proof and supervised the typing, sorting, and correcting of the references – errors there are my responsibility. Leslie Koechlin was a great help in editing the book, and Hans-Peter Thür provided administrative support of the project from start to finish.

In editing this book and writing part of it, I learned a lot and had a good time. I hope you share that experience and are stimulated to work further on these ideas. Have a pleasant journey.

Stephen C. Stearns  
Basel, Switzerland  
March, 1987

# Why sex evolved and the differences it makes

S. C. Stearns

## Introduction

"Nor do we know why nature should thus strive after the intercrossing of distinct individuals. We do not even in the least know the final cause of sexuality; why new beings should be produced by the union of the two sexual elements, instead of by a process of parthenogenesis . . . The whole subject is as yet hidden in darkness" (Darwin, 1862).

"One is left with the feeling that some essential feature of the situation is being overlooked" (Maynard Smith, 1976a).

"The apparent absence of parental control of progeny sex ratio is a serious theoretical difficulty" (Williams, 1979).

"Sex is the queen of problems in evolutionary biology. Perhaps no other natural phenomenon has aroused so much interest; certainly none has sowed as much confusion" (Bell, 1982).

Why sex evolved, how it is maintained by natural selection, and what consequences it has for the evolution of other traits are questions of central importance for all biologists. Sex has molecular, cellular, organismal, and populational consequences: its study integrates several fields. To understand the evolution of sex, we must understand natural selection itself. Answers to the questions, Why is there sex?, and, What differences does it make?, contain much of the answers to the more general questions: How does natural selection work?, and: What units does it work on?

This chapter gives an introductory overview as orientation to the main issues discussed in the book. Parts of it are taken from my introduction (Stearns, 1985) to the multi-author review on the evolution of sex published by *Experientia*, on which this book is based.

### Recombination, reproduction, and gender

Sex is ancient – about 1500 million years old by most estimates. It has been thoroughly integrated into the structure of the organisms that possess it. As a result, when we think of sex, we often confuse three things: recombination, reproduction, and gender (Ghiselin, 1974, 1987; Bell, 1982).

*Recombination* results in genetically diverse offspring. It is the direct consequence of Mendel's second law. The Law of Independence states that members of one pair of genes segregate independently of other pairs. Meiosis makes this true of genes located on different chromosomes, and crossing-over within chromosomes ensures that the Law of Independence is a good approximation so long as genes are not too close to each other on the same chromosome.

Naturally these statements hold primarily for eukaryotes. For many biologists the origin of sex is synonymous with the origin of the eukaryotic state – with a proper cell nucleus and complete meiotic apparatus – because that is the well-known, but derivative, mechanism used by all 'higher' organisms to produce genetically diverse offspring. The same effect can be achieved in bacteria through different mechanisms – conjugation or viral transduction, for example. Even viruses are capable of a type of recombination if two viral strains actively infect a single cell (e. g., Javier, Sedarati and Stevens, 1986). Thus the effects of recombination are mimicked by other mechanisms when recombination itself, as a fully developed eukaryotic mechanism, is not present. All living things appear to be capable of evolving mechanisms for the production of genetically diverse offspring.

The power of recombination to produce genetically diverse offspring is so great that no matter how many offspring a female produces, the likelihood that two of them will be genetically identical is virtually nil. A simple example shows why. Assume that an organism has 10,000 freely recombining loci – a large number for a bacterium, a small number for a eukaryote. If there are two alleles at each locus, and if all alleles are present in equal frequency, then the number of equally likely gametes is 2 to the power 10,000, or about 10 raised to the power 3000. The number of fundamental particles in the universe is estimated to be 10 raised to the power 130. In practice, this simply means that sexual reproduction ensures that each offspring has a different genotype no matter how many offspring there are.

*Reproduction* is not necessarily tied to recombination. The production of offspring can occur sexually or asexually, with or without recombination. The normal mode of reproduction in viruses, bacteria, blue-green algae, many angiosperms, rotifers, aphids, cladocera, flatworms, and some fish, frogs, and lizards is asexual, for example. The simplest form of asexual reproduction is fission. More complex, derived forms involve the production and development of uninseminated eggs, or parthenogenesis.

Some of the mechanisms of parthenogenesis do involve recombination. Called meiotic parthenogenesis, they result in genetically diverse progeny unless the mother is already homozygotic at all loci. Some of the mechanisms of meiotic parthenogenesis produce a genetically diverse array of offspring each one of which is a homozygote. When homozygotes are created in a population that is normally outbreeding, at a certain number of loci recessive deleterious alleles will be expressed. They can cause abnormal development, decreased physiological vigor and increased susceptibility to diseases, and even death. This condition is called 'inbreeding depression' by geneticists. In contrast, mechanisms that produce identical genetic copies of the mother are called *ameiotic* or *apomictic*. They are especially important in plants and entail none of the disadvantages of inbreeding depression.

*Gender,* indicating the 'maleness' or 'femaleness' of an organism, is the principal consequence of a history of sexual selection. It is technically defined as *anisogamy,* or the production of gametes of unequal sizes. By definition, males are the sex that produces small gametes (sperm), and females are the sex that produces large gametes (eggs). This distinction is most prominent in multicellular organisms: sexes cannot be distinguished in fungi and ciliates, for example, although mating types often can. Where mating types are found, only organisms with a certain type-relation can mate with each other. Type A might only be able to mate with Type A', B with B', and so forth. In many flowering plants, *incompatibility types* are found. This phrase describes the fact that genetically different pollen donors fertilize ova with different efficiencies. Whether this is simply a mechanism to promote outbreeding or a mechanism of female choice in sexual selection is an open question (see Charlesworth et al. in this book). The important distinction between incompatibility types in flowering plants and mating types is this: mating types evolved before gender, and incompatibility types, evolving later, have been superimposed on the distinction between maleness and femaleness.

*The historical sequence*

Most explanations of the evolution of recombination have been made with higher organisms in mind. In such organisms a wide range of sexual phenomena occurs; it obscures the effects that one would want to attribute solely to recombination. The historical sequence appears to have been:
1. asexual reproduction,
2. limited recombination (bacterial),
3. meiosis,
4. mating types,
5. anisogamy,
6. gender (products of sexual selection),
7. incompatibility types.
Hypotheses about the maintenance of sex are often tested in organisms that have passed through all steps in the sequence. To clarify the selection pressures their ancestors encountered, this book is organized so that the origin of mating types, of anisogamy and of gender-determining mechanisms are discussed before the major hypotheses for the maintenance of recombination are introduced.

## The traditional view

Until recently, the most widespread explanation for the evolution of recombination was that it gives an advantage to the *species* that possess it. The advantage consists of a smaller likelihood of extinction made possible by an increased rate of evolution. Sexual species are seen as adapting quickly to a changing environment; asexual species are seen as caught in an evolutionary dead-end. Recombination creates this difference through two effects: it increases the rate at which favorable combinations of alleles are brought together (Fisher, 1930), and it also increases the rate at which unfavorable combinations are discarded (Muller, 1932). The advantages of sex are seen as an advantage to the whole gene pool. All explanations claiming that sex evolved because sexual species could adapt more rapidly to changing environments and were therefore less likely to go extinct than asexual species rely on these features of recombination.

## Two problems with the traditional view

The first problem with the traditional explanation is that group selection – or in this case, selection of the attributes of *species* – does not work under most circumstances. The reasons for this are as follows:

(1) Selection pressure on a trait is proportional to the amount of variation in reproductive success among individuals that can be accounted for (a) by variation in the trait in question, divided (b) by the generation time. (In this case, the variation would be between sexual and asexual individuals.)

(1a) Generation times of individuals are much shorter than lifetimes of species – by a factor of 10,000 to 100,000 for most eukaryotes. Thus selection pressures on individuals are correspondingly much larger than selection pressures on species.

(1b) The response to selection depends on the amount of heritable variation available among the units selected. There is much more variation available for selection among individuals within a sexually reproducing species than there is among species within a lineage. Therefore the response to selection is much faster and larger for individuals than for species.

(2) Thus selfish individuals can always outcompete individuals that sacrifice their own interests to those of the species. If asexuality is an advantage to an individual, and sexuality an advantage to the species, then we should find that most organisms are asexual, for only rarely would the advantages of sexuality be so strong that species selection would overcome individual selection.

When Fisher (1930) und Muller (1932) first published their ideas on the adaptive advantages of sex, most biologists thought there was little genetic variation in natural populations, that mutants were rare, and that variation at most loci could be described by the contrast of 'wild type' and 'mutant'. In 1966, Lewontin and Hubby (1966) showed that there was much more variation in electrophoretically detectable isozymes than had been expected. Their discovery was followed by a deluge of papers applying electrophoresis and other techniques to the measurement of genetic variation (Lewontin, 1974; Nei and Koehn, 1983).

Recently there has been a comparable development in quantitative genetics: many more metrical traits than expected, including components of fitness, show measurable heritabilities in natural populations (Dingle and Hegmann, 1982; Clutton-Brock, 1987). As a result, an important and often implicit assumption about the genetic structure of natural populations has changed during the period in which hypotheses about the advantages of sex were multiplying. Since there is a great deal of genetic variability in natural populations, the impact of recombination in a single generation on the progeny of a single outcrossed female is now expected to be much greater than what one might have predicted in 1930 or even in 1965.

The importance of selection on individuals is also expected to be much greater. Fisher's and Muller's inclination to invoke species selection was probably influenced by their implicit assumption that there was not enough variation to generate some significant advantage to genetic diversity among the progeny of a single female (van Noordwijk, personal communication).

Group selection was common in biological thought before 1962, but it came under intense scrutiny after it was uncritically used by Wynne-Edwards (1962) to explain many puzzling patterns of animal behavior. It is no coincidence that three of the strongest critics of group selection also published important books explaining the evolution of sex through advantage to the individual: Ghiselin (1974), Williams (1975), and Maynard Smith (1978). They did so because sex was the outstanding example of a widespread adaptation of arguable importance that had until then *only* been explained through group selection, and which therefore came into direct conflict with the individual selection that they emphasized. (Wynne-Edwards (1986) has recently renewed his argument for the importance of a more sophisticated form of group selection. While it is possible in principle, I doubt that it has actually made much difference for most of the cases he discusses.)

The second problem with the traditional view is that sex does seem to cost a lot. It should pay the individual to be asexual. Lewis (1983 and this book) provides a good overview of the various costs of sex. For bacteria and single-celled organisms, the principal cost of sex is probably the time it takes to carry out recombination. This can lengthen the normal cell division time by a factor of two or more. All organisms encounter another cost of sex not related to anisogamy: recombination rearranges genotypes that have high fitness. Whenever selection is strong, this cost will be high, for by the fact of their survival all parents indicate their fitness in the local environment. In anisogamous species – all 'higher' plants and animals – the female provides cytoplasm to support the male genome. This results in the twofold cost of males, or cost of genome dilution, as Lewis describes it.

Compare a sexually reproducing female with an asexually reproducing female. Suppose that all offspring cost the same amount in both cases. The sexually reproducing female has offspring that are half male, half female. The asexually reproducing female has offspring that are all female, each of which also reproduces asexually and replicates her entire genome. Because she produces twice as many female offspring, after, for example, five generations she should have 32 times as many female descendents, each of which will also contain a complete copy of her genome. Not only does the sexually reproducing female make only half as many female offspring per generation; each of them contains only half of her genome.

It was this cost that first convinced Williams (1975) that the prevalence of sexual reproduction poses a serious problem for evolutionary theory. The advantages of sex to the individual have to be very large if sex is to be maintained by natural selection in any population in which parthenogenesis can arise.

In practice, the advantage of being asexual is rarely twofold. In most animals, the transition to asexuality is difficult. Asexual insects have on average only 67 % of the fecundity of sexual insects, for example (Lewis, this book). In plants the situation is reversed. Asexual plants often set more seed than sexual plants because they do not have to be pollinated. In such cases the advantage of asexuality is more than twofold. In small organisms, the actual costs of sex are reduced by intermittent sexuality – a series of asexual generations followed by an occasional sexual generation. In large organisms, such as mammals and birds, the realized costs of sex are quite small because sexuality is fixed in these lineages. Because the asexual option is simply unavailable, mutant asexual competitors cannot invade.

## The initial steps

*The evolution of mating types in isogamous organisms*

In fungi, ciliates, and some primitive algae there are no 'males' or 'females'. Instead, one finds mating types: only organisms in certain classes can mate with each other. Mating types are an innovation of single-celled organisms, whose isogamy implies no cost of genome dilution. These organisms provide a system free of one major cost of sex in which recombination normally occurs and genetically diverse progeny are regularly produced. Such organisms may offer convincing advantages in experimental attacks on the problem of sex.

An adaptive explanation for mating types involves more than the adaptive explanation for recombination (for which see below). It must explain why all individuals do not simply mate with any conspecific at random, instead restricting themselves to members of the complementary type. It is thus a theory about the number of 'sexes' that one expects to evolve *given* that there is an advantage to sex. In a highly original analysis, Hoekstra (1982 and this book) shows that the evolution of mating types may depend on the details of molecular recognition systems. Some of the mechanisms involved imply that only two types could co-exist. While this is clearly not the case for some incompatibility systems in plants, where a series of incompatibility types may be present, it may nevertheless provide a model for the origin of two types, a necessary precondition for the evolution of anisogamy.

*The evolution of anisogamy*

Hoekstra (this book) has also reviewed the adaptive explanations for the origin of anisogamy. Anisogamy evolves because small and large gametes have different advantages. Small gametes are cheap and can be produced in abundance; large gametes produce zygotes that grow rapidly and survive well. All theoretical analyses show that intermediate sizes are selected against. There should be only two types of gametes – as small as possible and as large as possible. A third, intermediate size of gamete apparently cannot coexist with the other two in an evolutionary equilibrium.

*The evolution of sex determination*

The most striking observation about primary sex ratios is that they do not vary from 50:50 in most outcrossed vertebrates (Williams, 1979). In some of these, such as Uganda kob, elephant seals, baboons, lions, bighorn sheep, and all others with polygynous or polyandrous social systems, it would be advantageous to evolve sequential hermaphroditism: to be born female, gain a certain amount of fitness through female function, then change sex and undertake the more risky but potentially more rewarding struggle for male dominance and access to many female mates.

To explain this invariance, we must look to constraints arising in the mechanisms that determine sex. This area of evolutionary biology is one in which the roles of history and constraint are particularly clear. Once a particular mechanism of sex determination has evolved, it becomes so thoroughly integrated into the machinery of the organism that subsequent change becomes very difficult, apparently so difficult that given sex-determining mechanisms usually characterize whole lineages. Bull (1983) recently published a book reviewing both theory and evidence on sex determination, and here he gives us his most recent views.

Sex-determining mechanisms can be classed as either genetic or environmental. There are two widespread types of genetic sex determination: either one sex is heterogametic and the other homogametic (XY and XX, for example), or females are diploid and males are haploid. Haplo-diploidy occurs in the Hymenoptera, in mites and ticks, white-flies, thrips, scale and some beetles. It has important consequences for sex-allocation and kin selection. The distribution of the heterogametic sex across lineages appears to be random, as though a coin were flipped at the origin of the lineage and one or the other sex chosen to be heterogametic. Thus among the flies (Diptera), the Anthomyids and Drosophilids, for example, have only heterogametic males, whereas among the Chironomids, Muscids and Tephritids the heterogametic sex is either male or female and varies among genera. Among the amphibians, the Bufonids, Pipids and Ambystomids have only heterogametic females, whereas the Discoglossids, Hylids, Leptodactylids and Proteids have heterogametic males. Among the Ranids, Plethodontids and Salamandrids either sex may be heterogametic. In birds the females are heterogametic, whereas in mammals the males are.

Environmental sex determination is also widespread and must have evolved independently several times. It is advantageous whenever the fitness of being a member of one or the other sex depends strongly on the particular environment one finds oneself in. Species in which sex determination depends on the environment include a number of reptiles, in which the sex of developing eggs depends on incubation temperature, mermithid nematodes, in which sex depends on larval nutrition (large larvae become females, small ones males), the echiurid *Boniella,* in which settling larvae develop into females in isolation and into dwarf, parasitic males if they encounter females, and the Atlantic silverside (a fish), in which sex determination is weakly temperature-dependent. Among haplo-diploid organisms, it is a mechanically simple matter for the female to alter the sex of her offspring as environmental conditions change. If she stores sperm in a spermatheca, then she can control the insemination of her eggs as she lays them. Fertilized eggs develop into females, unfertilized eggs into males. This is an example both of sex allocation and of environmental sex determination, but in this case the sex is determined by the mother, not by the developing offspring. Among plants, whether certain orchids develop as males or females depends on whether they grow in the sun (female) or in the shade (male). Some cucurbits show similiar patterns of variation in the percentage of male (shade) or female (sun) flowers on different branches of a vine (Charnov, 1982).

The advantages of environmental sex determination, like those of sequential hermaphroditism, are so convincing that it is genetic sex determination that calls out for explanation (cf. Williams 1979). Bull suggests that genetic sex determination (GSD) has two advantages over environmental sex determination (ESD). Under ESD, sex is determined late in development. This makes intersex with reduced function more likely and delays the onset of sexual dimorphisms. Secondly, ESD produces variation in sex ratios when there are systematic fluctuations in the environmental factors influencing sex. Temporal variation in the sex ratio selects for genetic sex determination because the geometric mean of the contributions of both sexes, taken over a series of generations, will be maximized when both sexes are produced in equal frequencies in each generation no matter what the environmental conditions. GSD with one sex heterogametic is probably the easiest and most effective way to ensure the production of two sexes with equal frequency. This explanation is a variant of that given by Gillespie (1974) of selection for minimizing variance in offspring number.

## The major hypotheses for the maintenance of recombination

*Saturated, heterogeneous environments*

A whole family of models embodies the notion that the production of genetically diverse offspring is advantageous in an environment that is saturated, heterogeneous, or both. Ghiselin (1974) appears to have priority for this idea, which was further developed by Williams (1975), Maynard Smith (1978) and Bell (1982). Bell named it the Tangled Bank hypothesis, after the closing paragraph in Darwin's *Origin*.

Just how effective such environmental conditions might be in selecting for recombination depends on the assumptions made. Williams and Maynard Smith found that the conditions under which one would expect sex to be maintained by this sort of selection were rather restrictive and unlikely. The models they investigated can be characterized as sib-competition models and lottery models. In sib-competition models, the offspring of a single female compete with each other for survival in patches in which only a few organisms can survive and reproduce. Genetically diverse progeny compete with each other less intensely than genetically identical progeny – hence, under certain conditions, the advantage of sex, which in this context is strongly density-dependent. Lottery models depend not on the density of competitors but on the frequency of various environmental conditions and genotypes. Williams likened sexual types to buyers of a certain number of lottery tickets with different numbers, and asexual types to buyers of twice as many tickets with the same number. Here the advantage of sex depends on the frequency with which the asexual type wins; in general, if rare types have a large enough advantage, sex will be favored.

Case and Taper (1986) found, however, that straightforward intraspecific competition with explicit resource dynamics is sufficient to maintain recombination if the niche width of the sexual population is larger than that of the asexual clones. Although Case and Taper qualified this result by noting that it could stem from the particular resource dynamics used, Koella (in press) has shown that their result holds for generalized competitive interactions under worst case conditions for sexuals. It appears that Ghiselin's idea, that sex is an advantage in a saturated environment, may well hold up under a rather broad range of assumptions.

In summary, one hypothesis on the adaptive value of recombination states that it is advantageous to be sexual when the environment is saturated and heterogeneous, in space or in time. The conditions that promote this advantage are density-dependent advantages of sexuals, frequency-dependence (the advantage of being rare), and genotype x environment interactions – meaning simply that a different genotype has an advantage under each different set of environmental conditions. Although these effects are normally based on resource-mediated intraspecific competition, they also occur when the determination of fitness is mediated by parasites, predators, or disease organisms. They are not restricted to environments that are heterogeneous in space; the heterogeneity may enter as a temporal fluctuation as well. In this book, Bell analyzes the Tangled Bank hypothesis and suggests an experimental approach to test it.

### Sex is an advantage in a coevolutionary arms race

The idea that recombination is an advantage in a coevolutionary race against competitors, predators, parasites and disease organisms has been brought up repeatedly since Levin (1975) suggested that it would plausibly explain geographic patterns in the genetic systems of plants. This notion has been used by Jaenike (1978), Hamilton (1980), Rice (1983b), and Bremermann (1980 and this book) to explain why sexual reproduction can have an immediate advantage to the individual if mortality from disease organisms is at all significant.

Disease organisms and parasites have a fundamental advantage in an evolutionary arms race. Their shorter generation time and high reproductive rate enable them to undergo significant evolutionary modification, within the lifetime of a single host, and thus to adapt themselves to a specific host genotype. This brings the host population under strong frequency-dependent selection, for it pays to have a rare genotype during an epidemic (Clarke, 1976), and it also creates the possibility that fitness will have negative heritability, i.e., the types that do well in one generation do poorly in the next because by then the parasites have 'evolved onto them'.

All the evidence at hand, especially that from plants (e.g. Klinkowski, 1970; Kulman, 1971; Newhook and Podger, 1972), indicates that herbivores, parasites and disease cause much of the mortality suffered by the world's organisms (Anderson and May, 1982). Even if the organism is not killed, its fecundity is often reduced (e.g. Rockwood, 1973; all cases of parasitic castration, common among invertebrates). Since the transmission of disease results in negative fitness correlations between parental and offspring genotypes, this mechanism deserves the closer experimental examination that both Bell and Bremermann recommend. Bremermann shows that sex can evolve as a defense against disease under quite general conditions, with no direct parent-offspring transmission. He also notes that recombination is essential for the efficient functioning of the immune system.

### Sex provides a mechanism for DNA repair

Most attempts to find a selective pressure strong enough to explain the evolutionary maintenance of sex have concentrated on the production of genetically diverse offspring. In contrast, this hypothesis, formulated by

Bernstein (1977) and Bernstein, Byers and Michod (1981), concentrates on fitness advantages accruing during the DNA replication step of the life cycle. They postulate that recombination evolved as a mechanism by which damage in one chromosome could be repaired by information in the homologous chromosome. This advantage was later followed by biochemical complementation between the homologous chromosomes that exploited the redundant information available in the diploid genome. As a result, the shift to diploidy became irreversible.

This hypothesis has the appeal of applying to all eukaryotes without any specification of particular ecological conditions; it is discussed at length in a recent book (Michod and Levin, 1987). Like all hypotheses for the maintenance of sex, this one too encounters certain problems. Although viruses and bacteria are not diploid and do not engage in meiosis, they do have efficient mechanisms for DNA repair, so efficient that despite their much shorter generation times, the apparent rate of molecular evolution at third codons is of the same order of magnitude in bacteria and mammals (Alan Wilson, personal communication). Further, the hypothesis is perhaps too powerful for its own good. If meiosis were necessary for DNA repair, which is arguably a universal problem, then it becomes more difficult to explain the persistence of ancient asexual lineages, like the bdelloid rotifers. It is also hard to see why one would observe geographical trends in asexuality, with more asexual types near the poles and at high elevations. The geographic trends in ionizing radiation are the opposite of what one would expect under the DNA repair hypothesis.

This is not to say that the hypothesis plays no role. It may very well contribute to the maintenance of sex. However, it does not appear to hold much hope for explaining either the taxonomic or the geographic distributions of sex.

**The evidence**

*Experimental evidence*

The best experimental evidence concerning the various hypotheses for the maintenance of sex has been gathered by people working on plants. This is no coincidence. Plants offer convincing technical advantages. They can easily be clonally propagated, the environment encountered by an individual can be clearly defined, and one can perform convincing manipulations in the field. The groups that have exploited these advantages include Antonovics and his students (notably Ellstrand and Kelley). Bierzychudek reviews their work in this book.

Most of the work has been done on a grass, *Anthoxanthum*, that reproduces both sexually through seed and asexually through tillers. Frequency-dependent selection has been demonstrated, i. e. minority types were fitter

*130,764*

**College of St. Francis Library**
Joliet, Illinois

than majority types in experiments with genotypes mixed at different frequencies. The advantage of sex was, however, not density-dependent. This confirms work on a small crustacean, *Triops,* in which sexual types did have an advantage over asexual, but the advantage decreased with density (Browne, 1980). The considerable evidence for frequency-dependent selection gathered by ecological geneticists, principally Clarke (e. g., 1976), underlines this point.

The experiments on *Anthoxanthum* also demonstrated that types grown in contact with 'alien' neighbors were fitter than types grown in contact with the same neighbors that the parents had had. This provides a straightforward ecological mechanism for generating negative heritabilities in fitness. However, genotype x environment interactions, while present, were not strong.

Experimental work in this area is making rapid progress and should lead to clarification of the relative importance of the major hypotheses within a few years.

*Comparative evidence*

Considerable work has been invested in the study of the geographical distribution and genetic characteristics of closely related sexual and asexual forms. In this book, Bierzychudek summarizes the evidence for plants and Hebert for cyclical parthenogens (rotifers, aphids, cladocerans). Their reviews complement the one undertaken by Bell (1982) for the animals as a whole.

While this sort of evidence cannot provide strong confirmation for any particular hypothesis because the number of confounding factors is always large, it does cast into doubt the idea that sex is an advantage in a physically harsh or unpredictable environment. In most groups studied, asexual clones do better than related sexual species in extreme, apparently unpredictable environments.

For example, in many plants asexual groups have larger ranges, reach into higher latitudes, and occur at higher altitudes than closely related sexual groups. The numerous clones and many fewer sexual species of dandelions, Genus *Taraxacum,* are an excellent case in point. A well-studied animal example is provided by the Genus *Daphnia,* in which two species are represented by obligatory asexual clones in northern Canada and sexual populations in the central United States: *D. capitata* and *D. pulex.*

*Summary*

No one has yet given a convincing, single-generation, microevolutionary and experimental demonstration of the advantages of sex, which must

nevertheless exist. Work on plants holds the most promise at present, but it is to be hoped that an increasing number of animal systems will also be exploited.

Although we now have good experimental evidence suggesting the advantages of sex in a spatially heterogeneous environment, most of the people performing these experiments believe that the interactions may well be mediated by microparasites (viruses, bacteria), parasites, herbivores, or predators. For example, in some of the experiments on *Anthoxanthum,* the plots were unexpectedly attacked by aphids. Damage was heaviest in the asexual clones, where local outbreaks built up rapidly.

The weight of the evidence suggests that biotic interactions are usually involved, and that in unsaturated environments clones do better. If this trend holds up, then recombination will probably come to be seen primarily as a response to the problems created by co-evolving organisms, be they competitors, predators, parasites or diseases. Sex appears to be an adaptation to the increasing diversity in and saturation of the biotic environment.

While the efforts to untangle the causes for the maintenance of sex have not yet reached their goal, they have clarified a number of important points along the way. Chief among them is the hierarchical nature of natural selection and the units that it works on. We now know that explanations for the maintenance of sex that are based strictly on selection of individuals do work in principle and are increasingly well-supported by experimental evidence. We do not yet know which of the many models to apply in most cases. Explanations based on the selection of groups or of species are not necessary. However, given that sex evolved because it conferred an advantage to a single reproducing female, it would still have long-term consequences that would appear as advantages to species (Stearns, 1986). Broad dispersal of offspring and a long life with many reproductive events are two other traits that, like sex, have good microevolutionary explanations and far-reaching macroevolutionary consequences. They all reduce the probability of extinction.

**The differences it makes**

*Sexual species live longer than clones*

Asexual lineages apparently do not persist as long as sexual ones – the bdelloid rotifers being the chief counterexample (Maynard Smith, 1978; Bell, 1982). Almost all eukaryotic asexual types are the recently derived descendents of sexual ancestors, and among eukaryotes the sexual species far outnumber the asexual clones. Bacteria and viruses have other mechanisms available for recombination, and therefore do not make good counterexamples to this pattern.

*Problems of sex-allocation arise*

Once sexual reproduction has been fixed within a lineage and anisogamy has evolved, then a reproducing organism is faced with a new set of problems. How should it allocate its efforts between male and female function? If it is a species with two seperate sexes, this question has to do with the sex ratio of the offspring. If the species is a sequential hermaphrodite, then the question has to do with the sex of birth and the age and size at which the organism changes into the other sex. If the species is a simultaneous hermaphrodite, then the question has to do with the relative production of male and female gametes.

In an incisive recent book on the subject, Charnov (1982) showed that these apparently different questions are essentially the same. The underlying principle is: *"Selection favors a mutant gene which alters various life history parameters if the percent gain in fitness through one sex function exceeds the percent loss through the other sex function"* (Charnov, (1982) p. 17, his italics). Where there is no within-sex frequency dependence, it is equivalent to say that selection maximizes the product of fitness gained through male function times fitness gained through female function. This idea was first formulated by Shaw and Mohler (1953) and is therefore called *the Shaw-Mohler theorem.*

One application of the theorem has motivated enough research to achieve its own label. *The Trivers-Willard Hypothesis* (Trivers and Willard, 1973) states that where the female can predict the difference in fitness of her sons and daughters, based, for example, on her own physiological condition or social rank, then she should alter the sex of her offspring to produce more of the sex predicted to have higher fitness. Thus in polygynous species, where males compete with each other for access to females, low-ranking females or females in poor condition should produce more female offspring, whereas high-ranking females or females in particularly good condition should produce males. This is, in fact, now known to be the case in red deer (Clutton-Brock, Albon and Guiness, 1986), in coypus (Gosling, 1986) and in wood rats (McClure, 1981), and is suggested for many other cases (see Clutton-Brock and Iason, 1986, for a review).

A second application is strikingly illustrated by the behavior of female parasitic wasps of the Genus *Nasonia* (Charnov, 1982). Because wasps are haplo-diploid and store sperm in a spermatheca, a female can determine the sex of her offspring simply by rotating the egg as it descends through the oviduct so that the fertilization pore is, or is not, opposite the opening of the spermatheca. If the egg is fertilized, it becomes female; if not, then male. *Nasonia* lays its eggs into the larvae of a grain-eating weevil. There is one weevil larva per grain, and the eventual size of the weevil larva, and hence of the emerging wasp, is determined by the size of the grain. If the grain is *relatively* small, then the wasp lays a male egg into the weevil larva; if it is *relatively* large, then she lays a female egg. Males lose less by being small than females, for a small male can still produce many sperm, whereas females gain fecundity more rapidly with size than males. This case, investigated in van den Assem's laboratory at Leiden, provides one of the loveliest experimental confirmations of a quantitative hypothesis available anywhere in evolutionary biology (Charnov, Los-den Hartogh, Jones and van den Assem, 1981).

An equally impressive piece of field work demonstrates the generality of the Shaw-Mohler theorem. Should a sequential hermaphrodite be born male or female? The blue-headed wrasse, Genus *Thalassoma,* is a prominent member of the fish community of Caribbean reefs. On small patch reefs, the species has a polygynous social structure, with a single dominant male defending the patch and holding a harem of several females. On such patches, all fish presumably start life as females, mature and function as females, and only the dominant female will turn into a male when the male dies (or is removed). When that happens, the dominant female is capable of changing her behavior and functioning as a male within one day, and the morphological change to the colorful, blue-headed male form is carried out in less than a week.

On larger reefs that a single male cannot defend, the social system is much more fluid, and two types of males are found: colorful, dominant males with elaborate courtship displays that cannot be distinguished from the males of small patches, and smaller, drab males that do not court but instead dash in to deposit their sperm in the cloud of sperm and eggs released by a courting pair. On such large reefs, the small drab males settle as males, whereas the large colorful males are born female and later change sex. Careful field work in Panama by Warner and his co-workers has shown, as expected from the Shaw-Mohler theorem, that the two male morphs on the large reefs have approximately equivalent lifetime fitness (Warner and Hoffman, 1980).

The success of the Shaw-Mohler theorem to date augurs for the distinguished future of work on sex-allocation. The current state of the field is represented in this book by two chapters. Herre, Leigh and Fischer discuss recent work on sex-allocation in animals, while Gouyon and Couvet review the evolution of gynodioecy in plants, in the process covering conflicts of evolutionary interest in some detail.

*Sexual selection begins with anisogamy*

In 1871 Darwin isolated one of the reproductive components of fitness and explained its evolution through a particular type of selection, sexual selection. In making this move he drew attention to the great strength of selection operating on traits that determine access to mates, in particular, male combat and female choice. He sought to explain the evolution of traits, such as the stag's antlers and the peacock's tail, that pretty clearly decrease the survival changes of the males that possess them. This type of selection begins as soon as anisogamy has evolved, for the consequence of anisogamy is different strategies of reproductive investment for male and female gametes. Sexual selection involves tight interactions between the male and the female genotypes. These lead to a particularly intimate kind of intra-specific coevolution between characters born by two different individuals requiring genetic and developmental coordination of genes working in different directions.

Ghiselin (1974) gave a good recent overview of sexual selection, and the two multi-author volumes edited by Campbell (1972) and Blum and Blum (1979) each contain excellent chapters. Fisher (1930, chapter 6) remains essential. The best single field study of sexual selection in a large mammal is in progress on the Isle of Rhum, where the red deer have been kept under close observation since 1973. Clutton-Brock, Guiness, and Albon (1982) report their progress to date in an excellent book combining ecological and behavioral approaches to sexual selection and life-history evolution.

Eberhard's (1985) recent book on the evolution of genitalia applies sexual selection theory to the evolution of morphological diversity in characters of taxonomic interest. I found it one of the most exciting books to come out in evolutionary biology in recent years. It synthesizes sexual selection theory and broad taxonomic experience, and its argument that female choice has driven the evolution of diversity in male genitalia is well argued and thoroughly substantiated. Read it – you won't regret it.

Much of the recent work on sexual selection has been based on the quantitative genetical approach initiated by Lande (1980a) and extended by Arnold and Wade (1984) and Kirkpatrick (1985). This approach is summarized here in a chapter by Arnold. The exciting possibility that sexual selection occurs in plants is given a critical evaluation by Charlesworth, Schemske and Sork. The current state of the field is described in greater detail in the proceedings of a recent Dahlem conference (Bradbury and Andersson, 1987).

*Selection arenas*

The process of editing these chapters suggested that many observations could be brought together under a single hypothesis. A talk by Hanne Ostergard on immunogenetics and reproduction in mice stimulated me to set these ideas down. I then discovered that the central idea, the selection-arena hypothesis, was not new at all, but could be traced back to Darwin, and that it had already been well investigated in plants, where Stephenson (1981) and his students have been particularly active.

The selection-arena hypothesis suggests that selection often takes place at an early stage of the life cycle when offspring have not yet cost very much. It makes four assumptions: zygotes are cheap; parental investment of time, energy or risk into offspring continues after conception; offspring vary in fitness; and this variation can be identified early on. Under these conditions, it should be advantageous to overproduce zygotes, identify those with lower expected fitness, and either kill and resorb them or at least stop investing in them so that only offspring with good prospects receive parental investment.

In the closing chapter, this idea is used to connect observations on polyovulation in mammals, flowering patterns in plants, superparasitism, recurrent consecutive spontaneous abortions in humans, the genetics of the major histocompatibility complex, kronism, and sibling cannibalism with the argument that sex evolved as an adaptation in a coevolutionary race with mircoparasites. While a limited version of the selection-arena hypothesis can work in an asexual clone, where variation among offspring is purely environmental, in its full generality it is definitely a consequence of the evolution of sex.

## Conclusion

Darwin thought the subject was shrouded in darkness. Bell suggested that no other subject in biology had produced so much confusion. Maynard Smith guessed that something essential was missing. Williams saw serious theoretical difficulties. Since they wrote, considerable progress has been made. Much of it is summarized in this book, in which I hope you find more light and less confusion than might have been expected just a few years ago.

Acknowledgments. Sabine Henrich and Jakob Koella read a draft. Their comments resulted in several clarifications. My thanks to both. The preparation of this chapter was supported by the Swiss National Science Foundation, Grant 3.642-0.84.

# The cost of sex

W. M. Lewis Jr

## Introduction

Between 1870 and 1935, sex presented three major problems for evolutionists to solve. First it was necessary to explain the widespread sexual differentiation of animals that extends beyond the primary requirements of reproduction. Darwin (1871) introduced sexual selection as the solution to this problem. Second, there was a need to explain stability of the sex ratio. This problem was resolved by Fisher (1930), who showed quantitatively how natural selection can stabilize the sex ratio. Third and most significant was the problem of explaining the value of sex in relation to its cost, and of showing by a favorable ratio of benefit to cost why sex should be widely maintained. This problem was dealt with by Fisher (1930) and Muller (1932), who resolved it by reference to group selection. Group selection, if accepted as sufficient to offset the cost of sex to individuals, would also account for the persistence of sexual selection.

Significant additions are still being made to our understanding of sexual selection and of sex ratios (Arnold, 1985; Charnov, 1982; Karlin and Lessard, 1986), but the general form of the original solutions to these problems remains intact. In contrast, the explanation proposed by Fisher and Muller for the maintenance of sex now seems of doubtful validity because of the high cost of sex to individuals and the weakness of the group selection mechanism in maintaining any trait that is accompanied by high cost to individuals (Williams, 1975).

Major costs of sex were evident to Darwin, Fisher, and Muller. Most obvious would have been loss of fitness associated with conflict or risk that occurs as a result of sex. Within a population that is exclusively sexual, this cost can be offset by sexual selection, as shown by Darwin. However, given that asexual reproduction may arise at some low but finite frequency in any group of organisms, sexual selection cannot explain why asexual forms, which would be free of such cost, would not have a great advantage over sexual forms. Also long recognized is the fragmentation of gene combinations of high fitness by sexual recombination, which would seem very costly to the individual, especially among organisms of low fecundity.

Interest in the cost-benefit analysis of sex was essentially dormant from 1935 until after 1960. Three factors seem to have caused a revival of interest in the subject. First, Crow and Kimura (1965) reopened Muller's work and, while not controverting it, introduced a basis for skepticism, as explained by Williams (1975). Next, widespread uncritical use of group-selectionist arguments by biologists stimulated Williams (1966) to stress the inadmissibility of group-selectionist explanations for most evolutionary phenomena. Shortly thereafter, Maynard Smith (1971a) identified a broadly applicable twofold cost of sex caused by the contribution of male genetic material to a zygote whose material existence is primarily an investment of the female organism. Given the increasingly critical atmosphere surrounding group-selectionist arguments and the stimulus of Maynard Smith's discovery of a major individual cost of sex that had not previously been identified, the costs and benefits of sex were examined rigorously for the first time, as summarized in reviews by Williams (1975), Maynard Smith (1978), and Bell (1982). As yet, there is no consensus of opinion on the manner in which sex has been widely maintained in the face of its high apparent cost.

Problems posed by the cost of sex can only be resolved by the discovery of major advantages of sex. For this reason, possible costs of sex have received less attention than possible advantages of sex. It is still useful, however, to consider the costs of sex. The costs of sex define the minimum opposing benefit that sex must offer individuals in order to be sustained by individual selection. Because the costs of sex vary among groups of organisms, it is also possible that the costs of sex will help define the advantages of sex.

My aim here is to present an overview of factors that are now known to contribute to the cost of sex, and to show how these may apply differentially across groups of organism. Most of these factors have been well explained elsewhere. However, the cost of sex has not been treated as a unified subject, and it may for this reason be useful to consider all potential costs together.

**Mode of selection for sex**

The advantages of sex are only peripheral here, but they cannot be completely ignored because they provide context for evaluating the costs of sex. It is impossible here to recapitulate the complex arguments that surround the advantages of sex, but a brief overview of the advantages is possible. A distinction between the group selection mode and the individual selection mode for sex is particularly important.

Williams (1975) searched for advantages of sex that would operate over the short term, and that would for this reason not require logic based on group selection. For organisms that have high fecundity, he showed that sex may offer a sustained advantage to individuals by maximizing the genotypic variety of offspring, and thus maximizing the probability that a parent will produce genotypes that are exceedingly fit for a specific but temporary suite of environmental conditions. According to this model, continual change in the environmental conditions (low heritability of fitness) prevents an organism from achieving an advantage from a reduced rate of recombination.

In analyzing William's arguments, Maynard Smith (1978) concluded that strong selection, when combined with sib competition, could support a sustained advantage for sex among individual organisms by reducing sib competition. At the same time, he expressed doubts about the magnitude of this benefit in relation to the cost of sex. Furthermore, in exploring the advantages of sex in a complex environment, Maynard Smith (1978) (cf. Felsenstein and Yokoyama, 1976) found that sex (recombination) offers clear advantages for individuals only if there is a continual reversal in the correlation of critical environmental variables, which he considered unlikely. Bell (1982) has called such an environment 'capricious'.

Maynard Smith may have taken an overly dim view of the likelihood that environments are capricious. Bell (1982) noted that the biotic component of the environment is likely to be capricious because of the homeostatic nature of selection among competitors or among predator and prey species. The genetic tracking of hosts by diseases may be an especially important example of this 'Red Queen' phenomenon (Bremermann, 1985).

While Williams and Maynard Smith agreed that individuals may derive advantages from sexuality, both concluded that the individual advantages for sustained sexuality, especially in organisms of low fecundity, are unconvincing as generalities. However, the possibility remains that a satisfactory explanation will yet be developed and widely accepted, and that sexuality will ultimately be understood as an adaptation that is maintained by individual selection.

Bell (1982) applied an hypothesis-testing technique to the entire suite of observations on the occurrence and distribution of sexuality among multicellular heterotrophs as a means of identifying the explanations of sexuality that are most consistent with the occurrence of sexuality. This approach is clearly preferable to the selection of examples that come to mind readily or that most clearly illustrate a theoretical principle. Bell, following principles suggested by Ghiselin (1974), Williams (1975), and others, concluded that environmental heterogeneity, and especially spatial heterogeneity, provides the most likely explanation for the maintenance of sex. Bell found the phylogenetic and ecological evidence in animals to be generally consistent with this interpretation, given two assumptions that seem correct on the basis of present knowledge: 1) The origin of parthenogenesis is assumed to be relatively rare, and 2) Parthenogens are assumed to be immediately reproductively isolated from the sexual forms that give rise to them. If environmental heterogeneity does indeed provide the principal individual advantage of sex, it still remains to be seen whether this advantage is large enough to offset the cost of sex to individuals.

Group advantages, the most extreme of which might be a lower extinction rate for sexual species, are the obvious alternative to individual advantages as an explanation for the maintenance of sex. An important role for group advantages has not been demonstrated, and any such demonstration would be difficult because it would have to be historical. In contrast, some empirical evidence suggests a sustained individual advantage for sex, even among organisms of relatively low fecundity. Most persuasive is the persistence of sources of genetic variation that are known to be responsive to individual selection. Both crossing-over and the division of the genome into multiple chromosomes contribute to genetic variation among progeny, and both are responsive to selection. Rates of crossing-over can be reduced by selection from their observed levels in natural populations (Bell, 1982; Williams, 1975). Multiple chromosomes are the rule rather than the exception, but the existence of a few species with very low chromosome numbers (*Haplopappus gracilis:* 2N = 2) illustrates the physiological and biochemical feasibility of low chromosome numbers; some species even show polymorphisms for chromosome number (Staiger, 1956). Consequently, it seems that individual selection maintains over the short term recombination rates significantly exceeding the minimum that would be expected to result from individual selection against recombination. This implies a continuing advantage for recombination. If this advantage exists, and yet is not understood in terms of population genetics, then our failure to understand the individual advantage of sex itself is not surprising.

A second line of empirical evidence for sustained individual advantage of sex is the persistence of sexuality in taxa that are also capable of asexual reproduction (Williams, 1975). Because individual advantage of sex in such species is difficult to illustrate in terms of quantitative population genetics, even though such an advantage must exist to prevent the asexual mode from taking over the sexual one, it seems reasonable to conclude that sex in species lacking asexual reproduction may yet be understood in terms of individual advantage.

The cost of sex is a more interesting subject if sex is maintained by individual advantage rather than group advantage. If the advantage is individual, then there must be constant interplay between cost and benefit that will determine the establishment and persistence of sexual life-history features. If sex is carried as a mortgage on phyletic persistence, the interplay will be duller and will have less explanatory power.

## Inventory of costs

The overall cost of sex can be divided into two groups of subsidiary costs: 1) costs not derived from anisogamy (broadly defined here as physical gametic differentiation of any kind, and thus including oogamy), and 2) costs derived from anisogamy. These two sets of costs are shown in outline form in Table 1. Beginning with the costs not derived from anisogamy, we have first the cost of recombination, which can be defined as the cost assignable to loss of fitness caused by recombination. As an extreme example, if heterozygotes for a particular locus have high fitness and homozygotes have low fitness, an asexual line can preserve heterozygosity at the locus with virtually 100 % fidelity, while a sexual line moves constantly toward a 50 % loss of heterozygosity.

Table 1. Inventory of the costs of sex

---

 I. Costs not derived from anisogamy
   1) Recombination
   2) Cellular-Mechanical costs
      a. Meiosis
      b. Syngamy
      c. Karyogamy
   3) Fertilization
      a. Exposure to risk
      b. Minimal density for reproduction
II. Costs derived from anisogamy
   1) Genome dilution (cost of males)
   2) Sexual selection
      a. Sexual competition (conflict, exposure)
      b. Dual phenotypic specialization

---

A second type of cost not derived from anisogamy has been called the cellular-mechanical cost of sex (Lewis, 1983). This includes any loss of synthetic potential in cells as a result of delays caused by the movement of supramolecular bodies such as chromosomes or gametic nuclei. Under this heading are three subsidiary costs for eukaryotic organisms: meiosis, syngamy, and karyogamy. The cellular-mechanical cost of meiosis should not to be confused with the 'cost of meiosis' identified by Maynard Smith (1971a), which is shown in Table 1 as the 'cost of genome dilution'. For reasons that are not entirely clear, meiosis requires much more time than double mitosis, to which it is superficially similar. The resultant delay in synthetic processes is the cause of the meiotic component of the cellular-mechanical cost. A time delay can also result from the union of gametes; this would be the syngamy component of the cellular-mechanical cost. Finally, a time delay can be associated with the union of the two gametic nuclei that make up a diploid cell; this would be the karyogamy component of the cellular-mechanical cost.

The third and final kind of cost not derived from anisogamy is shown in Table 1 as the cost of fertilization. This cost is caused by any loss of fitness derived from the peculiar necessity in outcrossing sexual organisms of arranging contact between gametes from two different individuals. The cost is broken into two large categories. The first of these is classified as exposure to risk, which refers to increased vulnerability to predation or disease transmission caused by contact between individuals or gametes from two different sources, or from waste of gametes or of effort expended in gamete transmission. The second is classified as the cost of minimum density for reproduction, and is associated with any delay or prevention of reproduction that may occur as a result of the unavailability of mates.

Anisogamy gives rise to a number of additional costs of sex. The first of these listed in Table 1 is the cost of genome dilution. This is new terminology, which I hope will lead to less confusion and point the way more specifically to the actual nature of the cost. This cost, which was first identified by Maynard Smith (1971a), has been given a number of names: cost of meiosis, cost of males, twofold cost, and cost of sex. The cost can be understood as the comparative success of amictic and mictic females in propagating their genomes. If the two produce eggs of the same size at the same rate, the sexual female contributes only half of the genetic material to each egg, and thus suffers a severe disadvantage. For reasons to be explained below, this cost applies only to anisogamous species, and is therefore derivative of anisogamy. It is misleading to call this the cost of meiosis, because meiosis imposes at least one other cost, as already mentioned above. It is also misleading to call this the twofold cost, because the cost may be less than twofold in some instances, and because other costs may be twofold under some circumstances. It is not so far wrong to call this the cost of males, but this term does not accurately represent the nature of the cost. Having synthesized the protoplasm and reserves essential for a zygote, and thus ensuring the propagation, of the genome, an anisogamous female suffers a large loss in propagation efficiency as a result of the male genetic contribution, which effectively dilutes her genes in relation to the cytoplasm and reserve materials that she has synthesized. Because genome dilution is the nature of the cost, it seems best to call this the cost of genome dilution.

In the last category, a variety of other factors are united as the cost of sexual selection. The common feature of these factors is that the environment extracts a penalty as a result of phenotypic manifestations of sex that are established and maintained by sexual selection. In one subcategory, we include increased vulnerability and loss of time associated with sexual competition. In another subcategory costs are associated with dual phenotypic specialization of a single species as a result of sexual dimorphism maintained by sexual selection.

## Context for evaluating costs

All of the individual costs listed in Table 1 vary widely according to the circumstances, and all can be negligible or inapplicable under at least some circumstances. The composite cost of sex for a given organism is thus likely to be a complex mixture of individual costs. A change in circumstances can alter the composite cost of sex, even within a single population.

The cost of sex can only be evaluated with reference to a specific alternative system. For present purposes, I follow the example of Maynard Smith

(1971 a) and use a hypothetical amictic organism as a reference point for evaluating the cost of sex in its mictic counterpart. This method fails to take into account historical factors related to the rate of origin of amixis from mixis, or of mixis from amixis, but these considerations can be added as needed.

It is useful to distinguish between two kinds of cost associated with each of the factors shown in Table 1. Potential cost is the maximum suppression of fitness that may arise from a given cost factor. Realized cost takes into account simultaneous advantages or modifying factors that reduce the potential cost. The benefit of sex need only balance the realized cost; sex may be made feasible by adaptations that minimize the realized cost.

**Categorizing organisms by frequency of sex**

In analyzing the cost of sex, a central consideration is the frequency of sex. From this viewpoint, it is possible to identify four categories of organisms. Category 1 can be called continuously asexual. Such organisms lack the capability for sex. Major groups of eukaryotic organisms that fall into this category are few. Examples include the Euglenophyta, numerous species of Chlorophyta and other algal taxa (Fritsch, 1965), and the bdelloid rotifers. Among multicellular organisms, continuous asexuality is principally limited to isolated species or to marginally distributed population of otherwise sexual species (Bell, 1982).

The second category of organisms can be called discontinuously sexual. Species of this category are composed of individuals that have sexual potential, but in which this potential is not expressed in all individuals. Examples include most species of Protozoa or unicellular algae, which grow extensively by vegetative or clonal reproduction, but that become sexual occasionally in response to environmental cues. Some multi-cellular organisms, such as monogonont rotifers and some species of Cladocera also belong to this category.

The third category consists of organisms that are continuously sexual. Among such species, each organism typically becomes sexual at some point in its life cycle, but each organism is also capable of reproducing vegetatively or by budding or fragmentation. Many invertebrates such as coelenterates and polychaetes belong to this group, as do many vascular plants.

The fourth and last category consists of organisms that are continuously and obligately sexual. Such organisms are individually committed to sexuality and lack the capability for reproduction by non-sexual means. In this category are the vertebrates, some of the more complex invertebrates, and some vascular plants.

This classification of sexuality glosses over many significant distinctions among sexual organisms, but in doing so allows us to focus on the frequency of sexuality, which is a key issue for evaluating the cost of sex. At one extreme of the classification organisms are exclusively committed to asexual reproduction. At the other extreme, organisms are exclusively committed to sexual reproduction. These can be regarded as evolutionary absorption states, within which the inertia favoring continuation of that state is likely to be high. Outside of these evolutionary absorption states, the frequency of sexuality is subject to adjustment in response to natural selection, and this is an important determinant of the realized cost of sex.

## Cost of recombination

Recombination has been explored quantitatively by simulation and by deterministic models (Maynard Smith, 1978). These approaches are useful because they yield tangible results. On the other hand, a hypothetical system that approaches intractability for quantitative study is often ridiculously simple by comparison with natural systems. Consequently, there are advantages to approaches that yield only qualitative results but that are capable of representing environmental complexities that are known to be important in relation to the costs and benefits of recombination. I use here one such approach in an attempt to illustrate the range of potential and realized costs that can be expected to result from recombination.

Figure 1 shows a generalized graphical model of a natural environment. The environment is represented as a gradient rather than as a series of alternate compartments, as must often be the case with quantitative models. Although the gradient is shown in only one dimension, it is actually multivariate. The gradient is divided into units, each of which represents the breadth of conditions corresponding to maximal fitness for the genotype that is best adapted to this position on the gradient. The fitness of a specific genotype decays from its optimum in either direction on the gradient. For convenience, I represent this decay function as a smooth curve with a well-defined peak. Although a specific example is shown in the figure, the breadth of the fitness decay curve in relation to the length of the gradient obviously varies according to the specific combinations of organism and environment. As shown in the literature review by Bell (1985b), there is little doubt that such fitness curves exist, even though their shapes and dimensions are seldom known.

Another feature of the generalized environment shown in Figure 1 is carrying capacity, which varies along the gradient. Carrying capacity is given graphically as distance above the line representing the environmental gradient. The irregular variation in carrying capacity along the gradient is explained by two factors. First, the environment is less favorable at some points than others, even for the best-adapted genotype, and the carrying capacity per unit area varies accordingly. Second, the area that is available for occupancy varies along the gradient.

The carrying capacity along the gradient is not stable through time. For

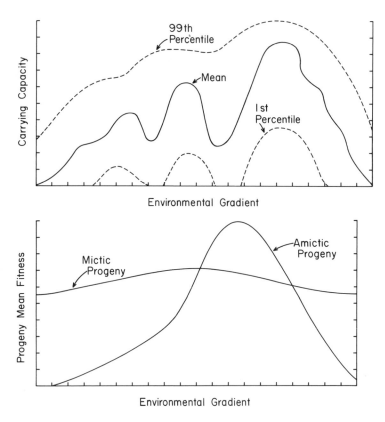

Figure 1. Schematic representation of a natural environment expressed as a linear gradient. The top panel shows the mean carrying capacity across the gradient and the 95 % confidence intervals for variations around the mean carrying capacity. The bottom panel shows mean progeny fitness for asexual organisms of a single genotype and for sexual organisms of similar genotype.

this reason, the graph indicates percentile bands around the mean carrying capacity. These bands correspond to some unspecified frequency distribution of carrying capacities through time for any given point on the gradient.

In considering the cost of recombination, the simplest way to use the gradient model of Figure 1 is by assuming the environment to be occupied by a single amictic genotype and a few mictic organisms of similiar genotype. The amictic population is fixed at a specific point on the gradient, where it consistently has maximum competitive ability. It is also an effective competitor at nearby points on the gradient. In contrast, the mictic population does not occupy a fixed position on the gradient. All offspring are genetically unique, and their fitnesses span the range of conditions on the gradient.

Three categories of potential cost for recombination can be recognized from this example. The first cost arises from the tendency of the mictic population to produce some genotypes that have negligible fitness under any of the conditions along the gradient. These may be regarded as lethal or sublethal genotypes. The second potential cost arises from the deviation of actual fitness from maximum fitness for any random pairing of a position on the gradient and a genotype produced by sexual processes. This potential cost is lowest if each of the offspring produced by sexual processes has the ability to seek out a position anywhere on the gradient, and is highest if each of the offspring finds a position on the gradient by random processes or by processes controlled mostly by the parent. The third potential cost is derived from the inability of the sexual population to avoid competition with the asexual population over that portion of the gradient where the asexual population is consistently superior.

Because the cost of sex is being measured in terms of the ability of an amictic population to replace a sexual one, most of the potential costs of recombination are realized in this example only as long as it takes for the amictic types to occupy that portion of the gradient where they are optimally fit. The disadvantages of recombination are not likely to lead to extinction of the sexual population because the amictic individuals are fit over only a portion of the gradient. After establishment of equilibrium, the mictic individuals suffer only a small disadvantage from production of some progeny whose genotypes are lethal, sublethal, or merely suboptimal on other parts of the gradient. The production of such genotypes, if accompanied by the production of at least some genotypes of high fitness, has much the same effect as a reduction in fecundity. Reduced fecundity of the mictic population would be important in determining the relative dominance of mictic and amictic types only over a narrow zone where the two types of progeny have nearly equal mean fitnesses; differences in fecundity would not be important over those portions of the environmental gradient where the amictic progeny are substantially superior or substantially inferior in mean fitness.

The realized cost of combination may be reversed by intense sib competition in the amictic population (Price and Wagner, 1982; Williams, 1975). The effect of environmental fluctuations causing changes in carrying capacity over the region of the environmental gradient dominated by the asexual population will be magnified by the phenotypic uniformity of the amictic organisms. If the environment is subject to resource depletion, or if the environment contains biotic challenges capable of focussing on particular phenotypes (e. g., disease organisms or predators), the effective carrying capacity of the environment may be lower for the amictic population than it would be for the mictic population, and this would be reflected in a higher probability of extinction (lower time to extinction (Cooper, 1984)) for the amictic organisms than might be implied by the carrying capacity of the environment for mictic organisms.

Over many generations, well after the establishment of short-term equilibrium, two additional factors come into play. First is a finite probability that random fluctuations in the carrying capacity of the environment will reach zero over that portion of the gradient occupied by the amictic population. The realized cost of recombination in this case will return to zero because of the elimination of the amictic population. The realized cost of recombination may also be modified by evolutionary response in the mictic population under certain circumstances. I have assumed that the position on the gradient where the amictic population has peak fitness is determined randomly by the change origin of amictic reproductive features from a mictic progenitor. If the amictic population happens to occupy the center portion of the gradient, it will probably be advantageous for individuals in the mictic population to produce offspring whose environmental optima span the entire range of the gradient, although some disruptive selection is also possible. However, if the amictic population occupies the edge of the gradient, the mictic individuals will experience directional selection leading to narrowing of the variance and shifts in the means of traits in such a way as to minimize overlap with the amictic population. This will reduce competition for both populations. It is a striking fact that amictic populations tend to occupy marginal or extreme environments over the range of conditions covered by their mictic counterparts (Bell, 1982; Bierzychudek, 1985). Numerous hypotheses have been proposed to explain this phenomenon, but none has yet been widely accepted (Bierzychudek, 1985). Directional selection in mictic populations because of the presence of derived amictic populations may be of unsuspected importance in explaining the association of amictic organisms with marginal environments, either directly or by reinforcement of other mechanisms that favor this bias in distribution. From the viewpoint of sexual organisms, the effect of reduced overlap would be to reduce the realized cost of recombination.

The realized cost of recombination is likely to increase in relation to degree of commitment to sexuality. Discontinuous sexuality, and to some degree continuous sexuality that is not obligate, would reduce the realized cost by reducing the frequency of sex and would provide a mechanism by which the realized cost could be adjusted over the short term. A population that has both sexual and asexual capabilities can develop increased emphasis on asexual reproduction in response to competition from a derived asexual line, but this is not possible in organisms that are continuously and obligately sexual.

In summary, the example of a sexual form competing with a corresponding amictic form shows that the realized cost of recombination will depend on the degree of commitment to sexuality and on the portion of an environmental gradient that can be dominanted by amictic individuals. If the breadth of conditions corresponding to high fitness in the amictic type is high in relation to the width of the environmental gradient, then the realized cost resulting from recombination will be high. However, a modification of the realized cost can occur if the amictic population occupies one extreme of the environmental gradient and thus causes directional selection in the sexual population in such a way as to reduce competitive overlap. In addition, realized costs can be nullified *a posteriori* by extinction of the amictic population, the possibility of which will be enhanced by lack of diversity in the amictic population as a result of (1) focussing by predators, parasites, and diseases on the minimally varying prey or host phenotype, (2) of random fluctuations in the carrying capacity of the environment over the range of high fitness, and (3) of reduction of effective carrying capacity by sib competition for resources.

A second example resembles the first one but is more complicated because it allows for the appearance of new amictic lines at some finite rate. The potential cost of recombination is the same as in the first example, but the realized cost is higher if the rate of origin of asexual lines exceeds the rate of their extinction by a sufficient margin to allow several amictic lines to exist at any one time. If this is the case, the realized cost of recombination could be expressed over the entire gradient, and could thus cause extinction of the mictic population.

The examples show that the potential costs of recombination can be identified in general terms, but that the realized costs are critically dependent on the rate at which amictic lines arise and the rate at which they become extinct. The inefficiency of recombination in preserving genotypes of high fitness has often been taken as evidence of a substantial cost of recombination. Flexibility in reducing this cost to a minimum by adjustments in the frequency of sex is available to all but the continuously obligate sexual organisms, which account for a very low percentage of all taxa. Even among continuously obligate sexual organisms, the cost of recombination is realized only in the presence of an alternate reproductive mode. Among such organisms, the factors that govern the frequency with which amictic strains arise and the probability that they will, once established, become extinct may be more important than short-term disadvantages of recombination. Among multicellular organisms, continuously asexual taxa are unusual (Bell, 1982). This suggests that amictic straits have low frequency of origin, low persistence, or both.

As long as other costs are not involved, it would appear that recombination can be preserved, even in the presence of amictic organisms, by individual advantage, so long as the entire environmental gradient is not occupied by amictic forms. If a variety of amictic forms were to accumulate in such a way as to occupy virtually the entire environmental gradient, it would seem possible that these amictic types could cause extinction of the sexual type. This change, which might qualify as group selection, would not be readily reversed because of the complexity of mechanisms that must lead to the evolution of sexual capabilities in organisms that lack these capabilities (Bull and Charnov, 1985). Thus a case can be made that recombination, in the absence of other costs, could be maintained by individual advantage but overcome by group disadvantage. In such a case, sexual individuals would continue to experience positive selection pressure for recombination as a means of exploiting gaps in the environment, even as these gaps were being filled by competitively superior amictic organisms. Conceptually speaking, this phenomenon would be the equivalent of group selection resulting from organisms 'running out of niche', as illustrated by Williams' example of a parasite that is optimally adapted to a host that is going extinct (Williams, 1975).

The possibility that recombination may be maintained by individual advantage but overcome by group disadvantage seems to run counter to a well-accepted line of reasoning that associates special group advantages with recombination (Maynard Smith, 1978). According to this line of reasoning, potential response to selection is directly related to recombination, which is in turn inversely related to expected time to extinction for a group of organisms. However, the two apparently contradictory conclusions really apply to different situations. The reasoning that associates possible group disadvantages with recombination applies to environmental conditions subject to stochastic but not secular change. If secular changes are occurring in the environment, there will be special advantages associated with recombination as a result of the greater adaptability of organisms that show recombination.

Given the general weakness of group selection (Wade, 1978) in relation to individual selection, the apparently slow origin rates of amictic lines in many sexual groups of organisms, and the prevalence of secular changes in the environment, it is not surprising that amictic lines seldom replace mictic ones (Bell, 1982). Where amictic lines have replaced mictic lines, not only the disadvantages of recombination, but also the numerous other disadvantages of sex may have been responsible.

## Cellular-mechanical costs

There is a trend toward increasing commitment to sexuality with increasing size among organisms. Although continuous asexuality is rare, it is much more common in small organisms than in large ones. Discontinuous sexuality is typical of small organisms; continuous sexuality and continuous obligate sexuality are much more likely among larger organisms. It would be difficult to make a general argument that sex has drastically different potential benefits among small organisms than among large organisms. Consequently, the most straightforward explanation for the increase in frequency of sexuality with size may lie in a higher potential cost of sex for small organisms than for large ones.

Sexuality in eukaryotes typically requires meiosis, syngamy, and karyogamy. The time taken up by these three processes is surprisingly great, and far exceeds the time required for two mitotic divisions (supporting primary literature for this section is given elsewhere (Lewis, 1983) and will not be cited here). In order to be retained by individual selection, continuous sexuality in small organisms, and especially in unicellular taxa, would have to confer impossibly high advantages in viability components of fitness to compensate for low reproductive rates. This explains why sexuality is not continuous in unicellular eukaryotic organisms, and may also explain the low frequency of sex in many other discontinuously sexual organisms. Because ameiotic sex in prokaryotes is apparently also time-consuming (90 min for genetic exchange vs 20 min for a full asexual cell cycle in *E. coli* (Bremermann, 1985)), some of the same reasoning may apply to prokaryotes.

The minimum mitotic cycle time for cells increases with the size of the cell, which is in turn directly proportional to the size of the nucleus and the amount of DNA (Cavalier-Smith, 1978; 1985). In eukaryotic unicells, or in the individual cells of multicellular organisms, mitosis requires as little as 15 min or as much as 3–4 h, depending on temperature and cell size. In contrast, meiosis seldom takes less than 10 h, and in multicellular organisms commonly takes more than 100 h. The time for meiosis, as for mitosis, increases with the size of the nucleus and amount of DNA. For example, Bennett (1973) found within a group of 15 diploid higher plant species a range for meiosis times of 25–20 h and a very tight linear relationship of meiosis time to the amount of nuclear DNA ($r = 0.96$).

Unicellular eukaryotes often experience high mortality rates and severe competitive pressure from other small organisms for the exploitation of resources. Even brief reductions in growth rate would seriously impair the fitness of these organisms. For most of the duration of meiosis, cells are not able to carry out any synthetic activities, and are, for practical purposes, growth-arrested. In fact, the cytoplasmic reserves of cells undergoing meiosis are subject to depletion by energy demands that are not offset by intake. The resulting growth handicap caused by meiosis can be very severe. In multicellular organisms, the burden caused by cessation of synthetic activity in meiotic tissue can be averaged across the somatic cells, so that the overall synthetic cost to the organism is smaller. Even so, meiosis has the potential to delay reproduction if meiosis takes more time than somatic maturation.

It is not clear why meiosis is so time-consuming, or what the minimum time for meiosis might be in organisms that have experienced steady and strong pressure for the reduction of meiosis times. However, a survey of the literature (Lewis, 1983) indicates that meiosis times cannot be reduced by selection to a range comparable to that of mitosis. The apparent resistance of meiosis to shortening implies mechanical limits, given that biochemical phenomena can typically be speeded up under strong selection pressure. Mechanical features of meiosis, including orientation, movement, and pairing of chromosomes, may set limits on speed. More detailed information on the specific reasons for differences in the duration of mitosis and meiosis would be useful, but apparently does not exist at present.

Syngamy may also consume time, although it is not likely to be so inefficient in this respect as meiosis. In some taxa, gametes unite in the environmental medium, and the delay associated with syngamy is limited to the time for gametes to find compatible mating types. Conjugation is an alternative in which the mother cells of sexually compatible strains unite and form a cytoplasmic bridge for nuclei or gametes (e. g., pennate diatoms, desmids (Brook, 1981; Geitler, 1932)). Conjugation might allow the overlap of syngamy with meiosis, which would reduce the overall time expenditure for the two processes. It seems likely, however, that the main advantage of conjugation is to raise the probability of contact between sexually compatible gametes. For multicellular organisms that make direct contact for gamete transmission, delay associated with syngamy would arise over mate location and gamete transmission.

Syngamy is not complete until gametes have actually joined. For oogamous taxa, the incorporation of the male gamete nucleus into the recipient cytoplasm seems to occur quickly. However, in isogamous species, delay is caused by the physical difficulties involved in the merger of two more or less equal cytoplasmic masses. This may be a significant evolutionary force favoring oogamy in small organisms.

Karyogamy, which follows syngamy, may also be important in delaying the resumption of significant synthetic processes. The two separate gametic nuclei must join inside the cytoplasm and become chemically integrated before the cell can resume function. More measurements of this time delay in eukaryotic unicells are needed.

Time requirements for syngamy and karyogamy are poorly documented. In small organisms, observations of the time for meiosis in relation to the time required for conjugation may indicate indirectly the magnitude of the delay caused by syngamy and karyogamy. In general, conjugation lasts significantly longer than meiosis proper.

It would be desirable to have a much broader statistical base than is now available, but it seems certain that the potential cellular-mechanical cost of sex is 5-fold to 100-fold for unicellular organisms. However, a number of mechanisms prevent this cost from being fully realized. The first and most important mechanism for minimizing the realized cost is intermittent sexuality. All other factors being equal, the realized cellular-mechanical cost of sex is reduced in direct proportion to the number of asexual generations that intervene between episodes of sexuality. In many unicellular organisms, sexuality is an annual occurrence. Under ideal conditions, a characteristic mitotic division interval for unicells would be of the order of one day. Given that ideal conditions do not persist over an entire annual cycle for most organisms, the relevant order of magnitude for the number of asexual generations separating annual sexuality for a hypothetical species would be approximately 100. The cellular-mechanical cost of sex would be reduced by this mechanism to a trivial level, but there might be a corresponding reduction in benefits derived from sexuality.

A second mechanism by which the realized cellular-mechanical cost of sex can be minimized is constraint of sexuality to a time during which cell growth is not occurring, or is occurring at minimal rates. If the instantaneous rate of increase for asexual cells is nearly zero, there is little or nothing to be lost by the cessation of synthetic processes over the interval required for meiosis, syngamy, and karyogamy. The cellular-mechanical cost of sex should thus force intermittent sexuality to occur toward the end of the growing season. This is commonly the case for unicellular eukaryotes. Traditionally, the occurrence of sexuality toward the end of growing season in such organisms has been interpreted in terms of genetic variance: low variance is supposedly advantageous during the growing season, and high variance is advantageous at the end of the growing season in anticipation of possibly changed conditions during the following year. Interestingly, this line of reasoning does not mesh very well with observations on the distribution of amictic populations in higher organisms, for which lack of sexuality is typically associated with marginal and uncertain environments (Bell, 1982). Perhaps the cellular-mechanical cost of sex is the key factor determining the timing of sexuality in small organisms (Lewis, 1983), in which case the distribution of amictic multicellular organisms is not inconsistent with factors governing the timing of sex in small organisms.

Still another mechanism by which the realized cost of sex can be reduced in unicellular organisms is incomplete commitment of cell lines to sexuality during the period of sexuality. The principle is best illustrated by diatoms, which appear to have evolved a mechanism specifically for this purpose (Lewis, 1984). In diatoms, sexuality requires a dual trigger: specific environmental conditions plus a specific minimum cell size (Drebes, 1977). Cell size, however, is controlled by an unusual division principle. The cells that are produced sexually have maximum size, as reflected by the size of the frustule (silica skeleton). The frustule consists of one upper and one lower component (valve). At each asexual cell division, the two valves of the mother cell become the upper valves of two new cells. For a given cell, the lower valve always fits inside the upper valve, and is thus smaller than the upper valve. Consequently, as time goes by, the mean valve size, and thus the mean cell size, decreases steadily, and the variance of valve sizes increases steadily. Only cells that have achieved a certain minimum valve size are responsive to the environmental cues that stimulate the occurrence of sexuality. Because of the variance in valve sizes for a given clone, this mechanism ensures that only a small proportion of the cells in a clone will be committed to sexuality in any one year. Thus the clone is able to follow a dual strategy by retaining good representation of the historically successful genotype, and at the same time creating an array of new genotypes that will occupy different portions of the environmental gradient or that may be more suited to novel conditions on the gradient in the future. This reduces not only the cellular-mechanical cost of sex by limiting the commitment to sexual processes, but also reduces the cost of recombination.

Another alternative for reducing the realized cellular-mechanical cost exists for unicellular organisms that consist of large cells. Such cells can have two nuclei, one of which can undergo sexual processes while the other continues with synthetic activity (Nanney, 1980). This may well be one of the major adaptive features of macronuclei in ciliates.

The potential cellular-mechanical costs of sex are so large that they may affect metazoans as well as unicellular organisms. Among small animals, the monogonont rotifers and the Cladocera are examples of organisms whose life histories might be strongly constrained by the cellular-mechanical cost of sex. In these groups of organisms, there is an extraordinary premium on rapid growth during brief favorable periods when resources are plentiful and before predators have responded to the availability of prey biomass. All monogonont rotifers and many species of cladocerans are discontinuously sexual; they typically experience a sexual phase when population growth has peaked (Bell, 1982). Individuals are reproductively competent shortly after hatching from the egg, and the size of the egg is high in relation to the body size of the reproductive individual. For these reasons, the time delay inherent in meiosis is likely to reduce the reproductive rate significantly because the synthesis of egg materials may be complete well before meiosis is complete. This will not be the case in organisms that require extended pre-reproductive development. For example, in human beings, meiosis requires a great deal of time (7500 h in males (Clermont, 1977)), but cannot delay reproduction because of the even greater time required for pre-reproductive development.

Another example of possible effects of the cellular-mechanical cost of sex in multicellular organisms is evident in the work of Bennett (1972), who showed that the time required for maturation in perennial flowering plants increases as a function of nuclear volume, DNA per cell, and meiosis time. The meiosis times are sufficiently long that they appear to determine the earliest date of fertilization in these plants.

Discontinuous sexuality in small organisms necessitated by the high cellular-mechanical cost of sex may be responsible for a higher probability that sexual capacity will be lost in these organisms than in larger organisms. Sex appears to be entirely lacking in the bdelloid rotifers, the Euglenophyta, and in many individual species of unicellular algae. Because mixis is much less likely to originate by chance than amixis, these lines may simply represent cumulative evolutionary accidents caused by the individual selection pressures minimizing the cellular-mechanical cost of sex.

## Fertilization

Physical contact between organisms or their gametes entails risks that are separate from those that are maintained by sexual competition. At least three kinds of such cost can be distinguished. First, increased risk of disease or parasite transmission accompanies contact between organisms, or even between gametes. This is the complement to a benefit of sex explained by Rice (1983a): asexual lines may be subject to genetic and environmental focussing by diseases and parasites. Second, contact between motile organisms typically entails loss or reduction of motility, and thus increased vulnerability to predation. Neither of these costs is derived from anisogamy, but anisogamy, operating through sexual selection, can increase the costs by modifying behavior or morphology. Risks associated with primary sexual structures or with a minimal component of sexual behavior can be considered a cost of fertilization, and not a product of anisogamy. A third type of cost associated with fertilization is waste of gametes, or waste of effort transmitting gametes. All three types of cost are reduced by intermittent sexuality.

A second cost of fertilization is the introduction, along with obligate continuous sexuality, of a minimum population size for reproduction. The potential cost could be quite high for continuously obligate sexual organisms that colonize empty habitats or that experience strong fluctuations in density. The realized costs are reduced to negligible levels by facultative sexuality, particularly if sexuality occurs in response to population density. The realized cost is also reduced by self-compatibility in hermaphrodites (Ghiselin, 1974), although suboptimal progeny fitness can be counted as a realized cost associated with the strategy.

Colonizing land plants are most frequently cited as examples of organisms in which continuous obligate sexuality would be disadvantageous (Stebbins, 1950). Planktonic environments provide another example that may be informative because it is different. Planktonic algae, microcrustaceans, and rotifers have only limited abilities of aggregation in the moving medium of a plankton environment. Furthermore, the organisms that occupy such environments are often dispersed very thinly through the environmental medium in relation to their body size. For example, a planktonic rotifer or diatom at a density of one organism per cubic centimeter occupies only approximately $10^{-7}$ of the environmental space and, because of poor aggregation capabilities, is seldom in close contact with other members of the same species at such densities. While planktonic organisms very commonly exceed such a density by several orders of magnitude, there is a continual and substantial possibility that in any given year any given species will not reach such a density, nor can any planktonic species colonize an empty environment without passing through a considerable amount of time at lower densities than this. Consequently, discontinuous sexuality plays a very important role in reducing the realized cost of sex in planktonic organisms, and has probably also influenced the timing of sexuality in planktonic species. Among planktonic metazoans, it is interesting that continuous sexuality appears first along the size spectrum at about 1 mm, among the Copepoda, which have sufficient motility to seek out mates even when the population densities are relatively low. Even so, it seems likely that obligate sexuality requires copepods to sustain higher birth rates than either rotifers or cladocerans (Lewis, 1979), both of which are typically smaller but are sexual only under a narrow set of circumstances.

For most organisms, significant cost associated with minimum population density for sexual reproduction seems unlikely. Small organisms are often not obligately sexual. Animals of moderate to large size have very high motility, and thus are capable of behaviorally offsetting very low population densities, and would only be handicapped by lack of asexual capabilities under very special circumstances such as colonization of remote islands. Large plants and sedentary invertebrates have limited vulnerability to low population density because of very different but equally effective adaptations. Many of these organisms are capable of producing vast numbers of microgametes, and thus occupy reproductively a very large area, despite their immobility.

In summary, the potential costs of sex associated with minimal density for sexuality are substantial, but the realized costs are probably small because of the large number of ways in which the potential costs can be reduced. Among large organisms, the ability of individuals or their gametes to saturate reproductively a large area offsets the problem of low population density, and in small organisms discontinuous sexuality allows the organism to forego sexuality or to develop through natural selection environmental cues that cause sexuality to occur when population density is highest.

**Cost of genome dilution**

Success for reproducing organisms has both a qualitative and quantitative component. The qualitative component is related to the fitness of progeny, whereas the quantitative component is related to the number of progeny, which is in turn a function of the synthesis capabilities of the parent and the optimum investment of cytoplasm and food reserve per gamete. In isogamous organisms, the amount of cytoplasm and food reserve per gamete takes into account the availability of an equal amount of cytoplasm and food reserve accompanying the second gamete that will contribute to the zygote. At the other extreme, in oogamous organisms, the amounts of cytoplasm and food reserve in the macrogamete must be adequate for all requirements of the zygote, since the microgamete will not contribute significantly in this respect. Consequently, the individual producing the macrogamete is providing twice the ratio of cytoplasm and food reserve to DNA as its isogamous counterpart, and suffers a 50 % cost in the efficiency of genome propagation as a result.

Numerous species of organisms occupy middle ground between isogamy and oogamy. These include particularly unicellular eukaryotes (Fritsch, 1965). It may seem surprising that the disruptive selection leading to anisogamy does not always carry through to the fullest extent in causing divergence between the sizes of the gametes. However, among unicellular organisms for which there is no supporting soma to assist in delivery of the gamete, even the microgamete may benefit from the availability of sufficient energy reserves to ensure its survival and activity for an extended interval. In such organisms, the realized cost of genome dilution is above zero, but is less than in oogamous species.

The realized cost of genome dilution need not always be 50 % for oogamous species. If the male parent enhances the fitness of his progeny after fertilization, the realized cost of genome dilution is reduced accordingly (Maynard Smith, 1978). Benefits to the offspring may be diffuse and subtle, as might be expected from certain types of territoriality, or may be definite and direct, as in the case of dual parental care of the offspring. It is easy to imagine examples in which the probability for survival of offspring is more than doubled by parental care from two parents. Sexuality will be reinforced by these advantages, as well as the advantages derived from recombination. Such a reinforcing effect may well account for almost complete absence of asexual reproduction in birds and mammals.

Oogamous organisms that are discontinuously sexual will also experience a reduced cost of genome dilution. Clonal and vegetative reproduction in oogamous organisms reduce the rate at which the disadvantage of genome dilution is expressed. Partial self-fertilization has the same effect. A reduction in rate at which the cost is expressed will in turn allow greater amount of time for the advantages of recombination to accumulate. If the advantages of recombination are not continuous in space or time, reduction in the rate at which the cost of genome dilution is expressed might well be critical to the preservation of sexuality.

Overall, it appears that the cost of genome dilution may not be so severe or widespread as it might at first seem to be. It does not apply to isogamous organisms, and applies only partially to organisms that are anisogamous but not oogamous. Among some large animals, biparental contributions to the probabilities of survival for offspring are sufficiently high to account for extensive reductions in the realized cost of genome dilution. Among organisms that are not obligately sexual, natural selection can reduce the cost of genome dilution per generation by reducing the frequency of sex.

## Sexual selection

Sexual selection can create novel risks in sexually reproducing organisms, or can exaggerate risks that are found in asexual organisms. It is convenient to group costs of sexual selection under two headings, as shown in Table 1: adaptations to sexual competition, and dual phenotypic specialization.

Sexual selection may enhance the performance of individual organisms by comparison with other organisms in the population, while at the same time lowering individual fitness with respect to a theoretical asexual organism not burdened with such competition. Thus sexual selection may work against natural selection of other types (Arnold, 1985; Darwin, 1871; Ghiselin, 1974). Combat and display behaviors as well as secondary sexual characteristics maintained by sexual selection almost inevitably handicap the organism to some degree, and thus constitute a potential cost of sex. Costs of this type are most obvious in vertebrates, but must also occur widely among invertebrates. For example, Saunders and Lewis (manuscript) documented a sudden increase in mortality between the subadult and the adult instar of freshwater diaptomid copepod, and associated this sudden increase with higher activity and changes in swimming behavior that facilitate mate location but that also make the animals more vulnerable to predators. Costs of this type for higher plants are low or negligible, as in unicellular organisms.

A more speculative and less easily quantifiable category of costs is associated with phenotypic differentiation of sexes. In sexually differentiated species, the male and female organisms essentially occupy differing niches. Separation of these niches is a matter of degree, but is often sufficiently pronounced to be measurable as a result of behavioral or physical specialization brought about by sexual selection. In such organisms, individual selection is optimizing the genotype along two separate lines at the same time. Although regulatory switching presumably directs the organism distinctively into one track or another, there would seem to be opportunity for substantial inefficiency as a result of opposing trends for the two sexes, either in dioecious or monoecious organisms. At the very least, the more extensive genetic information required to support distinct male and female phenotypes in dioecious organisms must be a measurable liability that can be classified as a cost of sex. This cost has evidently not been investigated.

### Phyletic distribution of costs

Both the potential and realized costs of sex are distributed unevenly across groups of organisms. Irregularities in cost correspond to differences in life histories, fecundities, environmental grain, and other factors that cannot be considered without a detailed inventory of groups of organisms. At the same time, however, there are certain general trends in potential and realized costs that run parallel to phylogeny. These trends, which are summarized in Figure 2, provide an overview of the cost of sex.

At the top of Figure 2 is a phylogenetic diagram that shows in very coarse terms the four eukaryotic kingdoms and the major groups within these kingdoms in relation to each other. The Protista from the broad base of stocks from which the other three kingdoms were derived. Passing from left to right, we find groups of more recent origin, beginning with weakly integrated multicellular or multinucleate organisms and culminating in the most complex and highly differentiated taxa on the right-hand side of the diagram. A non-linear scale of origins is indicated below the phylogenetic

54

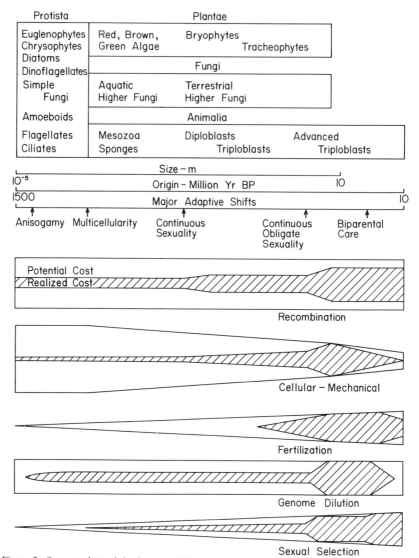

Figure 2. Suspected trends in the potential and realized costs of sex.

diagram. Also paralleling the diagram from left to right is a nonlinear size axis that is important to the phylogenetic trends in cost of sex.

Also shown in Figure 2 is a series of five adaptive shifts that have major significance for the cost of sex. The position of these shifts is indicated according to the approximate phylogenetic position at which they affect a large number of taxa. The shifts include the development of anisogamy from isogamy, the development of multicellularity from unicellularity, the development of continuous and of continuously obligate sexuality, and the development of biparental post-zygotic investment among animals. Each of these shifts has an identifiable influence on one or more of the costs of sex.

Figure 2 also shows the potential and realized costs of sex in diagrammatic form along the phylogenetic axis. The diagrams are based on suppositions and not measurements. Recombinational costs for sexual organisms are shown first. As the figure indicates, there is no basis for suggesting a phylogenetic trend in the potential cost of recombination. The potential cost of recombination may vary in relation to such factors as fecundity or environmental variability, but these bear no certain relationship to the phylogenetic axis. The realized costs, however, are likely to increase with increasing commitment to sexuality, as shown in the diagram, up to the point at which they are offset by biparental investments in progeny.

The cellular-mechanical cost is shown in the second diagram. The potential cost is high among unicellular organisms, for which the effective delay in synthesis of individual cells would be most costly, but the realized cost is far below the potential cost because of discontinuous sexuality and restriction of sex to periods of minimal growth. Beyond the boundary of multicellularity, the potential costs decrease steadily with the increase in organism size because of the progressively greater time required for the development of somatic tissues and the correspondingly lower potential for delay resulting from the creation of gametes. The realized cost increases, however, because of stronger commitment to sexuality in larger organisms. In some of the smallest organisms that show continuous obligate sexuality, the importance of cellular-mechanical costs is high because of the inability of these organisms to reduce costs by reducing the frequency of sexuality. In the largest organisms, the long pre-reproductive period reduces the cost.

Costs associated with fertilization probably increase with organism size. In animals, potential vulnerability arising from predation is likely to be highest in organisms of intermediate to large size that are adapted to avoid behaviorally complex predators. Vulnerability to parasites and diseases is likely to be an increasing function of size because larger organisms will support a broader diversity of such species. The realized cost is reduced, however, by the intermittent nature of sexuality in most organisms. Minimal population size for sexual reproduction, which is the second kind of cost associated with fertilization, is potentially applicable to all organisms, but is offset in small organisms by intermittent sexuality and in very large organisms by motility and high fecundity.

For genome dilution, the potential costs are essentially uniform, but the realized costs change drastically across the phylogenetic spectrum. For isogamous organisms, the cost is negligible, and for anisogamous organisms that are not oogamous, the cost is small. Beyond the transition to oogamy, there is little reason to suspect a phylogenetic trend until the shift occurs toward organisms that have continuous and continuously obligate sexuality. Up to this point the realized costs are reduced substantially below the potential costs of discontinuous sexuality. Organisms that are obligately and continuously sexual experience a drastically increased cost, which must approach the maximum potential cost. However, beyond the point at which biparental care becomes a common phenomenon, the realized cost is increasingly offset by increase in the probability of offspring survival through the male parent.

For sexual selection, the potential cost increases with size because of the general increase in complexity and lifespan with size, which allows more scope for sexual dimorphism. Realized costs are reduced by discontinuous sexuality. The realized costs, like the potential costs, increase steadily with increased organizational complexity in larger organisms. There is a shift toward higher costs with the abandonment of asexual reproduction. With the development of large body sizes in motile organisms that are behaviorally complex and sufficiently long-lived to be selected for elaborate strategies that maximize sexual fitness, the cost of sexual selection reaches a maximum.

**Conclusions**

Several general conclusions are possible from the schematic phylogenetic trend shown in Figure 2. First, there is a tendency for accumulation of larger and more diverse realized costs of sex in complex multicellular organisms, and particularly near the transition from taxa that are discontinuously sexual to taxa that are continuously sexual or continuously and obligately sexual. Asexual capabilities are extremely important in reducing the realized cost of sex, and it is therefore very important to understand what selective forces lead to the abandonment of asexual capabilities.

The phylogenetic trend toward increased commitment to sexuality may well be explained by organism size. As size increases, the time to maturity also increases. For a given level of commitment to sexuality, organisms with shorter times to maturity have the potential to evolve more rapidly than organisms that require longer to mature. Consequently, as organisms increase in size along the phylogenetic spectrum, they are increasingly subject to adaptive exploitation by smaller organisms that would use them as a resource (references in Bremermann, 1985), or to adaptive avoidance by smaller organisms that they would use as a resource. In other words, the Red Queen effect increases as size increases. This trend can be offset be increased recombination, which is manifested as an increased commitment to sexuality as size increases. A corresponding increase in the commitment to sexuality in smaller organisms is not feasible because of higher potential costs of sex that must be held in check by intermittent sexuality. Thus an increase in size allows greater commitment to sexuality in larger organisms, and a greater commitment to sexuality is forced on larger organisms by their slower adaptive responses. Phylogenetic trends in the frequency of sex may be to a large degree explained by these opposing size-related trends in the potential cost of sex and the rate of adaptive response.


## Summary

The potential costs of sex can be divided into two large categories. The first category consists of costs that are not derived from anisogamy, and includes costs associated with recombination, with delay of synthesis at the cellular level (cellular-mechanical cost), and with fertilization. The second large category of costs is derived from anisogamy. Specific costs include the cost of genome dilution (cost of males) and the cost of sexual selection. The cost of genome dilution applies to sexual females in anisogamous organisms because the female provides cytoplasm and reserve materials to support the male genome as well as her own genome. The cost of sexual selection is caused by loss of fitness through selection resulting from sexual competition, or through dual phenotypic specialization of a single species as a result of sexual selection.

All factors contributing to the cost of sex apply potentially to a broad spectrum of organisms. However, numerous mechanisms reduce the realized cost of sex to a level far below the potential cost. The potential cost is likely to be highest in small organisms, but the realized cost among these organisms is reduced greatly by intermittent sexuality. The realized cost of sex tends to increase with size and complexity among organisms because of increasing commitment to sex with increasing size. Increasing commitment to sex is required by the lower reproductive rates and consequently lower rates of adaptation in larger organisms, which would put such organisms at a disadvantage in relation to smaller organisms were it not for increasing commitment to sexual recombination in larger organisms. Comparable commitment to sex is not possible in smaller organisms because such organisms have a higher potential cost of sex, which maintains among them a lower limit on the frequency of sex.

Acknowledgments. I thank Dr Stephen Stearns for providing me with the opportunity to do this work and for his editorial services. I am fortunate that my office is located near an aggregation of helpful ecological geneticists who provided me with a steady flow of reading material and undoubtedly saved me from many errors; I especially thank J. Mitton and M. Grand. This work was supported by National Science Foundation Grand BSR 8604655 and by the University of Colorado Center for Limnology.

# The evolution of sexes

R. F. Hoekstra

## Introduction

*No practical biologist interested in sexual reproduction would be led to work out the detailed consequences experienced by organisms having three or more sexes; yet what else should he do if he wishes to understand why the sexes are, in fact, always two?*
(Fisher, 1930)

What else should he do? It is remarkable that Fisher did not consider the possibility of analyzing the logically simplest case of one sex (which could equally well be viewed as the case of no sexes, the absence of any differentiation in gender). Yet, this seems to be the most basic problem we face, when we want to understand the evolution and occurrence of different sexes: why is the simplest thinkable sexual system, namely a population in which all gametes are equivalent and every gamete can mate with any other gamete, not realized in Nature?

Such an imaginary system may seem absurd in organisms like mammals or higher plants, but it is not if applied to primitive organisms with a haplontic life cycle and external fertilization like many algae. Until recently, the problem of the evolution of different sexes was thought to be properly represented by the problem of the evolution of anisogamy from a more primitive condition of isogamy. On this view isogamy is equated with the absence of sexual differentiation. However, as will be discussed more fully later, this idea is almost certainly wrong. In isogamous species the two mating gametes which form a zygote cannot be distinguished morphologically, but they show characteristic physiological differences. These different types are called mating types, often designated as $(+)$ and $(-)$.

Thus it is clear that the evolution of different sexes in its most basic form is represented by the evolution of mating types in an isogamous population. It is desirable, then, for logical as well as for biological reasons, to start with reviewing the evolution of mating types, and to consider next the secondary problem of the evolution of gamete dimorphism as reflected in anisogamy.

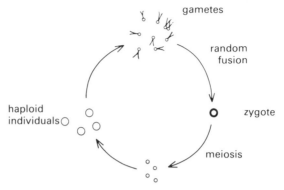

Figure 1. Life cycle of model organism.

## The model organism

All theories and models to be discussed in this paper apply to the following hypothetical organism, which could be viewed as an idealized representation of a unicellular green alga like *Chlamydomonas*. Consider a large population of vegetative haploid individuals living in water. Reproduction is asexual, but under suitable conditions the vegetative individuals produce gametes (by asexual division). The gametes fuse to form zygotes. When mating types are lacking, zygote formation is supposed to be random. The zygotes divide meiotically to produce a new generation of vegetative individuals (Fig. 1).

## The evolution of mating types in isogamous populations

### Terminology

In accordance with modern usage in the phycological literature, I shall use the term *'mating types'* to denote the two sexes in those cases where there is no morphological sex-differentiation.

There is some danger here of confusion, since the term *'mating types'* is also in use to denote incompatibility types in fungi and in ciliates. It is important to emphasize the distinction between sexuality and incompatibility. *'Incompatibility'* is a mechanism, superimposed on sexual differentiation, which interferes with the sexual differentiation process prior to fertilization (Raper, 1966; Van den Ende, 1976).

In higher plants the difference between sexuality and incompatibility is clear, but in isogamous species lacking morphological sex-differentiation, mating types and incompatibility types cannot be distinguished.

I will also conform to the usage in the phycological literature of the terms '*homothallism*' and '*heterothallism*' to denote respectively the possibility of successful sexual fusions within a clone, and the situation where only one sex is present in a clone. In this paper these terms are used as synonyms of respectively '*monoecy*' and '*dioecy*'. There are instances where the two pairs of terms are used in slightly different meanings (Starr, 1969), but that need not concern us here.

## History of the problem

Sexual differentiation without morphological differences between the sexes was first discovered by Blakeslee (1904) in his study of the sexuality of a group of fungal species belonging to the *Mucorales*. A few years later Hartmann (1909) formulated his theory of sexuality, elaborating on ideas expressed earlier by Bütschli (1887–1889) and Schaudinn (1905). According to this theory every gamete – even every cell – is in principle hermaphroditic in the sense that it contains the genetic potential for becoming male as well as female. Additional factors (environmental and/or genetic) determine the phenotypic sex. However, this phenotypic sexual differentiation is not necessarily into two discrete non-overlapping classes (male and female), but may result in a continuous distribution of sex phenotypes, ranging from extreme male via intermediate cases to extreme female. Thus in some species it is possible for a gamete to be male relative to one particular gamete, but female relative to another. There are in the literature a number of reports of experimental evidence for this principle of so-called '*relative sexuality*', for example Hartmann (1925) and Lerche (1937).

Hartmann's theory of sexuality (for a later account, see Hartmann, 1956) comprises three generalizations: the principle that every fusion is between unlike gametes, the principle of sexual bipotency, and the principle of relative sexuality. The ideas of Hartmann were influential mainly in the period between 1920 and 1940, and stimulated the search for sexual differentiation in species where no morphologically different sexes could be found.

The claimed general validity of the idea that without gamete differentiation no fertilization can take place has been questioned (Kalmus, 1932; Czurda, 1933; Pringsheim and Ondracek, 1939; Pringsheim, 1963).

According to Pringsheim, there is no gamete differentiation in the unicellular alga *Polytoma*, and sex can occur between two identical gametes.

Moreover, the credibility of Hartmann's theory has probably been damaged by the fact that the supportive evidence provided by Moewus, who worked on sex hormones and on the genetics of sex determination in *Chlamydomonas*, has to be discarded (Renner, 1958; Gowans, 1976).

With regard to the principle of relative sexuality, the experimental evidence from Hartmann's own work on the marine brown alga *Ectocarpus siliculosus* has convincingly been shown by Müller (1976) to be based on misinterpretations and errors. According to Müller (1976) there is now no significant data left in support of relative sexuality in plants. With regard to the other basic feature of Hartmann's theory, the existence of a sexual bipotency in all organisms (and therefore also in haploids), there is good evidence that in diploids the genes for both sexes are present in each sex genotype (Bacci, 1965). When sex determination occurs in the haploid phase, we must distinguish between homothallic and heterothallic systems. In homothallic systems, all individuals must necessarily possess the genes for both sexes, irrespective of whether sex determination is wholly dependent on physiological and/or environmental variation, or results from a switch mechanism as elucidated in yeast (Strathern and Herskowitz, 1979), and which very probably also operates in a homothallic *Chlamydomonas* species (Van Winkle-Swift and Aubert, 1983). In heterothallic species, there is ample evidence of genetic sex determination, where the sexes (or mating types) are inherited as if they were a single pair of alleles (Wiese, 1976).

Recent evidence on the genetic nature of the mating type locus in *Chlamydomonas reinhardii* suggests that the mating type locus may consist of a cluster of very tightly linked genes, all functionally involved in the mating process (Galloway and Goodenough, 1985). This is also expected on theoretical grounds (Hoekstra, 1982; 1984). On the other hand, several 'sex-limited' genes that specify gametic traits of (+) gametes are unlinked to the mating type locus, and yet are expressed only in gametes carrying the (+) mating type allele (Goodenough et al., 1978) and similarly with respect to the (−) allele (Forest and Togasaki, 1975).

This suggests a regulatory or 'switch' function of the mating type locus. In the latter case again the genetic information for both sexes is present in every genome.

Clearly, more work is needed to clarify in detail the genetics of sex determination in heterothallic haplonts, before more definite statements can be made concerning this matter.

Finally, it is of importance to note that Hartmann's theory is fundamentally non-evolutionary. I have been unable to find any indication in Hartmann's writings (Hartmann, 1909; 1956) of how selection could have produced sexual differentiation. Sexual bipolarity is just assumed to be a basic property of all organisms, and neither the problem of the origin nor that of its maintenance is considered.

The problem of the *evolution* of mating types has apparently not been considered in any detail until recently (see next section). I have only found some occasional remarks in the literature, suggesting that mating types as the first stage of sexual differentiation presumably evolved as blocks to inbreeding (for example, Ghiselin, 1974, p. 102).

*The theory of Hoekstra*

In this section I will summarize a group of related models in which I have attempted to analyze possibilities and difficulties for the evolution of mating types (Hoekstra, 1982; 1985; and unpublished).

As already mentioned in the introduction, sexual differentiation in its most basic form is presumably represented by the evolution of mating types in isogamous populations.

It may well be, however, that this view of mating type differentiation from a supposedly more primitive situation, where all gametes are equivalent and interchangeable in the sexual process, is erroneous, and that sex from its earliest appearance has been asymmetric in the sense that the partners showed characteristic differences. Under the latter assumption, the models discussed in the next sections can be interpreted to address the problem, by which evolutionary forces this sexual asymmetry is maintained.

*Mating type evolution based on gamete recognition.* Consider again as a starting point a population of organisms with a life cycle as shown in Figure 1, without pre-existing gamete dimorphism or mating types. First, we concentrate on the gamete recognition mechanism. Clearly, there has to be a specific recognition and adhesion mechanism preceding gamete fusion. In general, cell-cell recognition involves complementary interacting macromolecules at the cell surface. Various molecular models have been proposed for specific cell adhesion, based on unipolar and bipolar complementarity (Roth, 1973; Burger et al., 1975). There is good evidence that the specific recognition mechanism in microbial mating systems is of the unipolar type (Wiese et al., 1983; Musgrave and Van den Ende, pers. comm.). Bipolar complementarity is found, for example, in interactions between identical cell types in morphogenesis (Fig. 2).

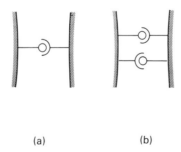

(a)                    (b)

Figure 2. Complementarity in gamete contact can be unipolar (a) or bipolar (b). (From Hoekstra, 1982)

If selection would favor a unipolar gamete recognition system (Fig. 2a), then two mating types would arise quite naturally. This possibility is analyzed in a model based on the following assumptions. There are two loci $A$ and $B$, each coding for one of the two complementary recognition molecules. The alleles $A_1$ and $B_1$ cause the respective molecules to be present at the cell surface, while $A_2$ and $B_2$ suppress their formation. The relation between genotypes and phenotypes which follows from these assumptions is shown in Figure 3. The same figure also shows the various types of matings with their relative mating efficiency, which is assumed to be proportional to the number of successful matings. The relative fitness of an $A_1B_1$ zygote is supposed to be lowered by a factor $1 - \delta$ due to the fact that a certain fraction of these zygotes results from fusions within the same clone. Thus, $\delta$ may be considered to be a measure for 'inbreeding depression', or alternatively as a measure for the disadvantage of not having sex with a genetically different individual. Furthermore, it is assumed that all genotypes have the same rate of asexual reproduction, and that $A_2B_2$ individuals cannot reproduce sexually. The recombination fraction for the two loci is represented by $R$, and gamete fusion is supposed to be random.

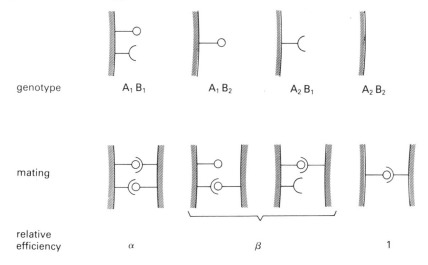

Figure 3. Specification of the notation used in the model in the section on 'Mating type evolution based on gamete recognition'.

Next, using standard population genetics methodology, recurrence equations connecting the genotype frequencies before and after sexual reproduction can be derived and analyzed. This leads to the following results (Hoekstra, 1982), which are graphically shown in Figure 4:
(i) A population consisting initially of $A_1B_1$ individuals can be invaded by $A_1B_2$ and $A_2B_1$ mutants if

$$\beta > (1-\delta)\alpha. \tag{1}$$

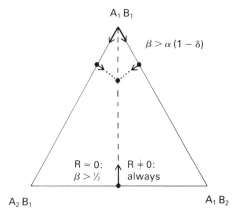

Figure 4. Results of the model described in 'Mating type evolution based on gamete recognition', represented in a de Finetti diagram (a particular population composition is represented by a point in the triangle, such that the frequencies of the three genotypes are proportional to the lengths of the perpendiculars to the three sides). (From Hoekstra, 1982)

(ii) The original $A_1B_1$ genotype will only disappear from the population if both

$$R = 0 \text{ and } \beta < \tfrac{1}{2}. \tag{2}$$

(iii) Selection will favor any reduction in recombination between the two loci, so that after a sufficiently long time $R = 0$ would be expected.

The main condition for the evolution of two mating types based on unipolar complementarity of gamete recognition is therefore $\beta < \tfrac{1}{2}$. This seems a very severe condition, since cellular interactions based on bipolar complementarity apparently are possible and function well – although in different biological systems. The fact that the conditions (2) are independent of the parameter $\delta$ for 'inbreeding depression' is a consequence of the assumption of random gamete fusion. When the bipolar $A_1B_1$ type is very rare, fusions between two $A_1B_1$ gametes occur at a negligible frequency.

This feature of the model is perhaps not quite satisfactory, since in natural situations different clones are presumably not very well mixed, so that mating within the clone will have a much higher probability than under the random fusion assumption. I will consider a model based on more realistic assumptions in this respect, but before doing so, I will first discuss a different version of the present model.

*Mating type evolution based on pheromonal attraction.* In this model the two complementary molecules of the previous model are interpreted as a diffusible pheromone and a corresponding receptor, which senses the presence of pheromone. It is assumed that once a gamete detects the

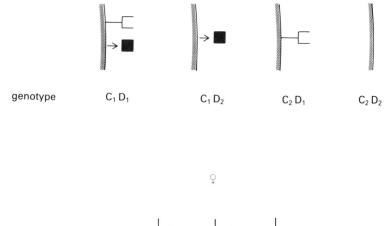

|  | genotype | $C_1D_1$ | $C_1D_2$ | $C_2D_1$ | $C_2D_2$ |

<table>
<tr><td></td><td></td><td colspan="2">♀</td></tr>
<tr><td></td><td></td><td>$C_1D_1$</td><td>$C_1D_2$</td></tr>
<tr><td></td><td>$C_1D_1$</td><td>$ml$</td><td>$m$</td></tr>
<tr><td></td><td>$C_2D_1$</td><td>$l$</td><td>1</td></tr>
</table>

Figure 5. Specification of the notation used in the model in the section 'Mating type evolution based on pheromonal attraction'.

pheromone, it will be able to adjust its swimming direction towards increasing pheromone concentrations and so find the pheromone producing gamete. Let us call the genetic loci in this model $C$ and $D$, to distinguish it from the previous model. $C_1$ individuals produce pheromone while $C_2$ individuals do not. $D_1$ individuals respond to pheromone while $D_2$ individuals cannot respond. Consequently, $C_2D_2$ genotypes will not participate in mating. Figure 5 specifies the notation of the model and also shows the relative mating efficiency of the possible types of mating: $C_1D_1$ is impaired in its 'male' function (chemotactic response) by a factor $m$ relative to $C_2D_1$, and is impaired in its 'female' function (pheromone production) by a factor $l$ relative to $C_1D_2$. The assumption of an impaired 'male' function is based on the idea that $C_1D_1$ individuals will to some extent suffer from self-saturation of their receptor sites; the impaired 'female' function results from the same phenomenon, since part of the pheromone production is captured by its own receptors.

A population genetic analysis based on these assumptions leads to the following results, illustrated in Figure 6:

(i) Suppose that the population originally consists exclusively of $C_1D_1$ individuals. Then a mutant $C_1D_2$ will invade if $l < \frac{1}{2}$; similarly, $C_2D_1$ will invade if $m < \frac{1}{2}$. In both cases a stable non-trivial equilibrium will be established between the two types.

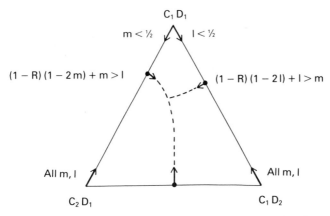

Figure 6. Results of the model described in 'Mating type evolution based on pheromonal attraction', represented in a de Finetti diagram. (From Hoekstra, 1982).

(ii) $C_2D_1$ will invade a $[C_1D_1,C_1D_2]$ population if $(1-R)(1-2l) + l > m$; similarly, $C_1D_2$ will invade a $[C_1D_1,C_2D_1]$ population if $(1-R)(1-2m)+m>l$.
(iii) The original $C_1D_1$ type will disappear from the population if both

$$R = 0 \text{ and } m + l < 1 \tag{3}$$

(iv) Selection will favor reduction in recombination frequency between the $C$ and $D$ locus, which implies that one may expect the condition $R = 0$ to be fulfilled.

Therefore, the main condition for the evolution of mating types based on a pheromonal gamete attraction system is $m + l < 1$. This seems more likely to be satisfied than the corresponding condition $\beta < \frac{1}{2}$ in the previous model; the self-saturation effect probably results in a very small value of $m$ (a poor 'male' function), so that a value of $l$ slightly lower than 1 will be sufficient. There is however a major drawback to this hypothesis of mating type evolution. This results from the fact that pheromonal gamete attraction systems are very rare in isogamous algae, which is precisely the group of organisms in which the occurrence of mating types is hardest to understand. But the model might provide a satisfactory explanation of the occurrence of mating types in isogamous haplonts with sexual chemotaxis, like yeast. The mating type locus in yeast is known to control the production of and reaction to pheromones (Wilkinson and Pringle, 1974).

*The effect of intraclonal fusion and parthenogenesis.* The model of equations (1) and (2) is perhaps unsatisfactory due to the assumption of random gamete fusion, implying a thorough mixing of gametes derived from different clones. Furthermore, in that model those gametes of 'true' (unipolar) mating types which happened to meet a gamete of their own type – so that no fusion could occur – were supposed to be genetically dead.

This is perhaps, too strong a disadvantage for the unipolar mating types, since the possibility of finding a suitable mating partner in a second or third encounter is ignored, and also the possibility of parthenogenesis. Parthenogenesis is in this context the development of a new vegetative structure from an unfertilized gamete. This is known to occur in at least some isogamous algae (Ettl et al., 1967), although I have been unable to find quantitative data for this phenomenon. Clearly, parthenogenesis would favor the establishment of mating types, since it alleviates the disadvantage resulting from their greater difficulty of finding a suitable mating partner. Here I will modify the model to incorporate these two aspects.

First, the genetics of mating type determination will be simplified by assuming that a single locus $M$ with three alleles determines mating type. $M_1$ (with frequency $x_1$) represents the bipolar type (corresponding to $A_1B_1$ in the earlier model) while $M_2$ and $M_3$ (frequencies $x_2$ and $x_3$) are the unipolar mating types (corresponding to $A_1B_2$ and $A_2B_1$). This simplification seems justified, since it was shown that selection is expected to create ultimately a 'supergene' consisting of the closely linked loci $A$ and $B$. A fraction $p$ of the gametes in the population is supposed to combine intraclonally. This results in $M_1$ clones in the intraclonal formation of zygotes with a relative fitness of $1-\delta$; in $M_2$ and $M_3$ clones this fraction $p$ is supposed to develop parthenogenetically with the relative fitness of the resulting vegetative individuals of $1-\theta$. The remaining fraction $1-p$ of the gametes is supposed to fuse at random, just as in the earlier model. The assumptions about possible differences in recognition or adhesion efficiency, as expressed in the parameters $\alpha$ and $\beta$, are retained. The following system of recurrence relations for the genotype frequencies is then obtained.

$$\overline{W}x_1' = p(1-\delta)\alpha x_1 + (1-p)[\alpha x_1^2 + \beta x_1(x_2+x_3)]$$
$$\overline{W}x_2' = p(1-\theta)x_2 + (1-p)[\beta x_1 x_2 + x_2 x_3]$$
$$\overline{W}x_3' = p(1-\theta)x_3 + (1-p)[\beta x_1 x_3 + x_2 x_3] \qquad (4)$$

The conditions for invasion of $M_2$ and/or $M_3$ into a $M_1$ population, and for removal of $M_1$ from a polymorphic population consisting of all three types, can be derived by determining the stability of the corresponding boundary equilibria of the system. This results in the following conditions.

$$\alpha - \varnothing < \beta < \tfrac{1}{2} + \varnothing,$$
where $\varnothing = p\,[1-\theta-\delta(1-\delta)]/(1-p)$. \qquad (5)

The left inequality in (5) guarantees the establishment of the unipolar mating types in an original bipolar population, and the right inequality guarantees that the bipolar type will disappear from a polymorphic population, so that finally a population results consisting solely of the unipolar mating types $M_2$ and $M_3$.

Are conditions (5) likely to be true? A difficulty is, that it is very hard (or not yet possible) to find data upon which estimates for the parameters can be based. First consider the case that differences in gamete recognition efficiency are absent or irrelevant, that is $\alpha = \beta = 1$. Then inequalities (5) reduce to

$$\delta - \theta > \frac{2(1-p)}{p}. \tag{6}$$

Therefore, mating types can evolve as a result of the effects of intraclonal fusion and gamete parthenogenesis only if $p$ and $\delta$ are high and $\theta$ small, that is if a very high fraction of the matings is intraclonal, the fitness of intraclonally produced zygotes is low, and if there is a very high rate of parthenogenesis with relatively little fitness loss. This combination of parameter values is clearly very special and does not seem likely on *a priori* grounds.

Next, let us try to make educated guesses with respect to realistic values of the parameters in this model. In the primitive situation we consider, sex is a rare event, which enhances the probability that individuals in each others vicinity belong to the same clone. It follows that a substantial proportion of matings will be intraclonal ($p$ high). Since the unipolar types are thought to arise by mutation from the bipolar types, the unipolar clones present in a particular locality will presumably not differ much genetically from the bipolar cells present. This suggests that the disadvantage of mating intraclonally will not be great ($\delta$ small). In many cases where parthenogenesis from gametes has been observed, the resulting vegetative individuals are smaller or have otherwise clearly a lowered fitness (Fritsch, 1965; Ettl et al., 1967). Therefore it seems reasonable to suppose that the parameter $\theta$ has an intermediate value.

Thus a 'typical' case might well be represented by the parameter combination $\{p=0.75, \delta=0.1, \theta=0.5\}$. For this case conditions (5) become:

$$-1.5 + 3.7\,\alpha < \beta < 2 - 2.7\,\alpha \tag{7}$$

Figure 7 shows a graphical representation of these conditions. Clearly, there are now no restrictions on the parameter $\beta$, but we still need to have $\alpha < 0.4$, so that in fact the condition that unipolar recognition must be at least twice as effective as bipolar recognition still remains, just as in the random gamete fusion model.

*The effect of segregation distortion and parthenogenesis on mating type evolution.* In this section and in the next one a short discussion will be devoted to two somewhat related ideas concerning mating type evolution. Both ideas have in common the feature that mating types may have evolved as an adaptation to alleviate the harmful effects of certain 'selfish' genetic elements. The first idea concerns the effects of a segregation distorting gene

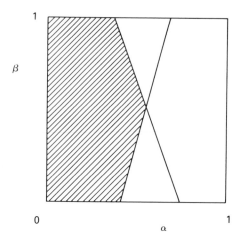

Figure 7. Conditions for mating type evolution in a 'typical case of the model' described in the section 'The effect of intraclonal fusion and parthenogenesis'. Mating types will evolve for ($\alpha$, $\beta$) values in the hatched area.

in conjunction with the possibility of parthenogenesis from unfertilized gametes (Hoekstra, 1985). Segregation distortion is the phenomenon that in a heterozygote a particular allele is favored at meiosis at the expense of the alternative allele, which results in an $Aa$ individual producing unequal numbers of $A$ and $a$ gametes.

This may lead to a stable polymorphism if the distorting gene that is favored at segregation (say $A$) causes a reduced fitness of $AA$ homozygotes (Hiraizumi et al., 1960). The idea of linking segregation distortion and parthenogenesis to the evolution of mating types is as follows. A distorting gene, if disadvantageous in homozygote condition, will lower the mean fitness of the sexual population. This will give a relative advantage to gametes developing asexually by parthenogenesis. Since gametes of bipolar mating type can mate with all types of gametes, while unipolar mating type gametes can only mate with gametes with a mating type different from their own, parthenogenesis will be more common among unipolar mating types. For this reason selection might favor unipolar mating types. This turns out to be the case.

When there is strong segregation distortion with a low fitness of individuals homozygous for the distorting gene, two unipolar mating types can evolve. There is however one snag to this model. For those parameter values which give rise to the evolution of two mating types, the mean fitness of a population in which there is *only* parthenogenesis is higher than that of a population with both sexual reproduction and parthenogenesis.

This implies that in fact this model leads to the prediction that when the bipolar mating type will be selected against, the population will ultimately become asexual.

*The effect of intragenomic conflict between cytoplasmic and nuclear genes on the evolution of mating types.* As will be discussed below, where the ideas of Cosmides and Tooby (1981) on anisogamy evolution are considered, the lack of a strict segregation mechanism for cytoplasmic genes will cause the cytoplasmic DNA from the two parental gametes to compete in the zygote for preferential transmission into the zygotic products. This implies that there is some scope for cytoplasmic genes combining an advantage at segregation with a fitness disadvantage to establish themselves in a population.

When there is transmission of cytoplasmic genes from both parents, such 'aggressive' cytoplasmic factors could spread through the population. When, however, there is uniparental inheritance of cytoplasmic genes (only through the mother) – which is the case in the great majority of organisms – the spread of such aggressive cytoplasmic genes is hampered, since their presence will be limited to the descendants along the female line of the female in which they originated. Natural selection will then tend to remove this line from the population because of the lowered fitness. Grun (1976) hypothesized that this beneficial effect of uniparental inheritance has been the selective force causing uniparental inheritance to evolve from biparental inheritance. This explanation clearly relies on the operation of group selection, and seems therefore not very attractive.

Charlesworth (1983) has suggested an explanation of uniparental inheritance as it occurs in *Chlamydomonas* with respect to chloroplast DNA (Sager and Ramanis, 1973). This mechanism is reminiscent of modification-restriction systems in bacteria: the chloroplast DNA from the mating type (−) parent is destroyed in the zygote; the mating type (+) derived chloroplast DNA escapes degradation since it is protected by methylation of its cytosine residues. Charlesworth assumes the occurrence of a mutation in the chloroplast DNA which interacts with the mating type locus such that it causes destruction of the mating type (−) derived DNA when transmitted via a mating type (+) gamete.

This scenario for the evolution of uniparental inheritance would work, but an unsatisfactory aspect is, that in fact two mutations are needed: one for destruction of unprotected DNA, and one for protection of mating type (+) derived DNA.

I have studied a model, details of which will be published elsewhere, of the evolution of uniparental inheritance *preceding* the evolution of mating types; after the establishment of uniparental inheritance, evolution of mating types appears to become possible. The scenario is as follows:

(i) Consider an isogamous population with biparental inheritance of cytoplasmic DNA. The majority of the gametes has a bipolar mating type ($M_1$ in the model of the previous section), but unipolar mating types ($M_2$ and $M_3$) occur at low frequencies (mutation selection balance). There is a gradual selective process of ever more 'aggressive' cytoplasmic variants spreading through the population.

(ii) A possible way out would be to make sex very rare: during intermittent periods of asexual reproduction the most harmful cytoplasmic variants would be removed or reduced in frequency.

Another possibility is formed by a chromosomal mutation which sacrifices its carrier's own cytoplasmic DNA, for example by greatly reducing the copy number in the gamete. This would save resources and give heterozygotes for this nuclear gene a selective advantage above the original homozygotes, but also above homozygotes for this mutant gene, which will suffer from their shortage of cytoplasmic DNA.

(iii) Next there will be selection favoring close linkage between the mating type locus and the locus interfering with the cytoplasmic DNA production, such that ultimately two types will remain in the population: one unipolar mating type ($-$, say) sacrificing its own cytoplasmic DNA in the gametes, and the other unipolar mating type ($+$) which exclusively transmits cytoplasmic DNA to the offspring. In this way both uniparental inheritance and unipolar mating types are established. However, this scenario appears to work only if the selective differences between the various genotypes are very great, which makes the hypothesis less plausible.

### Can there be more than two mating types?

If the basic sexual differentiation as manifested in mating types in isogamous species is indeed based on chemical complementation in recognition, adhesion or pheromonal attraction, as suggested in the models of the foregoing sections, then a system with more than two sexes becomes impossible. But of course, the evolution of an additional incompatibility system superimposed on the basic sexual differentiation *is* possible, and many instances of such systems can be found for example in the fungi and the higher plants. The evolution of such incompatibility systems is itself a fascinating area with very interesting ramifications to problems of self/non-self recognition (Burnet, 1971; Pandey, 1977a; Monroy and Rosati, 1979; Coombe et al., 1984; Heslop-Harrison and Linskens, 1984).

I will not discuss this matter, however, since it falls outside the scope of this article: there is a fundamental difference between incompatibility and sexual differentiation.

### Theories of mating type evolution: Conclusions

The discussion of the ideas about mating type evolution in the foregoing sections makes two things very clear.

First, the situation with regard to our understanding of the evolution of the basic sexual differentiation as reflected in mating types in isogamous species is comparable to the current situation with respect to our understanding of the evolution of sex itself. We can show that under particular circumstances mating types may evolve, but we do not have an 'easy' explanation which is both general and plausible.

Secondly, we need to know more facts. Are there indeed – as has been claimed in the past (Pringsheim and Ondracek, 1939) – isogamous species without (unipolar) mating types, so that all gametes are equivalent? How safe is the general conjecture (Kochert, 1978) that – with perhaps a few exceptions – isogamous species do not possess sexual pheromonal attraction mechanisms? The answer to this question has implications both for theories of mating type evolution and for theories of anisogamy evolution. What is the exact nature of gamete recognition and adhesion?

Can differences in effectiveness between unipolar and bipolar complementarity be determined? How general is the mating type switching, discovered in yeast, and in all probability also in a homothallic *Chlamydomonas* species? More questions could be added here, waiting for answers to be obtained from experimental work.

This list of questions however will suffice to demonstrate my conviction that at the present state of our understanding of sexual differentiation in isogamous species, there is a greater need to know crucial facts about these systems than to refine the theory further.

One of the aims of this review is to encourage experimentalists to attack the above-mentioned questions.

## Theories of the evolution of anisogamy

### The theory of Weismann et al.

In the last decade of the 19th century and the first of the 20th century a number of German biologists were interested in evolutionary explanations of sex and related phenomena. Among them were Weismann (1886, 1902), Hertwig (1906) and Bütschli (1889). Although divided in their explanations of sex, they were remarkably unanimous in their ideas about the evolution of sexual differentiation. The evolution of anisogamy from the more primitive condition of isogamy was thought to have resulted from two conflicting selection forces. First, there is selection for increasing efficiency of finding a mating partner. This favors the production of a large number of small motile gametes.

Secondly, there is the need of provisioning the zygote with a sufficient amount of reserve for development. This favors the production of big gametes, which are necessarily less motile. This conflict is resolved by a differentiation of two types of gametes, namely small, highly motile (male) gametes and big, less (or non-)motile (female) gametes.

Two aspects of this theory are important to notice. First, the two relevant selection forces figure in almost all the other models, as we will see. And second, there is no reference at all to a possible mechanism of evolution of the system.

*The theory of Kalmus*

The first mathematical model of the evolution of anisogamy is due to Kalmus (1932). He considered isogamy, where all gametes are alike phenotypically and interchangeable in the mating process, to be the primitive situation from which anisogamy has evolved. To understand the advantage of anisogamy, he used the same qualitative reasoning as Weismann et al.: there are two conflicting selection forces, namely the advantage of producing as many gametes as possible, and the advantage of producing sufficiently big gametes for provisioning the zygote.

In order to make more accurate quantitative statements about these matters, he considered a mathematical model based on the classical statistical mechanics approach to the collision of particles. In the following account I have modified his model slightly and simplified his notation.

Consider a population of primitive algae in a very large volume of water. There is a quantity $M$ of living substance available for the production of gametes. For successful development, a zygote needs to have a mass of at least $2m$. In the case of isogamy, the quantity $M$ is divided into $x$ gametes, each of mass $m$. In the case of anisogamy, the available material $M$ is divided in such a way, that from a fraction $p$ of it $x_1$ gametes of mass $m_1$ are formed, and from the remaining $(1-p)$ part $x_2$ gametes of mass $m_2$, such that

$$m_1 + m_2 = 2m \tag{8}$$

Without loss of generality we will assume $m_1 < m_2$.

It follows that

$$x_1 = \frac{Mp}{m_1}, \tag{9}$$

and

$$x_2 = \frac{M(1-p)}{m_2}. \tag{10}$$

It is supposed that gamete motility is random and independent of gamete size. Now the union of two gametes can be treated just as the elastic collision of two particles in statistical mechanics. From the assumptions it follows that the mean number of successful zygotes in a given, short time is in the isogamy case

$$N_1 = cx(x-1)/2, \tag{11}$$

where $c$ is a suitable constant, and in the anisogamy case

$$N_2 = c\{x_1x_2 + x_2(x_2-1)/2\} \tag{12}$$

Neglecting fusions between two macrogametes, eq. (12) becomes

$$N_2 \simeq cx_1x_2 \tag{13}$$

First we ask, which is the most favorable of all possible subdivisions in the anisogamy case. This will be that subdivision which gives the greatest number of successful zygotes. Therefore we have to maximize

$$N_2 = cx_1x_2 = \frac{cM^2p(1-p)}{m_1(2m-m_1)}.$$

Clearly, this expression is maximal for $p = \frac{1}{2}$, and for the smallest possible $m_1$. Thus the optimal subdivision of the material $M$ for gamete production is in an anisogamous population such that half of it is used for forming micro-gametes and the other half for macrogametes, and that the difference in size between micro- and macrogametes is as great as possible.

Next, we compare an isogamous population with an (optimal) anisogamous population with $p = \frac{1}{2}$. An anisogamous population will grow faster than an isogamous one *(ceteris paribus)* if $N_2 > N_1$, or approximately

$$\frac{M^2}{4m_1m_2} > \frac{M^2}{2m^2} \tag{14}$$

Writing $\theta = m_1/m_2$, inequality (14) reduces to

$$\theta^2 - 6\theta + 1 > 0, \tag{15}$$

from which it follows that we must have $\theta < 0.17$, that is at least a sixfold difference in size between micro- and macrogametes. Kalmus predicts from this result that as long as there is a weaker anisogamy ($0.17 < \theta < 1$) a whole range of gamete sizes will be found between the two extremes.

Only when the variation in gamete size has increased further so that $\theta < 0.17$, will there be selection for the strictly separated size-groups of gametes. He mentions two *Chlamydomonas* species which seem to conform to this prediction.

Kalmus further considered refinements of his model by allowing for differences betwen micro- and macrogametes in suspension ability and motility. These did not, however, change his conclusions in a qualitative way.

At first sight Kalmus' result, that in a random gamete fusion model more viable zygotes are produced with anisogamy than with isogamy, is counterintuitive. Clearly, the maximum number of viable zygotes in an anisogamous population is equal to the number of macrogametes (given by eq. (10)), since zygotes formed by the fusion of two microgametes are too small to be viable. In an isogamous population there is no wastage of gamete material, since all possible gamete pairs are viable; the maximum number of viable zygotes is therefore $x/2$. It follows that the maximum number of viable zygotes with isogamy is higher than the corresponding quantity with anisogamy, if

$$\frac{m_2}{m} > 2(1-p) \cdot \tag{16}$$

This inequality is always true if $p = \frac{1}{2}$, as in Kalmus' model. A closer inspection of Kalmus' reasoning reveals that his analysis is correct, but that the unexpected result is a consequence of the assumption that zygote formation takes place in a relatively short time interval, so that the number of viable zygotes formed is always much smaller than the number of macrogametes. This assumption is necessary for applying the statistical mechanics formalism which assumes *elastic* collisions between particles (implying that the number of gametes available for zygote formation does not change during the process). This unrealistic assumption of Kalmus (1932) has been removed by Scudo (1967b), who modified Kalmus' model by taking into account the actual density of gametes during the process of zygote formation. His conclusion is nevertheless qualitatively similar to the one reached by Kalmus: only a (very) high degree of anisogamy will give anisogamous populations an advantage over isogamous populations. Scudo's result becomes understandable by noting that he finds a different optimal allocation of material to the formation of micro- and macrogametes (he obtains a value for the parameter $p$ in eq. (16) of rather less than $\frac{1}{2}$).

Another difficulty with Kalmus' model is of course, that it relies on group selection: anisogamy is thought to have evolved as a consequence of the advantage of anisogamous populations over isogamous populations. Much attention has been given in recent years to the possible evolutionary consequences of group selection (Maynard Smith, 1976b; Wade, 1978; Wilson, 1983), and we now know that there are many difficulties in accepting an evolutionary explanation like that of Kalmus which is solely based on group selection. It is fair to say, however, that at the time Kalmus published his theory, group selectionist reasoning was not all uncommon, although already criticized by Fisher (1930, pp. 49–50).

*Early population genetic approaches*

The beginning of a new era in the theorizing about the evolution of sex-differentiation is marked by the first population genetic approaches to this problem by Kalmus and Smith (1960) and Parker et al. (1972). Kalmus and Smith (1960) were concerned with the question, through what mechanism the initial sexual differentiation may have evolved. They suggest the following possibility. A diploid population is initially homozygous for gene $a$, so that all individuals produce only small gametes. Then there arises a dominant mutation $A$ which results in larger gametes with more reserve material being formed. This gene $A$ is expected to spread through the population, since zygotes with more reserves have a higher probability of becoming a gamete producing adult. This spread will be checked by the difficulty the large gametes experience in finding each other for mating. Therefore, when $A$ reaches a high frequency, $a$ will be selectively favored. Hence a stable equilibrium is expected between dominant and recessive types. Natural selection may then be expected to develop from this system a more complete sexuality in which only disassortative gamete fusions occur.

Although Kalmus and Smith give no more than a sketchy outline of a model, their idea is basically correct in the sense that a stable equilibrium between microgamete- and macrogamete producers can exist under suitable conditions. Consider for example the following formalization of their idea in a (for the sake of simplicity haploid) population genetic model. In a very large population genotype $a$ with frequency $x_1$ specifies microgamete production, and $A$ with frequency $x_2 = 1 - x_1$ specifies macrogamete production. The selective advantage of a larger zygote size is expressed by the parameter $s$ (Table 1), and the advantage of microgametes in mating efficiency by the parameter $k$. Table 1 specifies the model.

Table 1. Population genetic model of the Kalmus and Smith (1960) idea

| Zygote | Frequency | Fitness | Offspring |
|--------|-----------|---------|-----------|
| aa | $\dfrac{k^2}{\overline{K}} x_1^2 *$ | 1 | a |
| Aa | $\dfrac{k}{\overline{K}} 2x_1x_2$ | $1 + s$ | $\tfrac{1}{2}A, \tfrac{1}{2}a$ |
| AA | $\dfrac{1}{\overline{K}} x_2^2$ | $1 + 2s$ | A |

$* \overline{K} = (kx_1 + x_2)^2$ is a normalizing factor.

It is elementary population genetics to write down the recurrence equations for $x_1$ and $x_2$ under the assumption of random gamete fusion in this model, and to derive the conditions for a stable polymorphism of $A$ and $a$. These conditions appear to be

$$\frac{1+2s}{1+s} < k < 1 + s, \tag{17}$$

which shows that the conjecture of Kalmus and Smith is correct for suitable values of $k$ and $s$.

It is of importance to note that this explanation of the evolution of anisogamy is based on individual selection within a population, and that the same model also explains the maintenance of anisogamy in a population, once evolved. It is therefore quite remarkable that Kalmus and Smith consider this explanation only to be relevant for the *origin* of anisogamy, and still adhere to the (group selectionist) model of Kalmus (1932) for the explanation of the selective advantage of anisogamy, once evolved – that is, for its maintenance.

A trace of this 'hybrid' position seems also to be present in the paper by Parker et al. (1972). These authors believe that Kalmus' (1932) theory indicates an important advantage of anisogamy once evolved, but is deficient in explaining the origin of anisogamy. They provide an alternative to the suggestion of Kalmus and Smith (1960) for the origin and evolution of anisogamy. Although much better elaborated, their model is not very different in its assumptions from the Kalmus and Smith (1960) idea. Parker et al. (1972) also start from the by now familiar two basic selection forces: selection for numerical gamete productivity and for large zygote size.

In particular, each individual is supposed to have an amount $M$ of material available for gamete formation. This amount $M$ can be subdivided into a variable number $n$ of gametes, such that for every individual $M = nm$, where $m$ is gamete mass. The population is supposed to consist of a number of genetical variants with different values for $n$ (and $m$). Zygote volume $m_{ij}$ is equal to the sum $m_i + m_j$ of the masses of the constituent gametes, and zygote fitness is supposed to be proportional to $m_{ij}k$, where $k$ determines the steepness of slope of the dependence of zygote fitness on size.

For two different modes of inheritance of gamete size, and assuming random gamete fusion, Parker et al. obtained the following results using computer simulation.

1) For values of $k$ lower than a critical value, $k_1$ say, which depends on the mode of inheritance and the variation in gamete sizes initially present, there is selection for decreasing gamete size: all variants except the smallest disappear.

2) For values of $k$ higher than another critical value $k_2$ ($k_2 > k_1$), there is selection for increasing gamete size: all variants except the largest disappear.

3) For intermediate values of $k$ ($k_1 < k < k_2$) there is disruptive selection on gamete size leading to a stable polymorphism between microgametes and macrogametes: all intermediate variants disappear leaving only the gametic extremes in the population.

This is the first model which is both based on individual selection and sufficiently detailed to allow quantitative conclusions.

This latter aspect has been improved in a number of more rigorous and elaborate models, based on the same assumptions as this model. These are discussed in the next section.

*Formalization and extensions of the theory of Parker et al. (1972)*

The theory of Parker et al. (1972) has been cast into formal – and therefore more rigorous – models by Bell (1978a), Charlesworth (1978), Maynard Smith (1978), and Hoekstra (1980). Bell (1978a) also reviewed the comparative evidence in a number of phyletic series among the algae, fungi and protozoa. This work I discuss below.

The above-mentioned models of Bell, Charlesworth and Maynard Smith differ in their formalism. Maynard Smith analyzes the evolution of anisogamy as a game, using a geometrical representation of the problem. Bell and Charlesworth both have a population genetic approach. Below I give an outline of Charlesworth's model because it offers the most suitable framework into which subsequent work can also be put.

Consider a haploid organism with the life cycle shown in Figure 1 and with random gamete fusion. There is a one locus genetic determination of the number and size of the gametes produced such that $A_i$ genotypes produce $n_i$ gametes of mass $m_i$. The product $m_i n_i$ is assumed to be constant and the same for all $i$. The mass of a zygote is equal to the sum of the masses of the constituent gametes, and the relative fitness of zygotes with genotype $A_i A_j$ is

$$w_{ij} = (m_i + m_j)^k \tag{18}$$

Let $x_i$ be the frequency of $A_i$ among the vegetative individuals and $y_i$ the $A_i$ frequency among the pool of gametes. Then

$$y_i = \frac{n_i x_i}{\sum_j n_j x_j} \tag{19}$$

The frequency of $A_i$ individuals in the next generation is then given by

$$\overline{W} x_i' = x_i \sum_j f_{ij} x_j \text{ where } f_{ij} = n_i n_j w_{ij} \tag{20}$$

Thus this model appears to be equivalent to a model of a random mating diploid population with genotype $A_i A_j$ having fitness $f_{ij}$.

The latter model is of course well known, and for the case of just two alleles, $A_1$ and $A_2$ a stable polymorphism will be maintained if

$$f_{12} > f_{11}, f_{22} \tag{21}$$

80

Using eq. (18), and writing $\theta = m_1/m_2 = n_2/n_1$ and assuming $\theta < 1$, conditions (21) become

$$\frac{\log\theta}{\log\{2\theta/(1+\theta)\}} < k < \frac{\log\theta}{\log\{(1+\theta)/2\}} \tag{22}$$

Figure 8 shows a graphical representation of these conditions. A necessary condition for stable anisogamy is that $k > 1$, implying that zygote fitness increases more steeply than linearly with zygote size. When $k$ is only slightly greater than unity (that is, when zygote fitness is relatively insensitive to zygote size), only a mutant which has a very large effect on gamete size can be selected. On the other hand, a slight degree of anisogamy ($\theta$ close to 1) can only be maintained for a very restricted range of $k$ values. For parameter values in the region above the upper curve there will be selection for larger gametes, and in the region below the lower curve selection will favor ever smaller gametes.

Charlesworth also analyzed a diploid genetic model, obtaining essentially the same conditions for stable anisogamy.

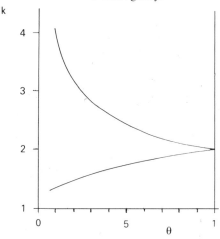

Figure 8. Graphical representation of conditions (eq. 22). Anisogamy is stable for (k, θ) values between the two curves.

In the same paper, Charlesworth considered the important case of anisogamy evolution in a population with mating types. As I have said already in the introduction, an isogamous population with mating types seems the correct starting point for models of the evolution of anisogamy (see for example Wiese, Wiese and Edwards, 1979), and all models discussed so far in this paper are unrealistic in their assumption of random gamete fusion. He obtained the following results:

(1) Pre-existence of mating types does not affect the conditions for the establishment of anisogamy given by (22).

(2) If linkage is sufficiently tight between the gamete size and the mating type loci, linkage disequilibrium tends to build up, so that gamete size alleles become associated with different mating types, thus creating disassortative fusion with respect to gamete size.

(3) If linkage is initially loose between the two loci, there will be anisogamy within each mating type. This situation has never been observed. However, selection will favor any mutant reducing recombination between the gamete size and mating type loci, leading ultimately to the situation described above under (2).

We see, then, that the analytical work described above confirms the results of Parker et al. (1972). There is, however, one aspect of the Parker et al. model which is not covered by the analytical models of Charlesworth (1978) and Bell (1978): this is the finding, that in the computer simulations under conditions for stable anisogamy all variants except the two extreme gamete sizes were selected against. Why are there only two types of gametes; can there be stable anisogamy with three (or more) different gamete sizes? This problem has been studied by Hoekstra (1980) in an extension of Charlesworth's (1978) model, as follows.

Consider an anisogamous population with gametes of two different sizes and random gamete fusion. A third allele $A_3$ is introduced into this population. For the moment we assume the mutant to be intermediate in size:

$m_1 < m_3 < m_2$.

The following notation is convenient:

$m_1/m_2 = n_2/n_2 = \theta_1$; $m_2/m_3 = n_3/n_2 = \theta_2$.

It can easily be deduced that $0 < \theta_1 < 1$ and $1 < \theta_2 < 1/\theta_1$. We now ask under what conditions $A_3$ will be protected.

Linearizing eq. (20) near $x_3 = 0$ gives

$$x_3' \simeq x_3(f_{31}\hat{x}_1 + f_{32}\hat{x}_2)/\hat{\hat{W}}, \tag{23}$$

Where $\hat{x}_1$ and $\hat{x}_2$ are given by

$$\frac{\hat{x}_1}{\hat{x}_2} = \frac{f_{12} - f_{22}}{f_{21} - f_{11}}, \tag{24}$$

and $\hat{\hat{W}} = \sum\limits_{i=1}^{2} \sum\limits_{j=1}^{2} f_{ij}x_ix_j$ is the mean fitness of the equilibrium population consisting only of $A_1$ and $A_2$ individuals. $A_3$ will be protected if

$$f_{31}\hat{x}_1 + f_{32}\hat{x}_2 > \hat{\hat{W}} \tag{25}$$

This inequality can be written in terms of $\theta_1$, $\theta_2$, $k$, $\hat{x}_1$ and $\hat{x}_2$. The lengthy expressions thus obtained are not reproduced here.

It is then possible to show that inequality (25) cannot be true, which implies that $A_3$ is not protected. Next, it can be demonstrated that in this model no stable equilibrium with all three types present will be possible. If the mutant gamete size is greater or smaller than the two already existing types, the allele coding for the then intermediate gamete size will be unprotected, and the same argument as above applies. It follows that since selection will always tend to eliminate the intermediate gamete size alleles, there will be selection for an increasing degree of anisogamy.

Hoekstra also considered a model where a mutant gamete size is introduced into an anisogamous population where the two gamete types are sexually distinct and gamete fusion is disassortative with respect to size. Also in this case a stable equilibrium with more than two gamete sizes does not exist, and the intermediate gamete size is selected against.

On the assumption of random gamete fusion, there is the need for explaining evolution towards disassortative gamete fusion, once anisogamy is established. This problem has been analyzed theoretically by Parker (1978). However, as pointed out earlier, the random gamete fusion assumption is unrealistic, and anisogamy is likely to have evolved in isogamous populations in which gamete fusion was already disassortative with respect to mating type (as has been modelled by Charlesworth (1978)). From this point of view the paper of Parker (1978) loses much of its relevance, and will not be discussed here. As Maynard Smith (1982) remarks, one still has to explain why in anisogamous populations disassortative fusion with respect to size (and mating type) does not break down. Charlesworth (1978) has shown that selection is expected to create close linkage between mating type and gamete size loci, so that the existence of different mating types within the gamete size classes is not likely. It follows that a breakdown of disassortative fusion with respect to gamete size would require a loss of the mating type dimorphism. This problem is of much wider significance than just within this anisogamy setting, and really requires an understanding of the evolution and maintenance of mating types, which was the main theme of the first part of this review.

*The comparative evidence*

The theories discussed in the foregoing sections are not easy to test. The assumption that a given amount of material can be subdivided into a variable number of gametes, such that gamete number times gamete size is constant, seems to be amenable to experimental verification in at least some systems (Wiese, Wiese and Edwards, 1979; Heimcke and Starr, 1979; Bell and Praiss, 1986). But the assumption that with anisogamy the fitness of the zygote increases (much) more with zygote size than with isogamy is difficult to test. Problematic for a direct study of the relationship between zygote fitness and zygote size is the absence of appreciable within species variation in zygote size. A possible indirect approach is to assume that zygote size is proportional to adult size and/or adult complexity, and then to relate the degree of gamete dimorphism with these adult characteristics. This has been done by Knowlton (1974) and Bell (1978a). Knowlton (1974) looked at two *Sporozoa* orders and at *Volvocaceae* (a family of green algae). However, the sporozoa data appeared to be unsuitable for this comparison, presumably because of the rather special life cycles of these organisms. In the *Volvocaceae* family she found isogamy generally associated with small colony size, and anisogamy with large colony size. Her statement – that fusion in a genus with intermediate sized colonies *(Pandorina)* is not disassortative – is questionable in the light of what is known about the occurrence of mating types in isogamous species (Wiese, 1976; Crandall, 1977).

Bell (1978a) made a more extensive study in a similar vein and found that the trend Knowlton (1974) observed was absent in other series of chlorophyte algae. His conclusion is that among the *Algae* as a whole there is a tendency for the transition from isogamy to anisogamy to be associated with the transition from simple to complex vegetative structure, but that there are many exceptions to this rule, especially in the *Chlorophyceae*. A difficulty is also formed by unicellular taxa containing both isogamous and anisogamous species, like *Chlamydomonas*.

Madsen and Waller (1983) noted that sexual reproduction in many algae is triggered by environmental deterioration and involves the production of a dormant zygospore. This could favor anisogamy irrespective of adult size, since large zygospore size could permit an extended period of dormant respiration or provide reserves for rapid growth after germination. Since situations that favor dormancy could also favor anisogamy, they investigated the relation between gamete dimorphism and characteristic habitat for a number of algal species in the *Chlorophycophyta* and the *Chrysophycophyta*. They found oogamy (anisogamy with immobile female gametes) to be more frequent in lakes. These results are in a sense similar to those obtained by Bell (1978a): the general trend is suggestive, but on the whole the data fail to convince because of the many exceptions. An additional difficulty is to explain why anisogamy is more common in lakes than in pools.

How could the situation with respect to the experimental and observational evidence be improved? The comparative approach as applied in the three studies discussed above seems to have reached its limits. I do not see how we can get much farther along these lines than the position we have arrived at, namely that there is a global trend that with increasing vegetative size and complexity anisogamy tends to replace isogamy, but that within many evolutionary lines this correlation is absent. Perhaps there is a possibility of a more direct experimental study of the relationship between zygote fitness and zygote size.

Apparently in some experimental systems (for example the four isogamous dioecious *Chlamydomonas* species studied by Wiese, Wiese and

Edwards, 1979), it is possible experimentally to induce variation in zygote size, which is normally absent. If this could also be done in a number of anisogamous species, there is in principle the opportunity to test directly the basic assumption of Parker et al. (1972) about zygote fitness in relation to zygote size.

There is one other aspect of this problem which seems worth signalling. In the older literature there are a number of reports saying that within a population a range of gamete sizes can be found and that both gametes of the same size, as well as others of very unequal sizes, are found to fuse. Kalmus (1932) cites the case of *Chlamydomonas multifilis,* where this situation has been found, and for example in the standard work by Fritsch (1965) this is claimed for the green algae *Dangeardinella* (p. 109), *Gonium* (p. 113), and a number of *Ulvaceae* (p. 216), among others. These reports are remarkable in two respects. First, such a range of gamete sizes (more than twofold) has to my knowledge not been mentioned in recent literature, and is also incompatible with the models of Parker et al. (1972) and Hoekstra (1980). Secondly, the existence of not strictly disassortative gamete fusion is not found in the modern literature and is also unexpected on the basis of recent evidence on mating types (Wiese, 1976; Charlesworth, 1978). Perhaps the older data on gamete fusion could be considered to be inaccurate due to less sophisticated research methods than are now in use. But it is not so easy to dismiss the gamete size data in the older literature. Moreover, variation in gamete size due to differences in vegetative growth conditions does not seem unlikely. A possible way to reconcile (at least some of) the old reports with modern insights is as follows: there can be some variation in gamete size around the mean standard size in basically isogamous populations; mating is disassortative with respect to the two (until recently not recognized) mating types; this may give the appearance of both isogamous and anisogamous fusions in the same population. The same interpretation may be obtained in a slightly anisogamous population, where there is some overlap between the size distributions of the two gamete types. Clearly, detailed observations on the species mentioned are needed to decide on this issue.

If it is true in some species that there can be considerable environmental variation in the size of gametes belonging to the same gamete type, then there is much more opportunity for a direct experimental approach to investigating the relationship between zygote fitness and zygote size, which plays such a decisive role in the theories described so far.

*Alternative theories*

In this section recent theoretical developments will be discussed that offer alternatives to (parts of) the Parker et al. (1972) theory.

Wiese, Wiese and Edwards (1979) point out that the random mating assumption in Parker et al. (1972) is unrealistic. They report experimental work on four dioecious isogamous *Chlamydomonas* species, showing that there is in these species a potential for a pronounced phenotypic anisogamy due to the existence of two different pathways of gametogenesis. In both the (+) and the (−) strain of each species, gametogenesis may occur by intracellular differentiation of each vegetative cell into one gamete, or by two or three mitotic divisions of each vegetative cell into four (or eight) smaller sized gametes. The switch to one or the other pathway depends on the physiological state of the vegetative cell at the time of induction of the sexual phase. Wiese et al. (1979) argue that a mutation blocking one of the two pathways would create a strain producing exclusively micro- or macrogametes. This scenario could provide the starting point for evolution towards anisogamy, for example along the lines of Charlesworth's (1978) model with pre-existent mating types. Thus, the hypothesis of Wiese et al. (1979) is not concerned with an explanation of the evolution of anisogamy, but provides a possible mechanism for how variation in gamete size may originate in an isogamous species.

In a later paper, Wiese (1981) considers the possibility of anisogamy evolving in an isogamous monoecious population, where mating is already disassortative with respect to mating type. Each asexually reproducing haploid cell possesses the genetic potential for (+) and (−) differentiation but realizes, upon sexual induction, only one of the two programs. This hypothetical situation is reminiscent of the recently elucidated mating type switch in the yeast *Saccharomyces* (Strathern and Herskowitz, 1979). Wiese furthermore assumes that cell size-associated factors influence the switch mechanism, such that for example cells smaller than a critical size $m_1$ develop into (+) gametes, cells larger than $m_2$ become (−) gametes, and in cells with size $m$, where $m_1 < m < m_2$, additional factors determine the mating type. Such a situation has been described by Von Stosch (1951; 1956), with respect to sex determination in two homothallic (monoecious) oogamous diatom species. Since small gametes will preferentially mate with large gametes of the other mating type, there will be a high degree of disassortative mating from the start of the process of the evolution of anisogamy, and there is no need for a theory like that of Parker (1978) to explain the evolution of disassortative gamete fusion with respect to gamete size. Like the hypothesis of Wiese et al. (1979) discussed above, this hypothesis also provides a scenario of how the various ingredients of anisogamy could arise in a homothallic isogamous population, but fails to indicate what selective forces could bring about the evolution of anisogamy. Thus both hypotheses are valuable in that they suggest possible (and perhaps plausible) mechanisms for the creation of the genetic variation which is a necessary prerequisite for the previously discussed theories to work. But they are not alternative theories for the evolution of anisogamy, since no evolutionary (selective) mechanism is specified.

In a remarkable and stimulating paper, Cosmides and Tooby (1981) discuss many possible consequences of what they call intragenomic conflict between cytoplasmic genes and nuclear genes. Their paper lacks, however, the rigor of mathematical modelling, and it remains to be seen how well their verbal reasoning will stand up against the scrutiny of population genetic models. They offer the following theory for the evolution of anisogamy as an alternative to the theory of Parker et al. (1972). Due to the lack of a strict segregation mechanism for cytoplasmic genes, there will be competition in a zygote between the cytoplasmic factors derived from the two parental gametes for preferential representation in the daughter cells (reviews of population genetic aspects of cytoplasmic inheritance are for example Grun, 1976, and Birky, 1978; 1983). In some systems, such as mitochondrial DNA transmission in yeast, minority alleles in the zygote appear to be selectively destroyed, thus creating a selective pressure for increasing the number of copies of mitochondrial DNA in gametes. Since there is a limit to the number of cytoplasmic DNA copies for a given gamete size, there will be a continuous selection pressure on cytoplasmic genes to make larger gametes in order to attain a larger input bias. A point will be reached where it becomes advantageous for nuclear genes to make many small gametes instead of a few large ones, since the mass of one large gamete is already sufficient for provisioning the zygote. In this way anisogamy will be established. Evolution towards disassortative gamete fusion with respect to gamete size will proceed by selection for cytoplasmic genes to resist fusion with larger gametes, and selection for nuclear genes to resist fusions between two microgametes.

Let us consider a population genetic formalization of these ideas, as follows. In a large population of haploid organisms with a life cycle as shown in Figure 1 there is genetic variation at a cytoplasmic locus $C$, such that $C_1$ codes for gamete size $m_1$ and $C_2$ for gamete size $m_2$, with $m_2 > m_1$. Furthermore, there is also variation at a nuclear locus $A$, such that $A_1$ does not interfere with the cytoplasmic gamete size determination while $A_2$ 'overrules' the cytoplasmic gene, and produces small gametes of size $m_3$ ($m_3 < m_1$). Because of the input bias in the number of cytoplasmic DNA copies, all offspring of a $C_1C_2$ zygote will be $C_2$. The reproductive resources for producing gametes are assumed to be fixed, so that the product of gamete number times gamete mass $n_i m_i$ is constant and the same for all $i$ ($i=1,2,3$). Denoting the relative frequencies of $A_1C_1$, $A_1C_2$, $A_2C_1$ and $A_2C_2$ by respectively $x_1$, $x_2$, $x_3$ and $x_4$, we obtain the following set of recurrence equations

$$\overline{W}x_1' = f_{11}x_1^2 + f_{13}x_1x_3$$

$$\overline{W}x_2' = 2f_{12}x_1x_2 + f_{13}x_1x_4 + f_{22}x_2^2 + f_{23}x_2(x_3+x_4)$$

(26)

$$\overline{W}x_3' = f_{13}x_1x_3 + f_{33}x_3^2$$

$$\overline{W}x_4' = f_{13}x_1x_4 + f_{23}x_2(x_3+x_4) + f_{33}x_4(2x_3+x_4)$$

Where $f_{ij} = n_i n_j w_{ij}$, as in eq. (20), and $w_{ij}$ is the fitness of a zygote composed of gametes with mass $m_i$ and $m_j$. The mean fitness $\overline{W}$ is a normalizing factor, so that $\Sigma x_i' = 1$. For the present purpose it is not necessary to analyze equations (26) exhaustively.

Which stable state the system will reach critically depends on the $f_{ij}$. First consider the case that zygote fitness is independent of zygote size, that is $f_{ij} = n_i n_j$. This assumption implies that the only selection forces involved are selection on cytoplasmic genes to increase gamete size and selection on nuclear genes to increase gamete productivity. When there is only cytoplasmic genetic variation, then it is easy to deduce that $C_2$ will replace $C_1$ provided the difference in gamete size between the two types is less than twofold; otherwise $C_2$ cannot invade, and a stable polymorphism is not possible.

On the other hand, nuclear genetic variation in only one type of cytoplasm leads to fixation of $A_2$. I will not consider here the case of joint cytoplasmic and nuclear variation, since this requires two mutations and is therefore very unlikely.

Next we add an assumption about the relation between zygote fitness and zygote size, so that $f_{ij} = n_i n_j (m_i + m_j)^k$, as in the model of Parker et al. (1972) (see equations (18) and (20)). Starting again with only cytoplasmic variation, it can be shown that $C_2$ will invade the population and replace $C_1$ if $2f_{12} > f_{11}$, or $2\theta > \{2\theta/(1+2\theta)\}^k$. Otherwise $C_2$ cannot invade; no stable polymorphism of $C_1$ and $C_2$ is possible. This brings us to the case of nuclear genetic variation in one cytoplasmic type, and this is identical to the model described above (Formalization and extensions . . .).

I conclude that the scenario for anisogamy evolution as proposed by Cosmides and Tooby (1981) does not work, unless an assumption about increasing zygote fitness with zygote size is added, but that then the model becomes virtually identical to that of Parker et al. (1972). The only difference is that in the Parker et al. scheme the starting point for anisogamy is formed by an isogamous population consisting of microgametes, and that in the Cosmides and Tooby scheme the starting point is an isogamous population of macrogametes.

Schuster and Sigmund (1982) have made the following refinement of the models discussed above. These models assume random gamete fusion, and suppose that the frequency of fusions between gametes of type $i$ and $j$ is proportional to the product of their frequencies $x_i x_j$. However, the number of random encounters also depends on the sizes of the gametes. Using the theory of the rate of diffusion controlled bimolecular chemical reactions, Schuster and Sigmund arrive at the result that the quantity $f_{ij}$ (eq. (20)) in a corrected form is equal to

$$\emptyset_{ij} = c_{ij} f_{ij}, \tag{27}$$

where

$$c_{ij} = [1 + (\frac{n_i}{n_j})^{1/3}][1 + (\frac{n_j}{n_i})^{1/3}] \tag{28}$$

The effect of this correction is that collisions (and hence fusions) between a micro- and a macrogamete are favored above those between two micro-gametes or two macrogametes. This effect is, however, not so strong as to maintain anisogamy on its own, i. e. without needing to assume a positive relationship between zygote fitness and zygote size.

In many algae and fungi the gametes do not meet by chance, but are aided by a sexual pheromone system. The literature on sexual pheromones in algae and fungi has been reviewed by Kochert (1978), and in a wider context by Machlis and Rawitscher-Kunkel (1963) and Van den Ende (1976). Although relatively few species have been inves-tigated, the observational evidence permits the generalization that in aquatic species with motile gametes a pheromonal gamete attraction system seems to be always present in anisogamous and oogamous species but absent in isogamous species. This remarkable correlation between anisogamy and sexual pheromones can be explained in two ways. Either isogamous populations will rather easily become anisoga-mous once a chemotactic gamete attraction mechanism has evolved, or chemotaxis will evolve easily in anisogamous populations but not in isogamous populations. In the first case sexual chemotaxis precedes and facilitates the evolution of anisogamy, in the latter case anisogamy pre-cedes and facilitates the evolution of sexual chemotaxis. This matter has been investigated theoretically by Hoekstra, Janz and Schilstra (1984) and Hoekstra (1984). Hoekstra et al. (1984) consider the evolution of gamete motility differences in the presence of a pheromonal gamete attraction system. The starting point is formed by the supposedly primi-tive situation of an isogamous population in which all gametes produce pheromone and are also able to respond to it. The pheromone diffuses into the water. There is a critical pheromone concentration which is the lowest concentration still perceptible by the gametes. The set of all points of this critical concentration forms a closed convex surface in three-dimensional space. The volume bounded by this attraction surface is called the attraction space.

The swimming paths of the gametes are supposed to be straight lines in random directions; when a gamete moves into the attraction space of another gamete, it will change direction so as to achieve gamete contact. A gamete's probability of finding another gamete in a given time interval is proportional to its swimming speed. A gamete's probability of being found by another gamete is assumed to be proportional to the area of its attraction surface.

Hoekstra et al. show that a stable swimming speed dimorphism may arise via disruptive selection on swimming speed, resulting from selection favoring a high efficiency of finding another gamete as well as a high probability of being found by another gamete. They also show that no more than two different swimming speeds can coexist in a stable polymorphism. Furthermore, a specific relation is deduced between gamete swimming speed and the surface area of the attraction space.

It is concluded that an initial difference in swimming speed of at least twofold will lead to a stable polymorphism, and that evolution towards one of the gamete types becoming non-motile (that is, towards oogamy) is very likely. In the second paper, Hoekstra (1984) analyzes the interaction between dimorphism in gamete swimming speed and anisogamy in a series of population genetic models. One interesting possibility is as follows. A mutation affecting gamete size will also result in a motility difference, since the resistance of the surrounding water is greater against large gametes than against small ones. Is it possible that anisogamy becomes established solely through its effect on gamete swimming speed? Assuming that the gametes have a spherical shape and that their swimming speed is inversely proportional to the area of a section through the center, the relation $V = \theta^{-2/3}$ is obtained, where $V = v_1/v_2$ is the ratio of the swimming speeds of the two types of gametes, and $\theta = m_1/m_2$ is the ratio of their volumes. Since a twofold difference in swimming speed is stable, as shown by Hoekstra et al. (1984), it follows that anisogamy with a gamete size difference of at least threefold can be established in a population without preexisting swimming speed dimorphism solely by its effect on gamete motility. Clearly, this is a scenario for the evolution of anisogamy alternative to that of Parker et al. (1972). No assumptions are needed concerning the relation between zygote fitness and zygote size or concerning the relation between number and size of the gametes. Moreover, if differences in gamete motility already have evolved in an isogamous population, then any mutation that effectively increases the difference in swimming speed between the gametes will be favored, as shown by Hoekstra et al. (1984). This implies that at this stage anisogamy becomes established much more easily, since even small differences in gamete size will be selected. Hoekstra (1984) shows that selection will favor close linkage among the loci affecting gamete motility, gamete size and mating type, so that ultimately the population can be expected to consist of two types in equal frequency: large slow-swimming (or non-motile) gametes producing pheromone, and small fast-swimming gametes responding to pheromone. Below I discuss how to discriminate between this theory and that of Parker et al. (1972) on the basis of empirical evidence.

Cox and Sethian (1985) considered effects of gamete size on rates of gamete encounter using theoretical results from the kinetic theory of gases and from search theory. Apart from this refinement they effectively use the assumptions of Parker et al. (1972). They find that isogamy is favored when gametes are small and the mutant introduced is not much larger; otherwise anisogamy is favored. They furthermore make the obvious point that gamete encounter based on chemotaxis is more efficient than encounter based on random motion, and they show that the particular search pattern shown by male gametes in the aquatic fungus *Allomyces* is optimal in some sense.

## Theories of anisogamy evolution: Conclusions

With the exception of Hoekstra (1984), all theories about the evolution of anisogamy assume two basic selection forces: Selection favoring increasing zygote size, and selection favoring numerical gamete productivity. The latter assumption is in some theories replaced by that of selection favoring increasing efficiency in finding a mating partner, but in these cases it is clear that the authors suppose this increasing efficiency to be realized by the production of more (and smaller) gametes.

The theories differ, however, in the evolutionary mechanism which is supposed to bring about stable anisogamy under these selection pressures, in various details, and in the degree of quantitative precision.

Alternatively, Hoekstra (1984) has suggested the possibility that anisogamy has evolved as a side-effect of selection for increasing differences in gamete motility favoring the efficiency of finding a mating partner through the use of sexual pheromones. This theory applies to species with aquatic mobile gametes and obviously cannot explain the evolution of anisogamy in the red algae *(Rhodophyta)*, which have immobile gametes and are oogamous.

What empirical evidence could discriminate between these alternatives? As discussed above, the comparative evidence concerning the relation between zygote fitness and zygote size in isogamous and anisogamous species is suggestive but not convincing, due to many exceptions to the general trend that isogamy is associated with small and simple vegetative structures, and anisogamy with large and complex structures. Perhaps some progress can be made by investigating this matter experimentally along the lines indicated.

Hoekstra's (1984) theory is in accordance with the general trend that in species with motile gametes anisogamy is associated with the presence of a sexual pheromonal system, and isogamy with its absence. To this trend very few exceptions are known, such as the isogamous *Chlamydomonas* species investigated by Tsubo (1961). There is, of course, also the possibility that the evolution of anisogamy precedes (and facilitates) the evolution of a gamete motility dimorphism, which would also explain the above-mentioned trend. Clearly, theory cannot decide between these alternatives, and what is needed is empirical evidence. As in the case of testing the assumptions of Parker et al. (1972), an experimental approach seems to me at least as valuable as collecting comparative data. For example, it would be very interesting to carry out selection experiments for a high efficiency among gametes of finding a mating partner in isogamous or slightly anisogamous species, and see whether (increased) anisogamy can be obtained as a correlated response.

**Summary**

It is very likely that sexual differentiation into two morphologically indistinguishable mating types has preceded the evolution of anisogamy. Therefore, the study of the evolution of mating types in an isogamous population is more informative for understanding the forces responsible for the evolution of different sexes than the study of the evolution of anisogamy; the latter represents the secondary problem of how, after the establishment of two sexes, an increasing degree of gamete dimorphism may evolve.

Mating type evolution has been analyzed theoretically in population genetic models. These explorations show that mating types may evolve as a consequence of selection for more efficient gamete recognition, and also as a result of intragenomic conflict between nuclear and cytoplasmic DNA. However, in both cases the selection forces have to be very strong, which makes these possible explanations less convincing.

Nearly all theories proposed for the evolution of anisogamy assume two conflicting selection forces to be relevant: selection for greater gamete productivity, and selection for greater zygote size. Although the explanation is intuitively plausible, the comparative evidence is a bit disappointing. Alternatively, anisogamy can be explained as a side-effect of selection for a greater efficiency in finding a mating partner by using sexual pheromones. Firm empirical evidence is lacking, however.

In both problem areas – mating type evolution and anisogamy evolution – experimental work is badly needed.

# Sex determining mechanisms: An evolutionary perspective

J. J. Bull

## Introduction

The beginning of this century marked the onset of studies on the inherited and environmental basis of sex, and a variety of sex determining mechanisms are now known in plants and animals. The two major categories of sex determining mechanisms are dioecy (separate sexes) and hermaphroditism (both sexes within the same individual), but there is further variety within both of these classes. Considering dioecious species, the most widespread mechanism is *heterogamety,* in which one sex is labeled XX and the other XY. Systems slightly more complicated than this, involving three or more factors, are also known. In addition, some species are haplo-diploid, with females arising from fertilized eggs, males from unfertilized eggs. In contrast to these systems, some dioecious species have environmental sex determination, whereby sex is determined in response to environmental effects experienced early in life. With hermaphroditic species, the distinction between male and female is not inherited as such, because both sex types occur in each individual (whether sequentially or simultaneously). However, the extent of male/female expression may be subject to genetic and environmental influences (Charnov, 1982).

The diversity of sex determining mechanisms poses an interesting set of questions to study from an evolutionary perspective (Bacci, 1965; Bull, 1983; Crew, 1965; Mittwoch, 1967; Ohno, 1967). As with any set of biological problems, three types of evolutionary questions may be addressed: what variety occurs, how the variety comes about, and why changes occur. All three levels of investigation are especially appropriate for sex determining mechanisms. For example, at the simplest level, we may consider *what* diversity of mechanisms is actually present. Understanding the existing diversity is chiefly a matter of observation, but more generally, to understand what mechanisms occur, it is often useful first to develop a conceptual framework for the larger set of theoretically possible mechanisms. Observations may then be made to reveal which mechanisms actually occur in nature, and which do not. Sex determining mechanisms are amenable to this analysis because they fall into discrete categories (heterogamety, haplodiploidy, and so forth), so the realm of possibilities is easily calculated.

The observed variety of sex determining mechanisms may then be studied to consider *how* the variety comes about. For example, how might male and female heterogamety arise in closely related species? Are the transitions straightforward or do they require fortuitous combinations of population structure and random effects? Along with investigating how different mechanisms evolve, it is rewarding to consider *why* they evolve: are some mechanisms selectively favored over others? The three evolutionary questions – what, how, and why – may be applied separately to the study of sex determination, but a perspective combining all three levels can lead to a particularly enriched paradigm of study.

This paper briefly reviews some recent work on the evolution of sex determining mechanisms along the lines of the above framework. This review aims to illustrate the different approaches outlined above, and is confined chiefly to two mechanisms, heterogamety and environmental sex determination. Some evolutionary problems concerning haplodiploidy and hermaphroditism will be considered briefly as well. Readers interested in further review and development of the ideas presented here are referred to my recent book on this subject (Bull, 1983). In addition, the question of *why* different sex determining mechanisms may evolve has been considered extensively for the different forms of hermaphroditism versus dioecy in Charnov (1982).

## Sex ratio evolution

A fundamental property of sex determining mechanisms, especially as regards how and why different mechanisms evolve, regards the sex ratio. As first proposed by Fisher (1930), there are a wide variety of conditions selecting a primary sex ratio of ½ in the population (primary sex ratio is the proportion male at conception). Consequently, sex determining mechanisms causing the population sex ratio to deviate from ½ will usually be selected against. This principle is fundamental to the following presentation and so will be elaborated here.

Fisher's argument may be explained as follows. Assume that a population mates randomly, and consider the number of offspring expected from a male zygote versus a female zygote. Suppose, for example, that a male zygote has a greater expected number of children than a female zygote: then selection will favor those genotypes producing an excess of males. Genetic variation permitting the sex ratio will evolve to the point that a male and a female zygote each have the same expected number of children. The relative reproductive success for a male versus a female depends on the population sex ratio. If males and females have identical survival rates and ages of maturity, then equivalence of male and female individual reproductive success is clearly achieved with equal numbers of male and female zygotes (a primary sex ratio of ½). This equivalence is true even with polygamy or any other factor that merely contributes variance to individual reproductive success, since the *average* number of children per male and per female will remain the same. (For example, if half the males acquire two mates and the other half acquire none, the average number of mates per male is ½ by 2, which equals unity, just as if each male acquired one mate.)

Fisher's result holds even with differential mortality of the sexes. Consider a male zygote in a population with a primary sex ratio of ½. If the probability of a male surviving to maturity relative to females is $s$, an *adult* male has an average of $1/s$ offspring times that of an adult female. Since the factors $s$ and $1/s$ cancel, a male and a female zygote have equal average reproductive success despite the differential mortality. Fisher's result does not apply in some special circumstances (Charnov, 1982), but his result appears to be general enough that sex determining mechanisms should often be selected accordingly.

### Evolution of heterogametic mechanisms

Heterogamety is the simplest sex determining mechanism with a consistent genetic difference between males and females: one sex is designated XX (homogametic) and the other XY (heterogametic). The X and Y influence the inheritance of sex in that any zygote inheriting Y develops as one sex, while a zygote wihtout Y develops as the other sex:

$$
\begin{array}{cc}
& \text{Sex} \\
\dfrac{1}{\text{XX}} \quad \dfrac{2}{\text{XY}} & \text{Parents} \\[4pt]
\text{XX} \quad\quad \text{XY} & \text{Offspring}
\end{array}
\tag{1}
$$

The X and Y are referred to as sex *factors*. (A sex factor is generally any gene or gene combination that influences the inheritance of sex.) Heterogamety is therefore more generally referred to as a two-factor mechanism. The notation ZZ/ZW is often used for female heterogamety, but in this review, XX/XY will apply to male and female heterogamety unless indicated otherwise. The symbol Y therefore also represents the W of female heterogamety, and X represents the Z. Species in which Y (or X) is absent are also consistent with this notation, since the labels X and Y serve only to represent the inheritance of sex.

Heterogamety is commonly identified by any of three means (Aida, 1936; Bacci, 1965; Crew, 1965): a) The sex factors may be inherited as part of *heteromorphic* sex chromosomes that are detectable cytologically (the sex chromosomes themselves also being designated as X and Y). b) Allelic variation at sex-linked loci segregates in the fashion depicted in (1), so if this variation is observed, its segregation implies heterogamety. c) Various means, artificial or natural, may be used to cause XY to develop into the sex that is commonly XX; XX may also be induced to develop into the sex that is commonly XY. The sex ratios from crosses of these 'sex-transformed' individuals with normal individuals distinguishes male heterogamety from female heterogamety and may be used in part to infer heterogamety. For example, XX males may be produced in a species with male heterogamety. The mating of these atypical XX males with normal, XX females produces all daughters and is consistent with male, but not female, heterogamety.

Heterogametic sex determination is widespread, occurring in some of the few dioecious flowering plants, in invertebrates such as insects, nematodes, and arachnids, and occurring in the vertebrate classes of fish, amphibians, reptiles, birds and mammals. The vast majority of cases have been identified only through cytological studies of sex chromosomes. Both male and female heterogamety occur in some groups (Diptera, lizards, amphibians; Table 1), only male heterogamety occurs in others (e. g. mammals, arachnids, and nematodes), and only female heterogamety occurs in yet others (birds, snakes, Trichoptera) (reviewed in Bull, 1983).

Table 1. Some groups exhibiting variety in the heterogametic sex

| Taxon | Heterogametic sex | |
| --- | --- | --- |
| | Male | Female |
| 1) Order Diptera (flies) | | |
| F. Anthomyidae | + | − |
| F. Calliphoridae | + | − |
| F. Chironomidae | + | + |
| F. Culicidae | + | − |
| F. Drosophilidae | + | − |
| F. Muscidae | + | + |
| F. Phoridae | + | − |
| F. Simulidae | + | − |
| F. Tephritidae | + | + |
| F. Tipulidae | + | − |
| | | |
| 2) Class Amphibia | | |
| Order Anura (frogs) | | |
| F. Bufonidae | − | + |
| F. Discoglossidae | + | − |
| F. Hylidae | + | − |
| F. Leptodactylidae | + | − |
| F. Pipidae | − | + |
| F. Ranidae | + | + |
| Order Urodele (salamanders) | | |
| F. Ambystomatidae | − | + |
| F. Plethodontidae | + | + |
| F. Proteidae | + | − |
| F. Salamandridae | + | + |

Table 1. Some groups exhibiting variety in the heterogametic sex

| Taxon | Heterogametic sex | |
| --- | --- | --- |
| | Male | Female |
| 3) Suborder Lacertilia (lizards) | | |
| Infraorder Gekkota | | |
| F. Gekkonidae | − | + |
| F. Pygopodidae | + | − |
| Infraorder Iguania | | |
| F. Iguanidae | + | − |
| Infraorder Scincomorpha | | |
| F. Lacertidae | − | + |
| F. Sincidae | + | − |
| F. Teidae | + | − |
| Infraorder Platynota | | |
| F. Varanidae | − | + |

Sources: Bull (1983), King (1977), Schmidt (1983), White (1973), F=family. + indicates that the indicated mechanism of heterogamety is known from at least one species in the taxon; − indicates that the indicated mechanism has not been observed.

## Variety and possible variety of mechanisms

If XX is strictly one sex and XY is strictly the other sex, only two heterogametic mechanisms are possible as regards these genotypes, male and female heterogamety. As noted above, both systems are known in animals, although some major taxonomic groups contain only male or only female heterogamety. Heterogamety has also been studied in other respects: the sex of YY individuals and the sex of aneuploid XO and XXY individuals, and these studies have revealed a wider variety of mechanisms than is evident from merely characterizing the heterogametic sex.

## The sex of YY

The sex of YY is a property of heterogametic sex determination that usually remains hidden because YY is not produced in the population: Y occurs only in the heterogametic sex, hence no zygote can inherit Y from both parents. However, various environmental effects such as temperature, hormones, or tissue transplants have been used to transform the phenotypic sex of XY, and YY can then be produced from matings of XY males with XY females. This analysis is limited to species in which the X and Y are not highly differentiated, since YY individuals are generally inviable in species with degenerate Y chromosomes.

There are two possible outcomes in this analysis: YY and XY are the same sex, or YY and XX are the same sex. Of the heterogametic systems studied in this respect (Table 2), YY and XY are invariably the same sex, as first discovered by Aida for the medaka fish (Aida, 1936). This conformity between different species is expected whether the X is a recessive female-determiner or the Y is a dominant male-determiner, or any intermediate scheme of this sort. However, there is no *a priori* basis for supposing that YY is never the same sex as XX, and there is no explanation of why nature consistently follows this one pattern. (A mechanism that operates on the principle that XX and YY are the same sex is in fact known in Hymenoptera; this mechanism operates with many sex factors under haplodiploidy, hence was not considered along with these heterogametic mechanisms.)

Table 2. The sex of YY individuals

| Taxon | Sex of | | |
|---|---|---|---|
| | XX | XY | YY |
| Fishes | | | |
| *Oryzias* | f | m | m |
| *Xiphophorus* | f | m | m |
| *Poecilia* | f | m | m |
| *Carassius* | f | m | m |
| Amphibians | | | |
| *Xenopus* | m | f | f |
| *Pleurodeles* | m | f | f |
| *Ambystoma* | m | f | f |
| *Hyla* | f | m | m |
| Dipterans | | | |
| *Musca* | f | m | m |
| Crustaceans | | | |
| *Orchestia* | f | m | m |
| *Armadillidium* | m | f | f |
| Plants | | | |
| *Thalictrum* | f | m | m |
| *Asparagus* | f | m | m |
| *Mercurialis* | f | m | m |

Source: Bull (1983), original references and additional details therein. m=male, f=female.

## The sex of XO and XXY

Two possible interpretations of heterogametic sex determination are a) that the Y is a dominant inducer of the heterogametic sex, or b) that the X is a recessive inducer of the homogametic sex (Beermann, 1955). Many other models are possible for the mode of action of X and Y, but these two models have had the appeal that they draw on genetic principles taken across a variety of characters from many species (recessivity, dominance, additivity). Both interpretations are further consistent with the observation that XY

Table 3. The sexes of XO and XXY aneuploids

| Taxon | Sex of | | | Interpretation |
| | XY | XO | XXY | |
| --- | --- | --- | --- | --- |
| **Mammals** | | | | |
| *Homo* (man) | m | f | m | Dom-Y |
| *Mus* (mouse) | m | f | m | Dom-Y |
| **Diptera** | | | | |
| *Drosophila* | m | m | f | Rec-X |
| *Musca* | m | f | m | Dom-Y |
| *Phormia* | m | f | m | Dom-Y |
| *Lucilia* | m | f | – | Dom-Y |
| *Pales* | m | – | m | Dom-Y |
| *Glossina* | m | m | f | Rec-X |
| **Lepidoptera** | | | | |
| *Bombyx* | f | m | f | Dom-Y |
| **Plants** | | | | |
| *Melandrium* | m | – | m | Dom-Y |
| *Rumex* | m | – | f | Rec-X |

Source: Bull (1983), original references and additional details therein. m=male, f=female, – =not studied.

and YY are the same sex. To distinguish these alternative models, individuals with the aneuploid genotypes XO and XXY have been studied (the individuals possessing a normal diploid complement of autosomes). Whereas the study of YY individuals was feasible only in species lacking heteromorphic sex chromosomes, the study of XO and XXY individuals is practical only with heteromorphic sex chromosomes, because the aneuploids are otherwise difficult to generate and to identify. The first such study was undertaken by Bridges on *Drosophila* (Bridges, 1916; 1925). Considering the sexes of XO and XXY, there are actually four possibilities (assume male heterogamety):

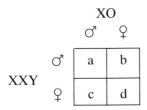

The observations, ranging from mammals to plants, are again confined to only half the possibilities (Table 3): a) recessive-X systems (c above): individuals with a single X are one sex, those with two XX's are the other sex, regardless of the Y. b) Dominant-Y systems (b above): individuals without a Y are the same sex, those with a Y are the other sex, regardless of the number of X's. Thus, even though not all possibilities have been observed, some variety among heterogametic mechanisms exists that is hidden from studies limited to those genotypes occurring in the population (XX, XY).

In summary, there are three characteristics of heterogametic systems that have been studied: the heterogametic sex, the sex of YY, and the sexes of aneuploids, XO and XXY. Only the heterogametic sex is evident in the context of the natural population; the other characteristics are observed only with special studies. These special studies have revealed a further variety of mechanisms than previously known, but some types remain to be discovered.

*How and why the variety evolves*

The study of variety, amid the larger set of possibilities, is interesting in itself, but it further inspires research into the processes influencing the observed variety. Because of their nature, sex determining mechanisms lend themselves to two questions regarding the evolution of variety: how does the change occur from one mechanism to another, and also, why does the change occur? In addressing the first question (how), there are obviously many possibilities that should be considered: a new heterogametic mechanism might evolve from a dioecious ancestor or might evolve from a hermaphroditic one (Charlesworth and Charlesworth, 1980; Westergaard, 1958). Both of these alternatives in turn offer many possible ancestral mechanisms. For example the dioecious ancestor of a novel heterogametic mechanism may have been characterized by environmental sex determination, polyfactorial sex determination, haplodiploidy, or even a different form of heterogamety.

All of these transitions are in fact theoretically possible, but the most common transition in many groups is probably from one heterogametic mechanism to another heterogametic mechanism. The transition, which in principle could involve a variety of intermediates, probably occurs most often through a simple intermediate involving few sex factors – a multiple-factor system. Furthermore, the transition is intrinsically easy (Bull and Charnov, 1977), as elaborated below.

A simple example is the multiple-factor system (2) first described by Beermann (1955) in the midge *Chironomus tentans,*

| Female | Male | |
| --- | --- | --- |
| bb XX | bb XY | (2) |
| | Mb XX | |

This system represents the segregation of sex factors at two regions in the genome (e. g. separate loci). The factors *b* and *M* segregate opposite each other, as do X and Y. If we imagine that the ancestral heterogametic mechanism was *bb XX/bb XY,* this multiple-factor system could have arisen in one step, by a mutation changing *b* into *M*.

The population genetics of this multiple-factor system is straightforward. Only two matings are possible. The mating of *bb XX* with *bb XY* behaves like male heterogamety XX/XY; the mating of *bb XX* with *Mb XX* behaves like male heterogamety bb/bM. If the two male genotypes have equal fitnesses (i. e. equal viabilities, fertilities, and attractiveness to females), both may coexist at any intermediate frequency. For example, if the frequency of *bb XY* males is designated as *p* (and the frequency of *Mb XX* males 1−*p*), the frequency of *BB XY* in the next generation is simply

$$p' = \frac{p/2}{p/2 + (1-p)/2} = p. \tag{3}$$

Hence any value of *p* between zero and unity is an equilibrium, and this multiple-factor system consequently has a continuous path of equilibria connecting a system of pure male heterogamety for XX/XY to a system of pure male heterogamety for bb/bM (Bull and Charnov, 1977).

The change from one heterogametic mechanism to another may occur for either of two reasons, chance and selection. Random changes in gene frequency may cause loss of Y or loss of *M*, for example. Alternatively, an advantage of one male genotype over the other – even a very slight advantage – will cause an increase in its frequency, and if the advantage persists, the favored male genotype will eliminate the other (subject to random effects). When fitness differences are introduced, there is no longer a path of equilibria (as occurred in the equal fitness case), and evolution proceeds toward the genotype of higher fitness. Thus the change from one heterogametic mechanism to another is not difficult in this system. It is easily noted that the change from *bb XY* males to *Mb XX* males involves a change in the locus of heterogamety and could also involve a change from a dominant-Y to a recessive-X system, or the reverse.

Other multiple factor systems are possible. An interesting one was discovered by Kallmann (1968; 1984) in the platyfish,

| Female | Male | |
| --- | --- | --- |
| XX | XY | |
| FX | | |
| FY | YY | (4) |

In this system, all three factors (F, X, and Y) assort in opposition; they are apparently undifferentiated sex chromosomes, because YY platys are viable and fertile. Again, starting with an ancestral system of XX/XY (or FY/YY), this multiple factor system could arise with a single mutation.

The population genetics of the platyfish system (4) is only slightly more complicated than in the midge multiple-factor system (2). With three female genotypes and two male genotypes, there are six possible matings. One of these matings constitutes male heterogamety, XX·YY, and another constitutes female heterogamety, FY·YY. Assuming equal fitnesses of all genotypes within a sex, there is again a continuous path of equilibria between the system of male heterogamety and that of female heterogamety. The sex ratio along this path is ½, even though one mating (FX·XY) produces an excess of daughters and another mating (XX·YY) produces all sons (Bull and Charnov, 1977). It seems remarkable that such a system can abide by the Fisherian sex ratio constraint along an entire path from male to female heterogamety.

There is no intrinsic problem in evolving from male heterogamety to female heterogamety or the reverse in the platy system. Random fluctuations in gene frequency may cause the loss of F or X and thereby establish heterogamety. In addition, certain combinations of fitness differences among the different genotypes also select pure heterogamety. The direction of change in these cases is not always from the ancestral mechanism to a new one; instead the multiple-factor system may degenerate to the ancestral mechanism, but many combinations do lead to the new system. However, the platy system has one property that was absent from the midge system: the possibility of an internal equilibrium in which F, X, and Y are all maintained by selection. Thus selection does not invariably lead to the degeneration of a multiple-factor system (Bull and Charnov, 1977; Orzack et al., 1980; Scudo, 1967a).

The platy and midge systems illustrate how one heterogametic mechanism may evolve to another. Starting with a single mutation, a multiple-factor system is generated in which a second heterogametic mechanism occurs. Evolution of this new mechanism depends on viability and fertility differences among genotypes and on random fluctuations in gene frequencies, but only slight fitness effects of the right combinations are sufficient to enable the new heterogametic mechanism to evolve. Both of the multiple-factor systems studied above have the property that, with equal fitnesses and no random effects, a path of equilibria connects the two heterogametic mechanisms. The points on this path are no longer equilibria when fitness differences or random effects are introduced, but the region of the gene-frequency space near this path serves as a 'corridor' along which evolution proceeds toward equilibrium. Thus, the existence of a path of equilibria connecting two heterogametic mechanisms in the equal fitness case seems to reflect a simple transition between them when fitness differences and random effects are introduced.

The question remains as to whether the platy and midge multiple-factor systems are typical of other possible multiple-factor systems. This problem is investigated in the following way. First, the set of multiple-factor systems that are theoretically possible is enumerated; this enumeration necessarily depends on the number of sex factors and 'loci' allowed. Second, each of these theoretically possible systems is studied with the methods of population genetics to determine whether a feasible transition exists from one heterogametic mechanism to another. A study of this sort has been conducted for some simple classes of multiple-factor systems, and the properties described for the platy and midge systems are common to many, but not all, of the possibilities (Bull, 1983; Scudo, 1964; 1967a). Cotterman (1953) offered the first enumeration of multiple-factor systems relevant to these problems, and Scudo (1964; 1967a) provided some of the first theoretical investigations of the evolutionary maintenance of multiple-factor systems; all three papers offer many important insights to the study of sex determining mechanisms. (Cotterman's enumerations were not cast specifically within the narrow framework of sex determination.)

*Evidence*

The foregoing text offered a framework in which to address the evolution of heterogametic sex determination: the transition from one mechanism to another proceeds through a multiple-factor intermediate. In principle, the transition could occur through other intermediates, such as polyfactorial sex determination, environmental sex determination, and haplodiploidy. Any theory for the evolution of heterogamety must therefore consider whether these other intermediates are important.

The present evidence suggests that changes in the heterogametic mechanism usually occur through multiple-factor intermediates, but the evidence is weak, due chiefly to a lack of knowledge about the transitions. Two types of evidence provide the most powerful tests of the theory: a) observed changes from one mechanism to another, and b) hybridization of closely related populations differing in the heterogametic mechanism (the hybrids may reveal the mechanism that was present during the transition; Avtalion and Hammermann, 1978). The actual transition from one form of male heterogamety to another was demonstrated in a laboratory population of houseflies by Herr (1970); the transition occurred via the midge multiple-factor system (2). This study incorporated selection of DDT resistance, and the linkage of a resistance factor with $M$ rendered the $Mb$ $XX$ males more fit than the ancestral, $bb$ $XY$ males. No other observations of a transition from one heterogametic mechanism to another have been reported.

Populations with different heterogametic mechanism have been hybridized to study sex determination in houseflies and midges (Martin et al., 1980; McDonald et al., 1978; Milani, 1975; Rubini et al., 1972). The observations are usually consistent with multiple-factor inheritance (i.e. only a few sex factors), but the multiple-factor systems have rarely been characterized in detail.

The present theory also predicts that multiple-factor systems should sometimes be observed in taxonomic groups exhibiting a variety of heterogametic mechanisms; alternative possible intermediates should be rare. Of the many theoretically possible multiple-factor systems, only two systems have been characterized in detail from natural populations, the midge system (2) and the platy system (4). The midge system, and simple extensions of it, have been reported in midges, mosquitoes, phorids, and houseflies (Beermann, 1955; Hiroyoshi, 1964; Martin et al., 1980; Milani, 1975; Thompson and Bowen, 1972; reviewed in Bull, 1983). The platy system is known in platyfish and in lemmings, except that YY is inviable in lemmings (Fredga et al., 1977; Gileva, 1980; Kallmann, 1968; 1984). Some other probable multiple-factor systems have been partially characterized from dipterans, but the systems seem to be somewhat more complicated than (2) and (4), and full characterizations have been difficult. The observations are therefore again consistent with the theory at the level of rigor permitted by the data, since changes in the heterogametic mechanism have occurred moderately often in *Xiphophorus* and Diptera (Kallmann, 1984; Table 1, above).

Studies on the nature of sex transformer mutations in nematodes *(Caenorhabditis)*, *Drosophila*, and mammals (man, mouse) add further credibility to the multiple-factor theory (Baker and Ridge, 1980; Cline, 1979; Eicher, 1982; Hodgkin, 1980; Ohno, 1979). In all taxa, presumed single-locus mutations are known with major effect on the sex phenotype, although the transformations are sterile in many cases. In *Drosophila* and the nematode, the mutations seem to comprise a hierarchy of major genes regulating the sex phenotype. The existence of such genes in these species strengthens the plausibility that mutations of major effect on sex phenotype occur in many other species. Furthermore, the failure to detect other forms of inherited effects on sex (i. e. polygenic influence or environmental effects, although a few sex-transforming mutations are temperature-sensitive) casts doubt on the plausibility that other intermediate systems commonly lead to new heterogametic mechanisms.

*Implications*

Several implications of the above theory are apparent. First, there do not appear to be predictable advantages of one heterogametic mechanism over another. For example, male heterogamety does not appear to be intrinsically superior or inferior to female heterogamety. The change from one heterogametic mechanism to another may be fortuitous on several levels: random changes in gene frequency may influence the outcome, as may selection on alleles closely linked to the sex factor – a sex factor may be favored because it happened to arise near a beneficial gene (as proposed by Kallmann, 1968; 1984; for platyfish).

One general prediction about the evolution of heterogamety is that systems with heteromorphic sex chromosomes (a degenerate Y) should be less likely to change than systems in which the X and Y chromosomes are similar. For example, the loss of an ancestral mechanism of male heterogamety requires creating XX males and/or YY males and females. YY will be inviable if Y is degenerate, and XX males may be sterile, as in mammals and *Drosophila*. As noted by Ohno, the X chromosome is apparently conserved throughout placental and marsupial mammals; the Z of snakes (female heterogamety) is also apparently conserved (Ohno, 1967). These conserved sex chromosomes are highly differentiated in both groups and are thus consistent with the above hypothesis, although alternative theories are also plausible (Bull, 1983).

An unsolved problem is the ubiquity of heterogamety in contrast to the rarity of multiple-factor systems and other systems with many sex factors (polyfactorial sex determination). When new sex factors arise repeatedly, heterogamety might be expected to yield to systems with more and more sex factors, yet this accumulation does not seem to occur. Some explanations have been proposed (Bull, 1983), but none are at present compelling, and there has been no careful study of this problem with respect to any of the hypotheses.

The present framework is useful in addressing correlations between the heterogametic sex and various other species' characteristics (e. g. the dominant gonad, viviparty, H-Y antigen – Mittwoch, 1973; Nakamura et al., 1984; Wachtel, 1983). When such a correlation can be demonstrated to be meaningful in an evolutionary sense (i. e. when the pattern has evolved independently too often to be ascribed by chance), two general explanations of the association may be considered. On the one hand, the association could be an invariant consequence of heterogamety (of heterozygosity per se); in this case the association with heterogamety should be preserved during the actual transition from one heterogametic system to another. In the platy system, for example, XY males should display the same characteristics as FY and FX females. An alternative explanation is that the association is not necessarily preserved during the evolution to a new mechanism, but that it arises after any new heterogametic mechanism evolves. These two hypotheses were considered in a nice paper on the association between H-Y antigen and the heterogametic sex, with special reference to platyfish: Nakamura, Wachtel and Kallmann (1984) found that H-Y expression was confined to males rather than to heterogamety; in this species the male was H-Y positive under both male heterogamety (XX/XY) as well as under female heterogamety (FY/YY).

## Environmental sex determination

At the other extreme from heterogametic sex determination is environmental sex determination: a zygote develops as male or female depending on an environmental effect experienced early in life. Whereas in heterogamety, a zygote's sex is determined by genotype virtually regardless of environmental conditions, here sex is determined by environmental conditions regardless of genotype. A necessary consequence of environmental sex determination (ESD henceforth) is that there are no consistent genetic differences between males and females. Thus, the term 'heterogametic sex' does not apply to ESD.

*Variety*

Mechanisms of ESD may be classified according to the environmental factors influencing sex and according to the relationship between those factors and sex ratio, although such categories necessarily lack sharp boundaries. A brief description of some known ESD mechanisms illustrates a considerable variety, however. In virtually all of these cases, the environmental effect on sex has been shown to be due to sex determination rather than to differential mortality of the sexes.

*Reptiles*

Some of the most recent discoveries of ESD have been in reptiles. The incubation temperature of the egg determines sex in various species of turtles, lizards, and crocodilians (reviewed in Bull, 1980; 1983 and Mrosovsky, 1980). The first discovery of this phenomenon was in a lizard by Charnier (1966), but the phenomenon was brought general recognition by the work of Pieau on two European turtles (Pieau, 1971; 1972; 1974; 1975a; 1982), and later by Yntema (1976) on another turtle. The temperature influences on sex occur under natural as well as laboratory conditions (Pieau, 1982; Vogt and Bull, 1984). More recently, the moisture available during development was also shown to influence sex determination in a species with temperature-dependent sex determination, so there may be a wide range of environmental effects on sex, with temperature the predominant one (Gutzke and Pankstis, 1983).

The magnitude and form of the temperature effect on sex varies between species (Figure). In many species a wide range of temperatures produces only one sex, another range produces only the other sex, and a narrow range produces both sexes. However, in some turtles, low temperatures produce males and high temperatures produce females, whereas the reverse pattern occurs in alligators and some lizards. Furthermore, in some other turtles and a crocodile, another pattern occurs in which females are produced at low and high temperatures, males in between. Finally, some species show no temperature effects on sex (which presumably applies to the many reptiles with sex chromosomes, although no direct study of temperature effects has been undertaken in most of these species). Further details of these systems are reviewed in Bull (1983).

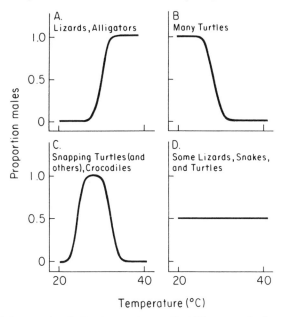

The relationship between incubation temperature and hatchling sex ratios for various reptiles. Four patterns are observed: *A* Females develop at low temperatures, males at high ones, as in two lizards and alligators. *B* males develop at low temperatures, females at high ones, as in many turtles. *C* Females develop at both high and low temperatures, males only at intermediate ones; this pattern is known in several turtles and a crocodile. *D* Incubation temperature does not significantly influence hatchling sex ratios in some turtles, snakes, and lizards. Redrawn from Bull, 1983.

### Mermithid nematodes

The second-oldest report of ESD in animals is from mermithid nematodes (Caullery and Comas, 1928; Christie, 1929). As adults, mermithids are free living, but as juveniles they are internal parasites of insects, deriving their nutrition from the insect's internal fluids. The parasitic phase of the nematode is very efficient at extracting resources from the host, and usurpation of the host is so extensive that the nematode's size (and presumably other measures of its quality) is strongly influenced by the size and nutrition of the host.

The nematode's sex is determined in response to the resources in extracts from the host: worms develop as female in the larger, uncrowded, and/or well-nourished hosts, and develop as male in the more poorly-nourished hosts (Bull, 1983; Caullery and Comas, 1928; Charnov, 1982; Christie, 1929; Petersen, 1977). Nematode mating takes place outside the host, so that there is no selection to produce both sexes within a host.

## Bonellia

Baltzer's report of ESD in the marine echiurid worm *Bonellia viridis* constituted the first valid case of ESD in animals (Baltzer, 1912). The larvae of *Bonellia* are initially planktonic but then settle on the marine substrate to become adults. Adult females are approximately the size of a walnut when contracted but have a long, extensible proboscis (1 m) used to filter-feed; males are tiny, living a parasitic existence in the reproductive tracts of females. Sex is determined when the larva settles from the water column: male development is induced by exposure to females (and also by exposure to female extracts and to various chemical effects); larvae settling in isolation develop as females (Baltzer, 1912; Leutert, 1975).

## Silverside fish

The Atlantic silverside, *Menidia menidia,* has temperature-dependent sex determination, although not to the extent of reptiles (Conover, 1983; Conover and Kynard, 1981). Eggs are spawned externally, and sex is determined by water temperature late in the larval phase: cool temperatures overproduce females, whereas warm temperatures produce males (as in lizards and alligators). The difference in sex ratio between the temperature extremes is small, only 0.2 to 0.4, and there appear to be large genetic effects on sex determination as well as the environmental ones.

## Others

Environmental effects on sex determination are also known in other species, although the resulting sex ratio variations are either small or their importance in natural populations has not been clarified. a) Bulnheim (1978) reported that the sex of the marine amphipod *Gammarus* was sensitive to photoperiod; major genetic effects on sex ratio were also evident. b) Laboratory-induced temperature effects on sex determination have been demonstrated in various amphibians that normally have heterogamety (Dournon and Houllion, 1983; Pieau, 1975b; Richards and Nace, 1978). c) The pH of the water was reported to influence sex determination in various fish (Rubin, 1985).

In summary, ESD as a natural mechanism or as a minor component of genotypic sex determining mechanisms seems to be a widespread phenomenon, but the environmental factors influencing sex vary between species. The following sections will consider why these various cues may be used for sex determination.

## How ESD may evolve from GSD

The evolution of ESD from GSD or the reverse poses no intrinsic problem. To illustrate, imagine that a species with male heterogamety experiences an environmental effect on sex such that some XX zygotes become male:

| Female | Male |
| --- | --- |
| XX | XY |
| | XX |

$$(5)$$

The presence of XX males selects a particular frequency of XX and XY zygotes, the actual value depending on the level of environmental influence. XY is selected to lower levels as the influence increases, and XY is selected out of the population if half or more the XX zygotes become male. The sex ratio is ½ at the equilibria, and once again, there is a continuous path of equilibria from one sex determining mechanism to another if fitnesses are equal for all genotypes within a sex. Here the path extends from strict GSD (no XX males occur) to strict ESD (all males are XX).

The path of equilibria is somewhat different here than in the multiple-factor systems above. The path of equilibria from heterogamety to ESD is a function of changes in the level of environmental influence, and it might seem that the level of environmental influence is imposed by external conditions and cannot evolve. However, there is in fact a direct parallel with the previous cases: the level of environmental influence on sex determination may evolve through changes in the frequencies of genes modifying and individual's susceptibility to environmental effects on sex. For example, in *Drosophila,* in the nematode *Caenorhabditis,* and in a mosquito, temperature-sensitive sex transformer genes are known that convert XX from male to female (Bull, 1983; Craig, 1965; Hodgkin, 1980). Increasing the frequencies of these genes increases the frequency of individuals that become XX male in response to environmental conditions. The response to environmental factors can be influenced genetically, and this model for the evolution of ESD from GSD is therefore similar to those above in that a path of equilibria exists from one mechanism to another through changes in the frequencies of appropriate genes.

*Why ESD?*

Whereas one type of heterogametic mechanism did not appear to have a systematic advantage over any other, there do appear to be relative advantages of ESD and GSD. A model in which ESD is favored over GSD was proposed by Charnov and Bull (1977) along the lines of a sex ratio model proposed by Trivers and Willard (1973). The model involves three parts. First, a species inhabits a 'patchy' environment, so that offspring reared in some patches have a different lifetime fitness than individuals reared in other patches. This patchiness further influences male fitness differently than female fitness, some patches being more beneficial to females than males, and other patches doing the opposite. Second, offspring cannot choose patch type, and parents cannot preferentially put offspring of one genotype in patches of one type. Third, mating takes place among individuals reared in different patches.

These three conditions select ESD over GSD (Bull, 1983; Bulmer and Bull, 1982; Charnov and Bull, 1977). An individual developing in a patch relatively beneficial to females leaves more offspring if it develops as female, whereas the converse applies to patches beneficial to males. Therefore, if an individual cannot choose patch type, it leaves more offspring by delaying sex determination until it has entered a patch, so that it may become the appropriate sex. Genotypic sex determination is selected against because it causes some individuals to develop as female in patches beneficial to males and vice versa.

Consider now whether GSD might be favored over ESD. The Charnov-Bull model does not offer any advantage for GSD, since the model was based on the premise that ESD allowed the evolution of sex ratios that GSD could not achieve. However, two possible advantages of GSD can be suggested on other grounds (described here as disadvantages of ESD). First, because sex is determined late in development and in response to environmental conditions, ESD might have two undesirable effects: intersexes might result, and the developmental onset of sexual dimorphism must be delayed until sex has been determined. A second disadvantage of ESD results from its susceptibility to sex ratio variation. Systematic environmental fluctuations from year to year may cause at least mild fluctuations in sex ratio under ESD. In the absence of the effects described in the Charnov-Bull model above, this sex ratio fluctuation selects GSD (Bull, 1981; Bulmer and Bull, 1982).

*Evidence*

Are the observations on ESD consistent with the Charnov-Bull model of a patchy environment? The model requires a) that the environmental determinant of sex correlate with some differential effect on male/female fitness, and b) that males (females) develop under conditions relatively beneficial to themselves.

In some cases the observations appear to be consistent with the model. In *Bonellia,* for example, one patch type consists of open substrate, and the other consists of adult females. A larva developing as male in open substrate would have negligable fitness (unless, perhaps, it was not dwarf); a larva developing as female inside of another female would also have low fitness (although a larva destined to be female might be able to avoid settling in another female). Thus, the patch types have differential benefits to males and females, and sex is determined accordingly. In mermithids, the observations are again potentially consistent with the model. Females develop under conditions of good nutrition. The model is consistent only if fitness as female is enhanced more by large size than is fitness as a male. These conditions seem to be met, although the evidence is circumstantial. (Despite the inadequacy of the data at present, it would be difficult to argue that males benefitted *more* from large size than females – Bull, 1983.)

Conover (1983) specifically investigated silverside ESD in the context of the Charnov-Bull model; the model was tentatively supported. Conover observed that the waters inhabited by silversides were cooler early in spring than late in spring, and fish born early grew to a larger size before winter. Thus water temperature during development correlated with adult size. Conover further showed that fitness as a female was enhanced by large size; various components of male fitness did not seem to be so enhanced by large size, although male fitness could not be estimated as directly as female fitness could. At the level of these observations, the model is supported, because females are overproduced at cool temperatures – the temperatures that enable them to reach the largest size as adults.

Reptiles offer the greatest enigma from this perspective. There is no *a priori* reason to suggest that incubation temperature has any effect that would persist into adult life, much less one that differentially affected fitness as male versus female. (Temperature affects embryo survival, but the effect probably does not differ between males and females.) Some recent work has shown that various environmental conditions during incubation affect the size and yolk reserves of surviving hatchlings (Packard et al., 1981), but again it remains to be demonstrated whether these effects translate into ultimate fitness differences and whether they have different consequences for males and females. At present, the null hypothesis that the Charnov-Bull model does not apply to reptiles stands unchallenged.

## Haplodiploidy

An interesting and common genetic system in arthropods is haplodiploidy: males develop from unfertilized eggs, females from fertilized eggs (also known as arrhenotoky). Haplodiploidy is known in all sexual Hymenoptera (ants, bees, wasps), in mites and ticks, whiteflies, thrips, scales, and some beetles (Bell, 1982; Bull, 1983; Hartl and Brown, 1970; Oliver, 1971; Schrader and Hughes-Schrader, 1931; White, 1973; Whiting, 1945). This section briefly outlines some major points about the possible evolution of haplodiploidy.

Consider heterogametic sex determination as a possible immediate ancestor of haplodiploidy. The origin of haplodiploidy requires that unfertilized eggs develop as male, and the heterogametic mechanisms in which all unfertilized eggs develop as male are as follows (Bull, 1983):

| 6 | | 7 | | 8 | |
|---|---|---|---|---|---|
| ♀ | ♂ | ♀ | ♂ | ♀ | ♂ |
| XX | XY | ZW | ZZ | ZW | ZZ |
|  | X |  | Z |  | Z |
|  |  |  | W | WW | W |
|  |  |  | WW |  |  |

Here, uniparental offspring are represented as haploid; XX/XY represents an ancestral mechanism of male heterogamety, ZZ/ZW female heterogamety.

All three systems violate any notion of recessive-X and dominant-Y principles, and they have the property that the sex factors causing males come exclusively from females of the previous generation. From studies on the sexes of YY individuals and of XO and XXY aneuploids, as described above, these three systems, 6, 7 and 8, are presumably rare among heterogametic species. Thus the very origin of haplodiploidy may be precluded by a lack of appropriate sex determination. (Several other requirements – viability and fertility of the haploid – may also prevent evolution of haplodiploidy – Hartl and Brown, 1970; Whiting, 1945).

Assume that a species has one of the three systems conducive to haplodiploidy, 6, 7, or 8, and consider the transition from heterogamety to haplodiploidy. If some eggs fail to be fertilized, the population contains haploid and diploid males (referred to as uniparental and biparental males, respectively). Hartl and Brown (1970), showed for system 6 that the equilibrium frequency of these two types of males is governed by two parameters, the fitness of uniparental males relative to biparental males, and the fraction of eggs developing without fertilization. In system 6, XY males are lost if enough of the eggs go unfertilized and if uniparental sons are sufficiently fit. With the loss of Y, all males are X (all of them then arise from unfertilized eggs), and haplodiploidy is established. An equivalent scenario applies to system 8, in which the loss of Z establishes haplodiploidy. (Obviously, systems 7 and 8 require that W and WW individuals be viable and fertile.)

Surprisingly, system 7 is only slightly different. Since all haploids and diploid homozygotes are male, diploid males cannot be lost (homozygotic offspring are unavoidable). However, suppose that additional factors with this type of sex determination can arise, so that this mechanism is more generally

$$
\begin{array}{cc}
\text{\Large ♀} & \text{\Large ♂} \\
\hline
A_iA_j & A_iA_i \\
 & A_jA_j \\
 & A_i \\
 & A_j \\
\end{array}
$$

for $i \neq j$ $(i, j = 1, \ldots, n)$.

This system is in fact the one known from several hymenopteran species (Bull, 1983; Crozier, 1977). If enough of these factors are present, the evolution of uniparental males obeys the same principles that Hartl and Brown discovered in system 6, and haplodiploidy can evolve (homozygous males become vanishingly rare when many factors occur in the population).

In conclusion, there are simple transitions from heterogamety to haplodiploidy, just as transitions were found between other sex determining mechanisms. In this case the transition depends on two parameters, the fraction of unfertilized eggs and the relative fitness of uniparental males. As in the model for the change from GSD to ESD, the evolution of haplodiploidy therefore appears to depend on a parameter that is largely environmental, in this case the fraction of unfertilized eggs. Once again, however, this parameter is subject to evolution. *A female may be selected to avoid fertilizing her eggs to produce uniparental sons.*

This last point may be the most important principle in the evolution of haplodiploidy, given that all the prerequisites have been met (sex determination, viability and fertility of uniparental males – Bull, 1979). Consider a mother producing a uniparental son versus a biparental son. When the uniparental son produces gametes, every gene locus in all of his gametes will carry her alleles. However, when the biparental son produces gametes, only half of the loci in his gametes will, on average carry her alleles; the other half will carry her mate's alleles.

Will selection therefore favor mothers who produce uniparental sons? The answer depends on the fitness of uniparental versus biparental sons. Denote $w$ as the fitness of a uniparental son relative to that of a biparental son. The uniparental son transmits $2w$ maternal alleles for every one transmitted by a biparental son. Thus if $2w > 1$, or $w > \frac{1}{2}$, selection favors mothers who avoid fertilizing eggs and produce uniparental sons. Given this condition, the evolution of haplodiploidy from diploidy automatically follows in all the above systems 6, 7, and 8 (Bull, 1983).

*Hermaphroditism*

The last case to be discussed here concerns hermaphroditism. The evolution of hermaphroditism from dioecy, or the reverse, has been addressed mostly from the perspective of *why* one system is favored over the other (Charnov, 1982). Briefly, the theory of sex allocation predicts that pure hermaphroditism is favored over dioecy when two conditions are met: a) the reproductive success realized by the hermaphrodite through male function is more than half that of a pure male, *and* b) the reproductive success realized by the hermaphrodite through female function is more than half that of a pure female. For example, if different resources are used to make male gametes than to make female gametes, a hermaphrodite may be able to make almost as many male gametes as a pure male and almost as many female gametes as a pure female; hermaphroditism would be favored in this case if reproductive success was limited chiefly by the number of gametes produced. Further details of these arguments are to be found in Charnov (1982).

The question of how (and why) the transition occurs between hermaphroditism and dioecy has been addressed in one elegant case that warrants description here. The transition from hermaphroditism to dioecy has occurred in several flowering plants, and the genetic basis has been worked out in some cases (Charlesworth and Charlesworth, 1980; Lloyd, 1974; Westergaard, 1958). According to Charlesworth and Charlesworth (1980), the most plausible selective force for the transition is inbreeding depression: the hermaphrodite may not be able to avoid selfing, and the selfed offspring may have depressed fitness from the high incidence of homozygosity for otherwise rare, deleterious genes (inbreeding depression). If the rate of selfing is high enough and the inbreeding depression severe enough, a pure-female mutant can invade – a plant that suppresses pollen production. As this female-type increases, the conditions relax for invasion of a pure male (a plant suppressing ovule production). However, to avoid generating plants that suppress both pollen and ovule production (which would be sterile), both factors must sort in opposition; heterogamety results. Charlesworth and Charlesworth's model further explains the preponderance of male heterogamety over female heterogamety in these plants.

## Summary

Theories on the evolution of sex determining mechanisms are reviewed for male and female heterogamety, environmental sex determination, and briefly, haplodiploidy and hermaphroditism. Because of their discrete and well-defined nature, sex determining mechanisms lend themselves to three types of evolutionary questions: *what* variety occurs and might be expected but does not occur, *how* do changes occur from one mechanism to another, and *why* do certain changes occur? All three approaches are illustrated for these different sex determining mechanisms. A generality emerging from these studies is that, at the level of selection on the sex ratio, there are no intrinsic problems in evolving from one sex determining mechanism to another: straightforward transitions between different mechanisms exist under various conditions.

# Two theories of sex and variation

G. Bell

For almost the whole history of evolutionary biology, sex has been thought of as creating preadaptation to an uncertain future, either permitting species to adapt more quickly or enabling individual females to produce a few unexpectedly fit offspring. The demise of this powerful idea dates from theoretical difficulties noted by Maynard Smith (1971 a) and was completed by the overwhelming hostility of the comparative evidence: it is parthenogenesis, and not outcrossed sexuality, that prevails in harsh, uncertain, disturbed and novel conditions (Bell, 1982). The comparative evidence points instead to a quite different role for sex, concerned with the efficient exploitation of the full range of possibilities presented by a diverse environment. Two theories of this sort have been especially prominent. The first is the Tangled Bank (Bell, 1982), which descends from the economic analogies of Ghiselin (1974) and is related to the sib-competition models introduced by Williams (1975) and elaborated by Maynard Smith (1976a) and Price and Waser (1982). In its simplest form, the Tangled Bank holds that the state of the environment varies widely from place to place on a very local scale, such that different genotypes are optimal at different sites. Since each site can support only a few individuals, the uniform progeny of an asexual female will compete intensely with one another for the same set of resources, while the progeny of a sexual female, which by virtue of their diversity will be able to exploit a much wider range of sites, will compete amongst themselves less intensely and thus achieve greater overall success. In this way, a narrowly specialized asexual clone cannot replace a diverse sexual lineage, despite its greater reproductive efficiency.

A formal population model (Bell, 1982; and Bell, in preparation) identifies two processes which are crucial to the maintenance of sexuality. Imagine a subdivided population with each local group contributing to a common pool of dispersive propagules. Each local group inhabits a diverse environment, in which the performance of a given genotype varies from site to site. Because the high local fitness of a successful genotype may not be reproducible when it is transferred to some other random site, it may be advantageous to break up even successful genotypes by recombination. At the same time, the total contribution made by each local group to the common pool will depend on the effectiveness with which the whole range of available sites is exploited, and will therefore increase as the genetic variance of the group increases. Mixed groups created by sexual diversification will therefore outyield genetically uniform clones. The Tangled Bank therefore involves both interaction between genotype and environment and interaction between genotypes.

The Red Queen, descending eventually from the theme introduced by van Valen (1973), was first adequately stated for the special case of sex by Jaenike (1978) and later formalized by Glesener (1979), Hamilton (1980), Hutson and Law (1981), and Bell (1982). It attributes the success of sexuality to the continual necessity to respond to the shifting challenge of antagonists such as predators and parasites. Since such antagonists will always be able to counteradapt eventually to any given genotype in the prey or host population, sex is necessary to recreate resistance in the progeny by producing new combinations of genes. Sex is equally necessary to the predators and parasites, which need to respond effectively to these counteradaptations, and the cycle of adaptation and counteradaptation continues indefinitely.

In formal population models, the crucial feature of the Red Queen is the negative correlation between the fitness of a genotype and its frequency at some time in the past. If the fitness of a genotype defined by the alleles present at two (or more) loci is a decreasing function of its past frequency, then not only are the allele frequencies at these loci likely to cycle through time, but the coefficient of linkage disequilibrium will also cycle, changing regularly from positive to negative and back again. This greatly favors the maintenance of sexuality, since the passage from coupling to repulsion genotypes, or vice versa, is greatly facilitated by meiotic recombination. We can validate this conclusion by building a model with three loci; current fitness is determined at two of the loci, and for any given genotype is a decreasing function of the past frequency of that genotype, while the third locus is segregating for a series of alleles whose only effect is to alter the rate of recombination between the two fitness loci. A model of this sort was analyzed by Bell (1982), who found, as expected, that alleles at the third locus – which suppressed recombination – were counterselected, and that in general alleles favoring intermediate rates of recombination were favored, confirming the analytical results of Hutson and Law (1981).

The Red Queen can be restated in terms which make it directly comparable with the Tangled Bank. There is an interaction between genotype and environment, though 'environment' is in this case a year (or some other appropriate measure of time) rather than a site. There is also an interaction between genotypes, since a population whose genetic composition is continually changing will outyield an invariant clone. The two theories thus equally involve complex genetic interactions, but differ in the source of these interactions. This perspective immediately suggests that the analysis of variance components in short-term experiments and observations may both show whether either theory is plausible, and may help to distinguish between them.

The Tangled Bank and the Red Queen are not merely two possibilities drawn by historical accident from a long list of possible theories. Rather, they represent two alternative opinions about variation, the former stressing spatial and the latter temporal heterogeneity in the environment. It will be a recurring theme in this review that the problem of how sex is maintained and the problem of how variation is maintained are, if not identical, at least very closely related. The dichotomy between these two theories is therefore of great interest. I have previously published an extensive comparative analysis of theories of sex (Bell, 1982), and shall not discuss the comparative evidence further here. Instead, I shall show how short-term experimental and observational techniques can be used to evaluate the worth of the Tangled Bank and Red Queen theories, independently of comparative predictions.

**Variation in space and time**

The basic concept invoked by both the Tangled Bank and the Red Queen is that the relative fitness of a genotype depends on the conditions in which it is reared. More precisely, when the relative fitnesses of a number of genotypes are measured under different experimental treatments there will be a substantial genotype x treatment interaction, showing that the ranking of relative fitnesses changes with treatment. This interaction is the familiar genotype x environment (GxE) interaction of quantitative genetics; I shall often refer to 'treatment' rather than 'environment' when the variance is created by some premeditated manipulation.

When a series of genotypes is raised under two treatments, their performances in the two treatments will be correlated to some extent. If there is no $G \times E$ interaction, this correlation will be large and positive, being less than $+1$ only because of measurement error. Theories of sex, whether Tangled Bank or Red Queen, require that the correlation should be zero or negative, at least for certain combinations of treatments, since it is only when the ranking of relative fitnesses tends to be reversed that there is a straightforward advantage for breaking up the parental genome.

This argument applies to the performance of genotypes when grown in pure culture, and deals with the interaction between genotype and treatment while ignoring any interaction between genotypes when grown together in mixed culture. If sex is to be favored, interactions between genotypes should occur, and must lead to the superiority of mixed over pure cultures with respect to yield, rate of increase, or some other measure of performance directly related to Darwinian fitness. This will be the case if different genotypes facilitate one another's growth, while individuals having the same genotype are antagonistic.

These two arguments suggest that there are two basic experimental procedures for detecting possible short-term advantages for sex and variability.

1) Transplantation. If there is a large $G \times E$ interaction then the relative

performance of a genotype will be altered substantially by moving it to a random site or by growing it under different random conditions. The Tangled Bank and the Red Queen differ with respect to the sort of transplant which is expected to produce this effect: essentially, the Tangled Bank predicts a large effect from transplanting between sites, and the Red Queen from transplanting between years. Another way of putting this is to say that if there is a large $G \times E$ interaction, the Tangled Bank predicts that the variance of treatments between sites will be much greater than that between years, while the Red Queen makes the reverse prediction. Naturally, the units of time involved are not necessarily years, and the definition of a site will also vary according to the organism involved.

2) Mixture. If there is a large positive interaction between genotypes then the combined yield of a mixture will exceed the average yield of its constituents in pure culture. The Tangled Bank and Red Queen differ in the type of mixture expected to be the more effective; the Tangled Bank predicts a large effect from a mixture of genotypes at a site or sites within a given year, and the Red Queen from a mixture over time, each site being occupied by a pure culture in any given year, the genotype of the culture being changed between years.

The literature of interactions between genotypes and between genotype and environment is very large, with papers scattered through the fields of population genetics, biometrical genetics, evolutionary biology, agronomy and stockbreeding. No critical review exists, and it would be impracticable to attempt an exhaustive treatment here; instead, I shall attempt to identify the major categories of evidence and survey some of the major empirical findings.

## Interaction between genotype and treatment

*The occurrence and magnitude of $G \times E$ interaction.* Most large-scale studies of variation involving many combinations of strains and treatments turn up substantial interactions. Some of the most extensive work has been done with crop plants and domestic animals. Blyth et al. (1976) analyzed data on 49 wheat cultivars grown in 63 environments and found that about 22 % of the total variance in yield was contributed by $G \times E$ interaction. In another very large experiment with cauliflower, involving 12 genotypes and 36 environments, about 10 % of the total variance in time to maturity and about 42 % of the variance in the weight of the flowering head was contributed by $G \times E$ interaction (Kesavan et al., 1976). A similar unpublished study on lettuce involving 13 varieties in eight field environments (Gray, cited by Kesavan et al., 1976) found that only about 2 % of the variance in time to maturity but about 24 % of variance in head weight was attributable to $G \times E$ interaction. Witcombe and Whittington (1971) found a substantial interaction for germination rate in *Brassica* seedlings grown at different temperatures.

In some cases, the results are less consistent. For characters of economic importance in poultry, for instance, some authors (e. g. Hill and Nordstrog, 1956; Nordstrog and Kempthorne, 1960; and Osborne, 1952) found rather large and consistent G × E interactions, but others found interactions for some characters and not for others (e. g. Gowe, 1956; and Gutteridge and O'Neill, 1942), while others found no interactions at all (Abplanalp, 1956; Gowe and Wakely, 1954; Lowry et al., 1956). In general, characters which are close to biological fitness (such as seed yield in cereals) usually seem to have substantial G × E interaction, while characters less highly correlated with fitness (e. g. fleece weight in sheep, King and Young, 1955) often have little or none. The retention of substantial interaction variance in many economically important characters, however, is doubly significant because one of the main aims of selective breeding is to remove this source of variance in order to create strains which perform uniformly well in a wide range of environments.

Extensive work on variation in plants under non-agricultural conditions has been done on *Nicotiana, Arabidopsis* and *Papaver.* Perkins and Jinks (1968) grew 10 inbred lines of *Nicotiana* in 16 soil environments and found small but significant line × soil type interactions, which accounted for about 4 % of the total variance in time to flowering and final height. In a large and carefully analyzed experiment, Pooni et al. (1978) found widespread genotype × microenvironment interactions in *N. rustica.* In *Arabidopsis,* large experiments involving inbred lines and their hybrids have shown substantial G × E interaction (Pederson, 1968; Westerman, 1971; Westerman and Lawrence, 1970). Westerman and Lawrence remark that although the G × E interaction is small relative to the total phenotypic variance, it is large relative to the genetic variance. Zuberi and Gale (1976) studied 11 metrical traits in 20 inbred lines of *Papaver dubium* grown under 16 combinations of fertilizer treatment, and found large G × E interactions for three of the traits, including fruit number, the trait closest to fitness; they comment that the relative fitness of different genotypes will probably change very markedly with changes in the natural environment.

Among fungi, Fripp and Caten (1971) found substantial G × E interactions for the growth-rate of *Schizophyllum commune* dikaryons, while Butcher et al. (1972) obtained a similar result for comparisons of growth-rate and cleistothecial production between compatibility groups of *Aspergillus nidulans.*

Relatively little work on animals has been directed to the measurement of G × E interaction, but it has usually been found when it has been looked for. Young (1953) found substantial interaction for body weight and litter size of three strains of mice reared under four combinations of temperature and food levels; Falconer and Latyszewski (1952) got a similar result for two strains and two food levels. Laboratory populations of fruitflies and flour beetles often show G × E interactions, which are described under the appropriate headings below.

The examples I have described above are only a small fraction of those it would be possible to extract from the literature. They suffice, however, to show that the G × E interaction is often substantial, and therefore that the performance of a genotype under one set of conditions will often be a poor predictor of its performance under different conditions.

*The correlation between genotype and environment.* If the relative fitness of genotypes varies between sites the result is likely to be local adaptation, with different genotypes prevailing in different types of site. Thus, G × E interaction is likely to lead to G × E correlation, and the existence of this correlation is indirect testimony to the existence, and the strength, of the interaction. Local adaptation has been the subject of some of the best empirical work in evolutionary biology, including the classical studies of heavy metal tolerance in *Agrostis,* industrial melanism in moths and crypsis in the snail *Cepaea.* These studies are especially valuable, both because they have identified the selective agents responsible for the G × E correlation and because they have shown that selection coefficients can change drastically over very small distances. The Park Grass Experiment at Rothamsted is particularly revealing, since this concerned differences in soil nutrient status which are directly relevant to the spatial heterogeneity commonly encountered by natural populations of plants. It has been described in a series of papers (Davies, 1975; Davies and Snaydon, 1973a; 1973b; 1974; 1976; Snaydon, 1970; Snaydon and Davies, 1972; 1976). They showed that the response of plants from different field plots to soil pH and nutrient status was related to the treatment the plots had received over the past 100 years. In general, plants from sites to which a given nutrient had been supplied could respond to high levels of that nutrient when grown in sand culture, whereas plants from unfertilized plots could not. The result was a pronounced physiological differentiation between plots, directly related to the local selective pressures exerted by the experimental treatments. It seems likely that these experimentally induced patterns are similar to the patterns observed in natural plants communities. Imam and Allard (1965) for instance, found that '. . . variability in wild oats in California takes the form of a mosaic in which highly localized differentiations are superimposed in a complicated way on patterns of differentiation associated with larger geographical areas' (p. 59). Although there are many such accounts of spatial differentiation on a very local scale in plants and sessile animals, I have found no corresponding accounts of temporal differentiation, such that particular genotypes are associated with particular types of year. The closest approach is made by seasonal fluctuations, such as those of inversion frequencies in natural populations of *Drosophila*.

*Selective response.* A general implication of the presence of substantial G × E interaction is that progress under selection will depend on the environment in which selection is practised. This is important in applied genetics, since it means that not only the pattern of selection but also the environment in which selection is applied must be specified to ensure high and consistent yield. From our point of view, it is another technique of demonstrating G × E interaction which highlights its role in producing local adaptation. Negative results, in which selective gains were independent of the environment in which selection was practised, have been reported for growth and size in mice (Dalton, 1967), rats (Park et al., 1966) and *Drosophila,* but similar experiments by other authors have given positive results for mice (Falconer, 1960; Falconer and Latyszewski, 1952) and flour beetles (Hardin and Bell, 1967; Yamada and Bell, 1969). Orozco (1976) gives a particularly careful account of G × E interaction for oviposition rate in *Tribolium.*

*Sites and years as sources of interaction.* The ubiquity of G × E interaction, while strengthening the case for a leading role of spatial or temporal heterogeneity in the maintenance of sex and genetic variation, does not help us to distinguish between the Tangled Bank and the Red Queen. In fact, I have been unable to find any extensive comparison of the relative contributions of sites and years to the interaction variance. Gooding et al. (1975) found that sites and years were about equally important for plant size, crown number, inflorescences per crown and fruits per inflorescence in strawberries. Killick and Simmonds (1974) give broader information, unfortunately for a character (specific gravity of potato tubers) which is of large economic but little biological importance; they conclude that while both years and localities may give rise to large interactions, years are generally the more important. Indeed, they refer (without authorities) to a general belief among agronomists that this is usually the case. But even if this were true, it must be remembered that differences between sites can be largely eliminated in agricultural practice by manuring, pesticide application and so forth, while differences between years are less easily dealt with. A more promising line of attack, for natural populations, is to argue that where G × E interaction is substantial, the relative importance of sites and years will depend on the variance of the treatment variates over sites and years. It is my strong impression, which I cannot substantiate, that variance between sites within years is usually much larger then variance between years within sites.

*Direct transplant experiments.* The crucial experiment is the transplanting of individuals from their home site to an alien site, their performance being compared with that of individuals having the same genotype which are transplanted back into their home site. While planting from natural sites into a common garden, with the object of demonstrating the existence of 'ecotypes', has often been done, reciprocal transplant and replant experiments are relatively rare. With the exception of the classical work on heavy-metal tolerance, the most extensive is probably that reported by Davies and

Snaydon (1976), using *Anthoxanthum* in the Rothamsted Park Grass Experiment. They found that plants survived longer, produced more tillers and produced more dry matter when planted back into their home plots than when transplanted to ecologically contrasting plots. The differences were large; the average half-life of plants put into alien plots, for example, was eight months, while that of plants put back into their home plots was two years. Davies and Snaydon calculated selection coefficients against the alien transplants as 1 − (performance on alien plot)/(performance on home plot); for 18-month survival these varied from 0.09 to 0.77 with a mean of 0.36, and similar values were found for tiller number and plant size. This demonstrates very powerful local selection though admittedly within a system deliberately engineered to provide sharp ecological contrasts.

Experiments in natural systems rather strongly support the Rothamsted result. Turkington and Harper (1979) transplanted *Trifolium* clones between and within different plant associations in a single old field. They found that shoots planted back into the same association usually had greater vegetative yield than those planted into different associations; even more strikingly, shoots planted into deliberately sown swards of the same grass with which they were associated in the field had almost twice as great a yield as those planted into an alien sward. Lovet-Doust (1981) performed reciprocal transplants of *Ranunculus* clones between adjacent woodland and grassland sites and found that those replanted in their home site generally grew faster and produced more leaves and ramets than those transplanted to an alien site. Antonovics and Primack (1982) made reciprocal transplants of *Plantago* between six field sites. They found that seedling survival was indifferent to the transplant site, but that growth and fecundity were generally greater in home than in alien sites. McGraw and Antonovics (1983) moved snowbed and fellfield ecotypes of *Dryas* over a distance of about 100 m and found that in 10/11 comparisons of fitness components (pollination success, seed germination, seedling and adult viability), seedlings replanted into their home site were superior to those moved to alien sites.

These experiments provide direct evidence of the site-specific adaptation required by the Tangled Bank. So far as I know, there are no comparable experiments in which genotypes are stored, either as seed or as vegetative material, and transplanted between years.

*The correlation between treatments as a function of their difference.* A conventional analysis will show whether or not a significant G × E interaction occurs in any particular case, but it cannot predict in advance whether or not an interaction will occur. We must be able to make such a prediction, however, if we wish to make general statements of how a given shift in place or time will affect the ranking of genotypic fitness.

Suppose that we have measured the fitness $w$ of a number of clones or inbred lines $i$ in each of two sets of conditions, $j$ and $k$. We wish to know the relationship between the genotypic fitnesses in the two treatments, $w_{ij}$ and $w_{ik}$.

If $j$ and $k$ are nearly the same, the $w_{ij}$ and $w_{ik}$ will be closely related, with a correlation coefficient of about unity. Conversely, if $j$ and $k$ are very different then we expect the correlation to be smaller. Indeed, if $j$ and $k$ are sufficiently different the correlation between $w_{ij}$ and $w_{ik}$ may drop to zero or even become negative.

This argument implies that the correlation between genotypic fitness under different conditions is a decreasing function of the difference between the conditions. There are two difficulties with this approach, which I have dealt with as follows.

1) The first concerns the array of environments, or treatments; the way in which the difference between treatments is measured has not been defined. The analysis must not depend on the particular units (degrees Centigrade, individuals per square meter, or whatever) used to express the difference between the treatments, which suggests the use of some dimensionless number. Moreover, we are not concerned with the difference between the treatments in terms of temperature, crowding or whatever, but only with the differences as expressed by the difference in performance of the strains under test. I propose that an appropriate statistic is the proportion of all the variance in performance over the two treatments being compared which is accounted for by the variance of the treatment means (i. e. the environmental variance, as a fraction of the whole). If the difference in treatment has little effect on performance, so that the variance of strains within a treatment is large relative to the difference between the treatment means, then this measure will be close to zero. On the other hand, if the difference in treatment has a very large effect on performance, so that the variance of strains within a treatment is small relative to the difference between the treatment means, then the measure will be close to unity. In this way, we can scale the effect of the treatment in a way which can be applied consistently to any particular sort of manipulation. This approach is similar to that taken by Hull and Gowe (1962), who suggested from poultry-breeding data that $G \times E$ interaction is likely to be large when the variation due to treatment is large relative to the nongenetic variation within treatments, or when the genetic variance between the groups which are subdivided and subjected to different treatments is large relative to the total phenotypic within-treatment variance.

2) The second difficulty concerns the array of genotypes. Whatever combination of treatments we choose, it is not unlikely that some genotypes will perform badly in both; this is especially likely when working with inbred lines. It is possible, though less likely, that a few types will be markedly superior in almost all conditions. This would tend to create large positive correlations for almost any combination of treatments, even if the bulk of the data, concerning more centrally located genotypes, showed negative correlation. Since this result would have little relevance to natural populations, in which unconditionally inferior genotypes are unlikely to occur, it is desirable to investigate the effect of removing extreme genotypes. I have

done this by calculating for each genotype the sum of squared deviations from the treatment means, and excluding the genotype with the greatest sum. If two genotypes are to be excluded, the procedure is repeated for the new data set created by the removal of the single most extreme genotype, and so forth, for any desired number of exclusions.

We expect to find that as the variance between treatment means becomes large relative to the variance between strains within treatments, the correlation between genotypic fitnesses under the two treatments will fall. By analyzing several combinations of treatments for a given array of genotypes, we can calculate the linear regression of the correlation between genotypic fitnesses, *Corr*, on the variance component attributable to the difference between treatment means, *PV*. Having estimated the parameters of $Corr = b_0 + b_1 PV$, it follows that the correlation becomes zero when $PV = -b_0/b_1$ (granted that $b_1$ is negative), and will be negative for greater values of *PV*. Thus, if $b_1 = -b_0$ the correlation between genotypes reaches zero when the variance within treatments is negligible compared with the variance between treatment means; if $b_1 = -2b_0$ then the correlation reaches zero when the variance is equally distributed within and between treatments; while if $b_1 > -b_0$ the correlation is always positive. The results from a number of studies involving fairly large numbers of genotypes under a variety of treatments are mapped in the Figure. (It should be emphasized this is a map rather than the regression of two variates, since neither correlations and variance components nor slopes and intercepts are statistically independent.) There is a tendency, though not a very strong one, for the slope $b_1$ to become more negative when extreme genotypes are excluded, as was anticipated. Leaving this aside, two major conclusions can be tentatively drawn. First, the regressions are all negative, showing that high fitness in one treatment becomes progressively less likely to be repeatable under a second treatment as the two treatments become more different. Secondly, in most cases the correlation remains positive no matter how extreme the difference between the treatments. This is by no means conclusive, since there may be manipulations which do induce negative correlations; indeed, the correlations are predominantly negative for rather large treatment effects on larval viability in *Drosophila,* while the only direct plot of genotypic fitness I have found in the literature (King, 1972) shows a striking negative correlation. However, what this small data set suggests is that if genotypes are moved from their home sites to alien sites within the range of tolerance of the species the correlation between their fitness in the two sites, though usually positive, will be small. More extensive experiments designed specifically to test this hypothesis would be of great interest.

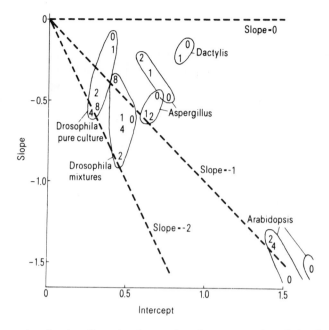

An attempt to describe the effect of a change of environment on the relative fitness of a genotype. Each data set analyzed consists of the performance of a number of clones or inbred lines under a number of different treatments. For each possible pairwise comparison between treatments I calculated the correlation between genotypic fitnesses, *Cor,* and the proportion of the variance contributed by the difference between treatment means, *PV.* I then estimated the intercept $b_0$ and the slope $b_1$ of the linear regression $Corr = b_0 + b_1 PV$. This procedure was repeated with the most extreme $n$ genotypes removed from each comparison, where the greatest value of $n$ used depended on the number of genotypes in the experiment. The results are mapped in the Figure; a solid line surrounds the estimates for each case, with a number indicating how many genotypes were removed from the data defore analysis. The broken lines indicate various slopes. If the data map below the line of zero slope, the correlation between genotypic fitnesses in different treatments declines as the treatments become more different. If they map below the line with slope $-1$, the correlation becomes negative for treatments which are sufficiently different. If they map below the line with slope $-2$, the correlation becomes negative when less than half the variance in the data is due to the difference between treatment means. Authorities are as follows. *Drosophila* pure strains (19 inbred lines, 6 density treatments) and mixtures (19 inbred lines, 5 density treatments, using larval viability – Lewontin, 1955). *Aspergillus* (5 compatibility groups 12 combinations of medium, pH and temperature), using cleistothecium production (upper) and colony growth rate (lower) – Butcher et al. (1972). *Dactylis* (5 strains, 10 combinations of site and harvest frequency), using yield of dry matter – Breese (1969). *Arabidopsis* (33 inbred lines, 3 temperatures), using flowering time (right) and fruit production (left) – Westerman and Lawrence (1970); doubtless because of the small number of treatments this analysis gave large positive values for the intercept and large negative values for the slope, but in this and the other cases intercept = $-$slope as a rough general rule.

## Interactions between genotypes

*The overall properties of mixtures.* Antagonism between individuals of the same genotype and facilitation between individuals with different genotypes will create greater rates of increase and higher asymptotic densities in mixtures than in pure cultures of a single strain. A common way of comparing mixtures with pure strains is to compare the yield of the mixture with the means of the yields of its components in pure culture; this is valid only when the strains are mixed in equal proportions and the composition of the mixture does not change between sowing and harvest. A more severe test is to compare the yield of the mixture with that of the component having the greater yield in pure culture; if the mixture is superior then facilitation has been proven, but a failure to obtain this result does not prove the absence of facilitation.

The bulk of the available information comes from inbred lines of crop plants, especially cereals. According to Simmonds (1962), 'the performance means of mixtures (of cereal varieties) are often equal to the means of the components but they sometimes exceed them, and occasionally even exceed the higher component; they are rarely inferior to the means of the components'. The impression of a general tendency to facilitation is strengthened by reviewing particular cases. Engelke (1935) combined five varieties of wheat in five combinations at various proportions and found that the yield of 11/13 mixtures exceeded that of the higher-yielding component (by up to nearly 20%). Nuding (1936) compared the performance of four wheat varieties with that of the six possible equal mixtures at seven sites over three years. He found that all six mixtures exceeded the component means, though not significantly for any individual case, while pooling the data showed that the mixtures had a slightly (about 3%) but significantly greater yield than the pure cultures. Frankel (1958) found little difference between mixtures and pure strains; Borlaug (1959) got mixed results form various mixtures of backcrosses from a standard wheat cultivar, 8/21 cases showing superiority of the mixture over the component means and 4/21 inferiority, with overall a 2.8% increase in yield being attributable to mixture. Jensen (1952) found that mixtures of oat varieties generally exceeded the component means by about 5% but were slightly inferior to the better pure strain. Sandfaer (1954) tested three oat mixtures and found that all were superior to the component mean and one exceeded the better component. Griffiths (1950–56) found that twelve mixtures all equalled or exceeded the better component. In experiments with barley, Sandfaer (1952) used six equal mixtures of four varieties and found that five exceeded the component mean and one the better component, giving an average gain in gross yield of about 5% overall. Gustaffson (1953b) conducted a careful study of three mixtures of barley cultivars and found them generally to produce more spikes, more grains and heavier grains than in pure culture; one mixture was intermediate between the two components, but the other two both exceeded the better component by about 2%. Roy (1960) grew mixtures of two high-yielding rice varieties and found that all exceeded the component mean while two exceeded the better component by 11–20%. In a larger experiment he compared 31 mixtures of varieties; only two mixtures were significantly different from their component means, both being greater. Allard (1961) reported a rare contradictory case in lima beans, where four equal mixtures of three varieties were each a little less productive than the component means. Hanson et al. (1952) found that mixtures of bluegrass clones exceeded component means by about 5%. England (1968) sowed six mixtures of four varieties of herbage grasses and found that they consistently yielded more than the pure strains, often exceeding the component means by more than 10%, all the significant differences (9/24 comparisons) being in this direction. In cotton, a widely-sown mixture of diploid and tetraploid strains performs better than either component over a wide variety of conditions (Hutchinson and Ghoose, 1937). According to Kahn et al. (1975), both flax and linseed often do better in mixture than in pure stand.

A few experiments have been done using insects in laboratory culture. Mixtures of karyotypes of *Drosophila pseudoobscura* generally yield more flies than pure cultures, whether the design enforces crowding among larvae (Beardmore et al., 1960) or among adults (Dobzhansky and Pavlovsky, 1961). These results might conceivably be due to the superiority of chromosomal heterozygotes, but by obtaining similar results at a temperature at which the heterozygotes are known not to be superior Battaglia and Smith (1961) showed that the increase in yield is an effect of mixture. In a further experiment, Dobzhansky et al. (1964) found that in uncrowded cultures the rate of increase of chromosomally polymorphic cultures exceeded that of pure cultures. This effect seems to disappear under very favorable conditions, when pure cultures attained very high rates of increase (Ohta, 1967). Kearsey (1965) found that the viability of *D. melanogaster* is greater in mixed than in pure culture at low or moderate density, but less at high density; an interaction with density was also found by Lewontin (1955), who found facilitation of larval viability in mixed culture only at intermediate densities. Sokal and Sullivan (1963) found that one strain of housefly did a little better and another strain a little worse in mixed than in pure culture. Bhalla and Sokal (1964), working with a different mutant, found that its performance in mixed culture 'furnishes a clear-cut demonstration of mutual facilitation among genotypes'. Finally, in a rare experiment with domestic animals, Moav and Wohlfarth (1974) found that the growth rate of carp is greater in mixed than in pure cultures, and showed that this was due to the greater genetic values of growth rate in mixed ponds.

A particularly interesting and relevant experiment was performed by Ellstrand and Antonovics (1985), who compared the success of blocks of clonally and sexually propagated tillers from semi-natural populations of the grass *Anthoxanthum.* The sexual tillers had greater survivorship and fecundity in each of two independent experiments, presumably because of their greater variability.

There are persistent suggestions that, aside from any difference in mean yield, mixtures are also less variable in performance than pure strains. This has been claimed, for instance, in cotton (Hutchinson and Ghose, 1937), wheat (Simmonds, 1962), corn (Sprague and Federer, 1951), lima beans (Allard, 1961) and fruitflies (Beardmore et al., 1960), but there seems to be too little hard evidence to permit a confident generalization.

The relevance of experiments involving arbitrary mixtures of pure strains can be questioned, on the grounds that facilitation is scarcely to be expected between genotypes which have never before interacted. The results with crop plants are, on this view, all the more convincing, since they concern genotypes which have actually been seen selected for their performance in pure culture. Seaton and Antonovics (1967) found that mixtures of wildtype and mutant *Drosophila* yielded more than pure strains only after a period of selection in mixture.

The bulk of the evidence, therefore, points unequivocally towards the general superiority of mixtures. All the examples cited above are mixtures sown in one place at one time, and are therefore consistent with the correlation between variation and yield required by the Tangled Bank. The practice of crop rotation shows that mixtures in time may also enhance yield, but I know of no experiments which use temporal mixtures of varieties rather than species, and the Red Queen therefore receives no direct experimental support.

*The effect of input frequency.* Mutual facilitation between genotypes implies that the fitness of a genotype will be greatest when it is rare, and when in consequence it interacts mostly with members of other genotypes. Thus, in the experiments on mixtures of *Linum* by Khan et al. (1975), the flax and linseed varieties were sown in three different proportions, allowing the authors to show that the superiority of mixtures was accompanied by frequency-dependence of fitness within the mixture. A particularly clear result was obtained by Birley and Beardmore (1977), who varied temperature and density as well as the input frequency of genotypes at an esterase locus in *Drosophila.* They found not only significant effects of input frequency on fitness, but also significant interactions between input frequency and treatment.

While instances of frequency dependence are too numerous to review here, the relevance of natural mixtures of clones in asexual organisms must be pointed out. It is difficult to see how such mixtures could persist, unless a clone has greater than average fitness when it is rare. In fact, the number of clones in grass communities often declines with time, perhaps eventually leaving only the single best genotype (Charles, 1964; Harberd, 1962; Kays and Harper, 1968). However in fertile lowland habitats populations are often highly polyclonal (Cahn and Harper, 1976; Harberd and Owen, 1969). Bell (1982) concludes that populations of asexual animals such as anemones, rotifers and cladocerans are usually, though not invariably, polyclonal.

*Proximal causes of the superiority of mixtures.* The mechanism of facilitation is often unknown. In some crop plants, differential resource utilization has been suggested; in Gustafsson's barley experiments, for instance, the superiority of mixtures was attributed to differential exploitation of the soil, one type having a shorter but more branched root system than the other (Wettstein, cited by Gustafsson, 1953a). In bluegrass, Hanson et al. (1952) suggested that the components of high-yielding mixtures have complementary morphologies, e. g. sod formers and taller grasses.

A much more common explanation of the superiority of mixtures is that they resist pests and pathogens more effectively than pure stands. Besides a great deal of more or less anecdotal evidence for the sensitivity of pure stands (e. g. Simmonds (1959), for banana; Borlaug (1959) for maize; Borlaug (1959) and Thorpe (1959) for wheat; review in Adams et al. (1971)) there is some quantitative evidence that both the incidence of pathogen damage (Summer and Littrell, 1974; Wolfe et al., 1981) and the rate of increase of pathogens (Berger, 1973; Burdon and Chilvers, 1976; Burdon and Whitbread, 1979; Leonard, 1969) is lower in mixed than in pure stands. The properties of monocultures, with special reference to the susceptibility of genetically uniform strains of crop plants and domestic animals to damage by pathogens, has been brillantly reviewed by Barrett (1981). From the opposite perspective, it is generally true that biological control (usually by insects) is more likely to be effective with asexual than with sexual weeds (Burdon and Marshall, 1981).

A particularly extensive and important experiment was conducted by Wolfe and Barrett (1980; 1981), who compared 37 different three-cultivar mixtures of spring barley, involving 25 cultivars in all, with the performance of their components when grown alone. Of 47 possible comparisons, 39 showed the mixtures to equal or exceed the component mean, while in 26 cases the yield of the mixture exceeded that of the highest-yielding component when grown alone. Overall, the effect of mixture was an increase in yield of 6.5%. However, the effectiveness of mixture varied with the degree of pathogen damage: at seven sites with little or no mildew infection the superiority of mixtures was about 3%, while in 10 sites with heavy mildewing it was about 9%. This suggests that, out of a total yield increase of about 10% associated with these rather simple mixtures, about two-thirds is attributable to increased resistance to pathogens and about one-third to other factors, including enhanced resource utilization. The implication of pathogens in the superiority of mixtures immediately suggests the Red Queen as a possible explanation.

It must be emphasized, however, that the Red Queen is not necessarily a theory about pathogens, nor is pathogen resistance necessarily evidence for the Red Queen. What the Red Queen states is that the function of sex and variability is to enable a response in time to pathogen pressure to be made. The Tangled Bank can refer to pathogens as well as to any other ecological variate, but predicts that current diversity is in itself a sufficient protection against pathogens. The operational difference between the two hypotheses is as follows. If we compare a pure strain with a mechanical mixture of genotypes, we expect the mixture to do better, perhaps because of resistance to pathogen attack. The Tangled Bank predicts that a mixture sown in one place at one time should be effective; the Red Queen predicts that a temporal mixture should be effective, a different genotype being sown in each year. The crucial experiment would be the comparison of a mechanical mixture of clonal genotypes with a sexual population; the Tangled Bank predicts that both will do equally well, whereas the Red Queen predicts that the sexual population, with its greater capacity for adaptive change, will do better. In fact, most of the observations refer to mechanical mixtures sown in a single season, and therefore support the Tangled Bank as much as the Red Queen. The only observations that directly suggest a role for the Red Queen are the greater effectiveness of biological control for apomictic weeds and the fact that parasites almost always manage eventually to overcome disease resistance based on one or two loci, so that the only effective agronomic strategy is then to introduce a new resistant variety. In neither case, however, it is clear that coevolutionary response rather than high current variation is crucial for success.

*The effect of neighbors.* A particularly elegant and vulnerable experiment to test for mutual facilitation between different genotypes is to measure the performance of a test plant when surrounded by related or unrelated neighbors. Any increase in yield associated with the presence of unrelated neighbors is a conclusive demonstration of facilitation. The classical experiment was done by Allard and Adams (1969), using a $3 \times 3$ planting design to evaluate the effect on the central test plant of its eight neighbors. Their results are summarized in Table 1. When cultivars of barley or wheat are used there is a small (1–4%) but significant and highly consistent increase in the yield of the test plant when its neighbors belong to a different variety or varieties. An even clearer result, giving a 5–6% increase in yield, was obtained by drawing neighbors from the genotypes produced by intercrossing varieties of wheat. A similiar result was obtained by Antonovics and Ellstrand (1984) for *Anthoxanthum* under semi-natural conditions. Tillers were transplanted into blocks or rows in such a way that each was surrounded by genetically similar or dissimilar neighbors. Plants with dissimilar neighbors had greater reproductive output, the effect being much greater than that in Allard and Adams' cereals. Although there were statistical difficulties arising from highly skewed fitness distributions, substitution of unlike for like neighbor appeared to be associated with a doubling of fitness, an astonishing result.

Table 1. Summary of the results obtained by Allard and Adams (1969) from an experiment to measure the effect of genetically similar and dissimilar neighbors on a target plant

| Neighbors | Barley vars | | Wheat vars | | Wheat genotypes | |
|---|---|---|---|---|---|---|
| | Mean | SD | Mean | SD | Mean | SD |
| 1) All eight of a single different type | 101.54* n = 24 | 2.529 | 102.10** n = 24 | 2.598 | 104.5** n = 16 | 4.379 |
| 2) Equal numbers (four each) of same type and another type | 101.16* n = 24 | 1.581 | 101.36** n = 24 | 1.517 | – | |
| 3) Two of same type; two each of three different types | 101.95* n = 8 | 1.258 | 103.78** n = 8 | 1.436 | – | |
| 4) One of same type; one each of seven different types | – | | – | | 106.31** n = 16 | 4.600 |

The mean and SD are estimated from 25 replicates of each design, and refer to yield of seed relative to yield = 100 when surrounded by eight neighbors of the same type. On this basis, t-tests for significance of the excess of yield over 100 are shown by $*p < 0.01$, $**p < 0.001$.

## Conclusions and suggestions

The evidence that I have very briefly reviewed above strongly suggests the general occurrence of substantial interactions among genotypes and between genotype and environment. These short-term observations and experiments therefore run parallel to comparative work showing that sex is associated with stable heterogeneous environments, and the confluence of these two independent streams of evidence argues very strongly that either the Tangled Bank or the Red Queen, or both, are crucial to the maintenance of sex and variation. Unfortunately, the data do not seem sufficiently decisive to eliminate one of these two possibilities; genotypic fitness may interact substantially with time or place or both. Indeed, both hypotheses may be required for a satisfactory solution to the problem, and experiments such as those conducted by Wolfe and Barrett (1980) show how the variance in yield between mixtures and pure lines might be partitioned between the two effects.

The most effective designs are the reciprocal transplant-replant and the effect-of-neighbor experiments I have described above, and it is very desirable not only that more of these should be performed, but also that they should be capable of detecting between-year effects. A different type of experiment, however, has not to my knowledge been attempted: the comparison of mechanical mixtures with sexual populations within a single site, to evaluate the relative importance of current variation and selective response. It would be necessary to use a short-lived outbreeding organism which can be propagated vegetatively or by apomictic seed. From a diverse initial sample from natural populations, a large number of genotypes are extracted and propagated clonally. Each site within a block of sites is then stocked at the beginning of each generation in one of five ways:

1) same clone;
2) different random clone;
3) same mixture of clones;
4) different random mixture of clones;
5) the indefinite mixture generated by random-mating the surviving individuals in proportion to their fecundity.

From the mean performance of individuals from each of these five treatments, we can then identify four types of effect.

    A) The effect of temporal mixture, estimated by the excess of 2 over 1.

    B) The effect of contemporary mixture, estimated by the excess of 3 over 1.

    C) The interaction of A with B, estimated by the excess of 4 over 1 not explained by the excess of 2 or 3 over 1.

    D) The effect of appropriate genetic response, generated by selection, estimated by the excess of 5 over 4.

The main effects A and B identify any advantage to the random dispersal of genotypes in time or space; if both are substantial then it is not unlikely that the interaction C will also be important. The final effect D is intended to show whether outbred sexuality is more or less effective in enhancing the properties of mixtures than deliberate randomization. The results of such an experiment would be interesting on general grounds, showing whether the Red Queen (effects A and D) or the Tangled Bank (effect B) or both (effect C) were supported by the properties of different kinds of mixture. They might also have some practical utility in suggesting the best ways of constructing mixed crops. It would obviously be very instructive to run such experiments so that spatial heterogeneity within sites or exposure to an appropriate pathogen were deliberately contrived. It is, I believe, very desirable that the extensive comparative work on sexuality should be supplemented by increased effort to devise short-term experimental tests of the rival theories.

# The adaptive significance of sexuality

H. J. Bremermann

## Introduction

Birds do it, bees do it, plants, algae, protozoans, bacteria, and last but not least: mammals and *Homo sapiens*. Much of the entire biosphere is engaged in sexual recombination. The near universality of sexual recombination lets one suspect a common cause or function.

It has been argued that sex speeds up evolution by bringing together rare advantageous mutations (Felsenstein, 1974; Fisher, 1930; Muller, 1932). Kimura and Ohta (1971) have stated this effect dramatically: 'Sexual reproduction has played a very important role in speeding up evolution in the past, helping to produce man before the sun in our solar system burns out.'

Lewontin (1974) said that sexual recombination creates the variety which constitutes the potential of further evolution, while E. O. Wilson (1975; 1978) describes its role as creating variety as insurance against changing environments.

Until the mid-sixties these effects of sexual recombination were often taken as an explanation: it is good for evolution, hence nature invented sex (Felsenstein, 1974). The only remaining question was: when? Thus one may find in many works on the origin of life speculations about when sexual recombination was invented. Once this happened evolution could proceed to create man.

Sex thus is explained through its purpose. The end explains the means. Biological methodology quite generally observes a taboo against explaining biological facts by the ends which they serve (Fraenkel and Gunn, 1961). However, biologists often have not hesitated to enshrine evolution itself as the ultimate and final cause of all things biological.

The trouble is that what is good for evolution need not be good for the individual. If sex is bad for the individual but good for evolution in the long run, what prevents individual mutants from pursuing their own fitness advantage by reproducing asexually? Many microorganisms can reproduce both asexually and sexually. Plants can spread by vegetative propagation and many species can produce seed without being fertilized by pollen from other plants. Mutations that turn off the complicated machinery of sex are possible, and many eggs can be made to develop without fertilization by sperm. There are parthenogenetic species of salamanders and fish. Parthenogenetic plant species are frequent and often successfully invade the territories of their sexually reproducing ancestors and relatives. (For details on parthenogenesis see Maynard Smith, 1978, and Cuellar, 1977).

Maynard Smith has pointed out that a parthenogenetic mutant producing only females has an initial 2:1 fitness advantage over its sexually reproducing competitors. Why then do parthenogenetic varieties not replace sexual ones? This is the fundamental problem. (Actually, the 2:1 advantage holds under *ceteris paribus* assumptions. It is assumed that lifetime fitnesses of sexual and asexual types are unaltered except with respect to the sex of the offspring. This is often not the case – see the chapter by Lewis.) If the assumption is satisfied, then the percentage of parthenogenetic females in a population initially about doubles with each generation. Subsequently the mutant allele keeps on spreading (though at reduced rates) until it asymptotically approaches 100% (Maynard Smith, 1978). If parthenogenesis leads to reduced fecundity, parthenogenetic mutants would still take over as long as the advantage is better than 1:1. (A further complication is the phenomenon that in contrast to parthenogenetic lizards, fish, insects, and numerous plant species there are no parthenogenetic birds or mammals among 12,000 species. Why?)

If the benefit to evolution at large (to evolve man before the sun burns out) were the driving force that maintains sex against individual fitness advantage, it would be a case of group selection: the group being the entire biosphere. What mechanisms would prevent individual parthenogenetic mutants from spreading? Unless we believe in replacing divine intervention by an equally mystical force of intervention guided by the presumed long-term goals of evolution, we must account for the persistence of sex in terms of individual fitness advantages or in terms of mechanisms that prevent parthenogenesis from taking hold. These mechanisms in turn would have to be explained as evolutionarily stable, or as being stable over long periods of time.

Many attempts at understanding sex have been made (Williams, 1975; Shields, 1982; Bell, 1982). In his monograph, 'The Evolution of Sex', Maynard Smith (1978) writes in the preface: 'I am under no illusion that I have solved all the problems which I raise. Indeed on the most fundamental question – the nature of the forces responsible for the maintenance of sexual reproduction and genetic recombination – my mind is not made up . . . it is not clear to me whether the short-term selective forces I discuss are sufficient to account for the facts, or whether models of a qualitatively different kind are needed.'

The models discussed by Maynard Smith are based either on temporal changes in the environment or on spatial heterogeneity (unpredictability). He concludes (p. 89): 'At equilibrium in a uniform environment if there is any selection on recombination it will be for reduction.'

This suggests that species in environments that have remained temporally and spatially homogeneous for extended periods of time should have a preponderance of asexual and parthenogenetic species.

It appears that off-shore marine environments, especially deep-sea environments, are highly uniform. Yet sexually reproducing species flourish there as elsewhere. (A careful tabulation of data on sexuality in exceptionally stable environments would be in order.)

Hamilton (1980; 1982), Jaenike (1978) and Bremermann (1980) (independently) have pointed to another force that could favor sexual recombination in multicellular organisms: microparasites that cause disease. (Rice (1983a) has made a similar suggestion; he emphasizes the transmission of parental pathogens.)

Microparasites (broadly defined to include viruses, fungi, bacteria and protozoans) have shorter generation times, smaller genome sizes and have greater speed of adaptability than their multicellular hosts. These parasites depend for their own survival and propagation upon their hosts. For example, viruses cease to replicate when the host dies. This creates a feedback between host and microparasites with a population dynamics of its own (Anderson, 1979; Anderson and May, 1979; 1981; 1982; May and Anderson, 1979). We propose that the key to understanding sexuality, at least in multicellular organisms, lies in this feedback between microparasites and their hosts and in the different rates with which hosts and parasites can adapt through mutations and other genetic changes. We discuss this proposal further in the following sections. (For another discussion see B. Levin et al., 1982.)

A different mechanism may be responsible for the maintenance of sexual recombination in single-celled organism: Some microorganisms live indeed in unpredictable environments and environments that change with the seasons. Bacteria repress genes that are not used. For example, *E. coli* represses transcription of the beta-galactocidase gene when there is no lactose in the growth medium. The gene can remain repressed for many generations when no lactose is present. Similarly other genes remain repressed for generations until their function is needed again by a change in the environment.

While genes are repressed they are not subject to purifying selection. Thus even when the mutation rate per cell division is low, the cumulative rate in an unused gene can be high. It is proportional to the number of generations that have passed since last used.

*E. coli* can go through three generations per hour, 72 generations per day. After several days mutations have accumulated in the repressed and unused genes. When the environment suddenly shifts and repressed genes are derepressed, they may no longer function. At this point there is an advantage in exchanging genes with another individual. For example, if each donor carries one nonfunctional gene (at different loci, and assuming that the loci are not too closely linked) then there is a chance, up to 25 %, that the recombinant genome is fully functional. A 25 % chance of producing fully competitive descendants is better than producing no competitive descendents at all. Thus for microorganisms sexual recombinations may be something like a repair mechanism.

Note that *E. coli* reproduces normally through asexual division. Only occasionally will it shift to sexual reproduction where a copy of the genome is transferred to another bacterium through a sex-pylum. Transfer of the genome takes 90 min (versus 20 min generation time for asexual reproduction under favorable conditions – Lewin, 1977).

Yeast normally is diploid and reproduces asexually through cell division. On occasion yeast produces haploid cells of two different mating types (a and alpha). Cells of different mating types are chemotactically attracted to each other and fuse, forming diploid cells which then divide asexually for many generations (Lewin, 1983).

The sex-as-repair model for single-celled organisms was proposed in an appendix to Bremermann (1979). (Note that portions of that paper, notably the section on the generation of antibody diversity, have been made obsolete by recent developments in immunology. The appendix on sex as a repair mechanism remains valid.)

Our model could be tested by subjecting continuous colonies to environmental shifts and by manipulating mutation rates through mutagens or UV. Also, sexual mechanisms require resources. In the absence of any benefits from sexual recombination one would expect that asexually reproducing mutants would prevail. Hence, in a steady environment that is spatially and temporally homogeneous, and in the absence of parasites (such as phages) one would expect asexual mutants to take over. Besides the exchange or diploid association of entire genomes there is a lively trade of smaller pieces of DNA. In addition to the main chromosome many bacteria contain plasmids which are small, independently replicating circular pieces of DNA. Plasmids are also found in eukaryotic cells like yeast. They are easily acquired and lost. Plasmids are essentially parasitic since they are replicated at the expense of their hosts. Often they also carry genes that are beneficial to their hosts like resistance factors, for example, genes for endonucleases that provide protection against invading phages. Like phages, plasmids may on occasion become integrated into the genome. Lysogenic versus lytic infections are advantageous to the phage, depending upon circumstances (Bremermann, 1983b). The population dynamics of plasmids has been discussed by Stewart and Levin (1977), Novick and Hoppenstaedt (1978), Van der Hoeven (1984), and Hockstra and Van der Hoeven (1984). For the molecular biology of plasmids and phages see Lewin (1977).

## Pathogens are not just another factor in the environment

Pathogenic parasites are a powerful factor of host survival and fitness. For example, prior to the advent of modern hygiene families typically lost half their children to intestinal and other diseases before the onset of puberty. Today, malaria is endemic in many countries. Trypanosomiasis excludes cattle from wide stretches of Africa (Anderson and May, 1982). The American chestnut *(Castanea dentata),* which at the beginning of this century constituted 25 % of Eastern forests, has been virtually wiped out by a fungus (Van Alfen et al., 1975). The list could be continued indefinitely.

Most pathogenic microparasites are species specific and are thus produced by the very population which they attack.

While multicellular predator-prey species evolve on comparable timescales, microparasites do not. Many microparasites have an advantage in speed of adaptability by orders of magnitude. They have shorter generation times, smaller genomes and greater genetic flexibility than their hosts. For example, viruses can acquire host genes, bacteria acquire, maintain, insert, and delete plasmids beyond and above the regular mechanisms of mutation and recombination.

The speed of adaptability of microparasites is difficult to compute theoretically. It may be seen, however, in involuntary experiments with agricultural crops such as wheat and corn. When a new variety has been created by plant breeders which is resistant to all existing strains of rust or leaf blight fungus, a mutant strain of fungus tends to appear after a few years that spreads like wildfire and may ruin an entire crop. This has happened repeatedly (National Academy of Sciences Committee Report, 1972). A natural host population that is infected by a pathogen would fall to a low density until infective and susceptible individuals are at an equilibrium. If the host population falls below the equilibrium, fewer parasite propagules are produced and fewer individuals are infected. Hence the host population increases. When the host population exceeds the equilibrium, the percentage of infected individuals rises – with the effect that the host population decreases. In this way hosts and parasites can fluctuate stably around an equilibrium. (For a mathematical analysis see Anderson, 1979; Anderson and May, 1979; 1981 and May and Anderson, 1979).

With time the parasite may evolve towards an optimal level of virulence that balances the inevitable damage to the host resulting from the parasite's replication and shedding of propagules (damage to tissues, leaves, etc.) against the cumulative advantage of shedding propagules throughout the host's lifespan. The longer the host lives, the more propagules are shed (Bremermann and Thieme, 1987).

Parasitism, even highly coevolved parasitism, is never beneficial from the host's vantage point. An infected host suffers excess morbidity and/or reduced fecundity and thus a reduction of fitness.

A mutant host that is not subject to infection by the parasite thus has higher fitness and would increase to levels determined by the carrying capacity of the environment. If the disease-free host strain is genetically homogeneous (which would be the case with an asexually reproducing clone) then it could become a target for a mutant parasite strain that would neutralize the host mutation.

This process may continue until a polymorphic host population and corresponding polymorphic collection of parasite strains has evolved. *This phenomenon should be observable experimentally in E. coli populations grown from a clone and infected with a clone of phages.*

Host resistance against parasites can be cellular and systemic. The vertebrate T cells and B cells constitute a systemic defense. At the cellular level it is molecule against molecule. We show below that the sexual combination of alleles from a polymorphic host population are an essential ingredient of both cellular resistance and systemic immune defenses.

## Host-parasite interactions at the molecular level

Both systemic and cellular defenses involve molecular interactions of great specificity. Currently many studies are underway that identify the effects of amino acid substitutions upon antibody binding and virus neutralization. Monoclonal antibodies have made possible serological specificities that one could only dream of with multiclonal sera.

Single amino acid substitutions in viral proteins can change the serological identity of a virus strain (Knossow et al., 1984). Conformational changes that are induced by amino acid substitutions are being studied by X-ray diffraction analysis and computer studies of variations in amino acid sequence. A picture of several critical sites on viral proteins emerges. At the critical sites substitutions affect serological specificity while other portions of the polypeptide chain are less sensitive (Wiley at al., 1981).

Epidemiological studies trace the variety and antigenic shifts and drifts of common infectious diseases such as influenza, hepatitis, polio, measles, etc. The variety of viral strains and their genetic drifts and shifts are of prime importance for recurrent epidemics as well as for the preparation of vaccines for immunization. For a recent survey see Chanock and Lerner (1984).

The processes that bring about antigenic drift and shifts are very complex. In the case of influenza, human populations throughout the world as well as animal hosts participate in this process. It would be unreasonable to model such a complex process. Instead it seems appropriate to accept the appearance of new strains as an empirical fact, to determine the time it takes for new strains to appear, and to model their effect upon host populations. This is the approach which will be taken in the mathematical models that are discussed below.

The molecular battles between hosts and parasites are not limited to systemic immune defenses but take place at the cellular level as well. The principles of host-parasite dynamics have been studied theoretically and experimentally in cultures of bacteria and bacterial viruses. Here the binding of a bacteriophage to a receptor site is highly strain specific. The battle between host and phage involves host mutations that abolish binding of a phage strain and mutations in the viral capsid protein that restore it (Lewin, 1977; Stent and Calendar, 1978). Once a virus particle has been absorbed, its

RNA or DNA enters the cell where it faces nucleases (Arber and Dussoix, 1962). The restriction nucleases of DNA are site specific, and restriction sites are characterized by short palindromic DNA sequences (Lewin, 1983). Any invading DNA that contains a restriction site gets degraded. Mutant viral strains that do not contain the restriction site can infect the host. However, hosts can acquire new genes for different restriction nucleases. The evolution of bacterial host-parasite systems can be observed in chemostat experiments. It happens quickly and would lend itself to the testing of theoretical predictions, such as the emergence of polymorphic populations with different strains of hosts and phages.

Similar specificities affect the binding of fungal toxins to specific targets in the host cell (Scheffer and Livingston, 1984; Strobel, 1973; 1975). Plant molecular genetics is a very young field. Advances in molecular biology, especially gene cloning and sequencing of parasite genomes, have made it possible to study host-parasite interaction at the molecular level. The polymorphism of both hosts and parasites is becoming apparent (Verma and Hohn, 1984). The full range of host-parasite interactions at the molecular level is rather complex (Gracen, 1983); however, in some cases susceptibility and resistance depends only upon a single locus in the host and a corresponding single locus in the parasite. Such correspondence is known as gene-on-gene resistance.

## Gene-on-gene resistance, sex and polymorphism

Gene-on-gene resistance is a widespread phenomenon for cereal grasses and their pathogens. Mode (1958; 1960) pioneered models of such coevolving systems. Numerous similar models have been studied subsequently (Jayakar, 1970; Levin, 1976; Levin and Udovic, 1977; Yu, 1972).

Flor (1955; 1956) investigated the genetics of the interaction between flax *(Linum usitissimum)* and rust *(Melampsoa lini)*. He found that there were 27 genes for resistance (R genes), distributed as multiple alleles at five loci. Resistance is inherited as a dominant character. Virulence in rust is controlled by a complementary system with a one-to-one relationship of each R-gene in the host with a corresponding gene in the parasite. Virulence is recessive. (This is a somewhat idealized picture. For a recent review of the genetic and biochemical basis of virulence see Panopoulos et al., 1984.)

Such a gene-on-gene relationship has been found in many other host-parasite systems (Person, 1966; Pimentel, 1982). Pimentel (1982) and Bremermann (1983) have argued that the evolution of gene-on-gene relationships resembles the response of parasites to other single environmental stress factors, such as pesticides or antibiotics.

Gene-on-gene resistance is frequently encountered in fungal plant diseases. For example, cereal rusts damage their hosts by injecting hyphae to collect nutrients. They kill host cells for food. Damage to the host may be severe (Van der Plank, 1975). A virus, in contrast, requires live host cells for its own reproduction.

Fungi often produce toxins that disrupt host cells. They may interfere with membrane transport or with enzymes in the metabolic machinery of the cell. It is a battle of molecules against molecules. For example, Strobel (1973; 1975) showed that susceptibility of sugarcane to the fungus *Helminthosporium sacchari* involves the binding of a toxin, helminthosporoside, to a membrane protein. A resistant variety had a protein that differed by four amino acids.

For diseases with gene-on-gene resistance we thus envision a battle of adaptation between host and parasite. The parasite evolves molecular species that bind, the host evolves proteins that do not bind. Here is an analogy to the immune system except that the roles are reversed: In the immune response the host produces molecules that bind and the pathogen loses if it is bound. In gene-on-gene resistance there is the added constraint that host molecules still must perform their function as enzymes or membrane proteins. The selective pressure creates *isozymes which are neutral mutations with respect to the metabolic or membrane function which they serve.*

A sexually reproducing host, drawing upon pollen from a polymorphic host population, produces seed with a variety of allele combinations. The risk that a particular allele will become susceptible through pathogen adaptation is thus spread widely. Fungal strains may evolve that overcome resistance for some of these alleles. The majority of seed, however, will carry alleles that remain resistant.

As fungal strains evolve that overcome resistance for some alleles, these alleles will become rarer. On the other hand new mutant isozymes that arise in the host population which confer resistance will spread. (For a discussion of mathematical models of this type of dynamics see the section entitled 'New mathematical models'.)

In contrast, without sexual recombination, a host variety creates a genetically homogeneous clone. This clone is a target for pathogen adaptation, especially if it successfully proliferates (Rice, 1983a). The situation resembles that of a cereal grain variety that at the beginning is resistant to all strains of rust and which is planted in large populations. The length of time till resistance is lost may vary. For highly homogeneous crops like wheat (Apple, 1977) and tobacco (Lucas, 1980) the period of grace is only a few years. The parthenogenetic mutant thus will enjoy a period of reproductive success after which most of its descendents are wiped out by disease.

In contrast, the plant that reproduces sexually in a sufficiently polymorphic population is hedging its bets: Some seeds may be worse off with foreign alleles, but the majority will be resistant to all fungal strains, and so on for subsequent generations. Bet-hedging is not limited to the uncertainties of encountering different strains of parasites. It is a wider phenomenon in the context of strategies under uncertainty (Stearns, 1976). The concept of 'strategy' is well defined in the context of game theory.

Sexual recombination thus is a strategy in a game of molecular pursuit and evasion. The game puts a premium on speed of adaptation. Here the microparasites are at a definite advantage. It means that parasites must be genetically flexible. Some pathogenic fungi have indeed been described as 'genetically highly unstable' (Lucas, 1980).

The host, in contrast, is at a disadvantage in speed of adaptation. It compensates by existing as a polymorphic population, polymorphic at loci whose products are directly affected by corresponding gene products of the parasites. The contest between parasites and hosts thus can be described as a game of pursuit and evasion between two fundamentally different contestants.

Sexual recombination and polymorphism thus must be seen together. Separately they pose seemingly unresolvable problems. Together they are a mechanism by which multicellular species hold their own in their battle with microparasites.

The fertile crescent in the Middle East is the origin of many of our cereal grasses. Despite favorable conditions for pathogen growth the crops there have been relatively stable and free of epidemics (Browning, 1974; Dinoor, 1974). In contrast American cereal crops have been beset by periodic crop failures and are subject to genetic management that attempts to stay one step ahead of disastrous outbreaks of cereal rust and similar epidemics (Apple, 1977). Several authors have argued that this disparity is due to the genetic oversimplification of American agriculture (Bremermann, 1980; 1983; Browning, 1974; Pimentel, 1982). It is important, however, to see that this phenomenon is not an isolated one but part of a general principle. The game of pursuit and evasion is general.

## Game models of pursuit and evasion

Mathematical genetics has had unusual difficulties explaining polymorphism in natural populations. In a recent survey on the evolutionary dynamics of genetic diversity (Mani, 1984), Nevo et al. (1984) have compiled a vast amount of data on polymorphisms in natural populations of plants and animals, vertebrates and invertebrates. They have correlated the measures of heterozygosity with all kinds of important biotic variables with one exception: pathogens.

In the same survey, Cook (1984) describes the phenomenon of genetic diversity as 'The Problem'. It has resisted understanding within the framework of mathematical genetics for some time. Solbrig et al. (1979) write on page 46: 'One of the unresolved paradoxes of modern population biology is the apparent high level of heterozygosity in populations which is impossible to account for under present single gene models with selection.'

Several theories have been proposed, most notably the balance theory (Lewontin, 1974), the neutral theory (Kimura, 1979; 1983b; Kimura and

Ohta, 1971; Nei, 1984) and various models of frequency dependent selection (Clarke, 1979).

Our model combines features of all three: Since gene-on-gene resistance tends to be dominant, heterozygotes have a selective advantage. Secondly, alleles that represent different conformations of an enzyme or membrane protein and which differ in their resistance/susceptibility to pathogen strains should be selectively neutral with respect to their metabolic function. Thirdly, the probability of appearance of a parasite mutant to which a previously resistant allele is susceptible rises when the allele is common. Hence, in the long run selection is frequency dependent.

Our model differs fundamentally from conventional models by including feedback between host and parasite. In classical models the selective value (usually denoted by W) is static or varies randomly in space or time. In our model the host population itself causes delayed changes in W.

Our model is a game model: Both host and parasite must optimize their fitness. Fitness of both host and parasite, other things being equal, is proportional to the host's lifespan. The parasite also must optimize the rate at which it sheds propagules (spores, virions, etc.). This objective conflicts with longevity: the greater the rate of parasite reproduction the greater the damage to the host. A high rate exacts a toll from the host and shortens its lifespan. This game model was first proposed by Bremermann (1980). For a further discussion see Bremermann and Pickering (1983).

May and Anderson (1983a; 1983b) and Levin (1983) have pointed out the importance of combining population dynamics with genetics. Unfortunately the monster of intractibility raises its ugly head as soon as these models are written down. *To remedy the situation we propose that host and parasite genetics do not deserve to be treated on the same footing.* Parasites have short generation times, large populations, produce vast numbers of spores and have unusual modes of mutation and genetic modification (Schimke, 1982). Once a mutant parasite appears that can attack a frequent host genotype it spreads rapidly. This is demonstrated by the numerous instances of catastrophic diseases in artificially homogenized agricultural crops and garden plants (Bremermann, 1983a; Klinkowski, 1970).

Instead of keeping track of all individual genetic events in the parasites, we treat mutants as singular perturbations of the systems equations that occur from time to time. We characterize mutants by their phenotypic ability to attack one of the existing host genotypes. What matters is the frequency with which such mutants appear and how their appearance depends upon allele frequencies of the host population. Once such a mutant appears the system goes through a transient and may settle down at a new equilibrium.

Formulating these models is fairly straightforward. They can be simulated on moderately fast computers (for specific values of the parameters). Analytical results, however, due to the nonlinearities, are difficult to obtain and are in any case beyond the scope of this paper. (Results for special cases have been obtained by Bremermann and Fiedler, 1985 and Bremermann and Thieme, 1987).

We have so far focussed mainly on fungal diseases in plants and have demonstrated how sexual recombination in a polymorphic population is a strategy by which hosts can compensate the advantage of faster adaptation enjoyed but the parasites. Polymorphism and recombination are also crucial in other host-parasite encounters.

Vertebrates have a lymphocyte immune system which fights viruses, fungi, bacteria, and protozoans. In the following we will show that recombination of alleles from a polymorphic population is indispensible for the functioning of the vertebrate immune system. The alleles in question are the alleles of the major histocompatibility complex (MHC). Their combination gives individuals an immunological identity of self versus non-self. This identity is necessary since in a homogeneous host population parasites could adapt to camouflage. Again, the disparate speed of adaptation between microorganisms and metazoans is the crux of the matter.

### The vertebrate immune system responds to rapid pathogen evolution

Vertebrates possess in addition to cellular defenses a systemic immunological defense: the lymphocytes. They are a specialized cell population, in humans, that consists initially of at least $10^6$–$10^7$ distinctive subclones. The lymphocytes that circulate in the blood stream are screened to be self-compatible: They do not attack the organism's own cells, while they respond to foreign antigens of all kinds: viruses, bacteria, fungi, as well as cells of another individual from the same species. To escape attack from the lymphocytes, a cell must carry a specific combination of histocompatibility antigens. These differ from individual to individual (except for homozygous twins).

The total number of cells in a human organism is of the order of $3 \times 10^{14}$. If we divide this number by $10^7$ (the approximate number of lymphocyte clones) we obtain an upper bound of $3 \times 10^7$ cells per clone. Since lymphocytes amount to only a fraction of the cells of the organism, the actual number is more like $10^4$–$10^5$.

When an antigen invades, two things can happen, depending upon the relative numbers of invaders and defenders: If the number of invading antigens is small, they are bound by lymphocytes and eliminated. When the number exceeds the number of lymphocytes and the immunoglobulin molecules that they can produce, then the invader has a chance to multiply. At the same time the specialized clone whose antibody binds to the invader is stimulated to proliferate. Initially the responding clone cannot produce enough antibody to bind all invading pathogens, and the pathogen population expands. (In other words there is a threshold effect, depending upon the size of the inoculum.)

Lymphocytes that bind antigen are stimulated to proliferate: a competitive proliferation results. The pathogen population expands until so much antibody is produced that more pathogens are destroyed than are being born. At this point the balance tips in favor of the organism: the pathogen population declines and the organism can recover (Perelson et al., 1976; 1978).

The invading pathogens can be compared with an invading army. If the invading army is small, it is eliminated before it can proliferate. Otherwise, the battle becomes one of competitive proliferation. One may ask: Why are infections necessary in the first place? Why aren't organisms adapted to have sufficiently large lymphocyte clones to begin with?

This is a logistic problem: it depends upon the number of different kinds of pathogens that can assault the organism. For each potential pathogen the organism must have an 'army' of specialized lymphocytes. The number of different armies, that is, the number of different kinds of lymphocyte clones, each producing one kind of antibody, has been estimated for humans as $10^7$ to $5 \times 10^7$ (Klinman et al., 1977). For small tadpoles the number is $10^5$ (Perelson and Oster, 1979).

Using no more than some basic assumptions about the size of recognition sites and some reliability theory, Perelson and Oster (1979) have analyzed 'minimal antibody repertoire size and reliability of self-non-self-discrimination'. They come to the conclusion that the lower limit of the repertoire size is around $10^5$: below this number a lymphocyte system could not cope with all potential antibodies and function reliably. For an upper limit Perelson and Oster show that repertoire sizes of $10^6$–$10^7$ are fully sufficient. These figures are derived from assumptions about the size of the recognition site of antibody molecules and from logistic requirements alone and are valid for all species with antibody-producing monospecific lymphocytes.

If all possible pathogens are equally probable to attack, then the 'standing armies' that can be maintained to absorb the initial attack are limited to $10^4$–$10^5$ cells, as we pointed out before. Or in other words: The organism cannot maintain $10^7$ different standing armies of large size (the number of lymphocyte clones producing different kinds of immunoglobulins). One could object that the actual number of infectious diseases that afflict man is much smaller than $10^7$. Why hasn't man evolved hereditary immunity against the diseases that actually afflict him, rather that keeping clones against millions of potential diseases, most of which do not exist and have never existed? If a species would concentrate on maintaining lymphocyte clones against the few hundred or thousand infectious diseases that actually occur, then it could raise the threshold of infectivity – the size of the inoculum required to produce disease – correspondingly. Since infectious diseases greatly reduce the fitness of afflicted individuals, one would expect the human species to have evolved hereditary immunity. There appears to be a paradox.

The apparent paradox of the maintenance of defenses against nonexisting diseases can easily be resolved if one takes into consideration the speed at which new pathogens can evolve. An illustrative example is the influenza virus which causes world wide flu epidemics in man. After a pandemic (such as the worldwide flu epidemic in 1919 which claimed millions of lives), there is widespread immunity, preventing further infections. The worldwide immunity reduces the flu virus to very low residual levels. Nevertheless, in the course of a few years, a new worldwide flu epidemic appears. What happens is that the flu virus changes its antigenic characteristics to such an extent that the existing immunity (from the previous epidemic) is no longer effective. In other words the standing army of flu specific lymphocytes (acquired after considerable suffering) cannot cope with the new flu virus. A new, specialized defensive clone is needed to combat the changed virus.

The flu virus is an extreme case: it is capable of major antigenic shifts within a few years. These shifts involve animal intermediate hosts ('swine flu') and the usual genetic recombination processes (Kaplan and Webster, 1977). In addition there is ordinary antigenic drift of the virus. Residual amounts of virus are attacked by people's immune system with an intensity that varies according to the antigenic characteristics of the virus. Therefore, mutants are selected that have altered characteristics. This antigenic drift is responsible for minor outbreaks of flu epidemics between the major pandemics every ten or twenty years, which are due to antigenic shifts (involving unusual genetic recombinations). Thus the flu virus can evolve significant antigenic changes well within a person's lifetime. This discrepancy in the speed of evolution is by no means unique to the flu virus.

*The vertebrate immune defense system is a general purpose system. It can cope with all potential antigens, not only with already existing antigens. Without this capability to respond to ever novel antigens, it would be useless because of the speed with which new antigens evolve.*

## Antigen recognition depends upon recombination of MHC alleles from a polymorphic population

In the battle between pathogen and organism it is imperative that the organism be able to recognize the pathogen as such. Conversely, if a pathogen can avoid recognition, it will not be attacked by lymphocytes or inhibited by antibodies. Thus pathogens should camouflage themselves, taking on the characteristics of the organism's own cells.

Cercariae of *Schistosoma mansoni* use camouflage. After a cercarium penetrates the skin of its human (or mammalian) host, it sheds its tail (which is attacked by the immune system) and covers itself with glycoproteins from host cells (Haas, 1976), thus acquiring the 'self'-defining histocompatibility antigens of its host. Exempt from attack, the cercarium can travel to the host's liver without being destroyed. After finding and joining a mate, the cercarium develops into a schistosoma, capable of shedding hundreds of eggs a day for years. Cercariae are quite large in comparison with bacteria and rely on a complicated life cycle for the transmission of the disease (schistosomiasis) involving an intermediate host (snails). Fortunately, bacteria seem to be less capable of adopting this strategy of 'stealing' the host's self-defining tissue antigens. (For further details see the survey article by Bloom, 1979.)

Bacteria and viruses are small enough for airborne transmission from host to host, and they enjoy an advantage over larger parasites that do not reach new hosts in this way. However, they also carry less genetic information than cercariae or protozoan parasites. They do not play and probably are unable to play the complicated games of camouflage or sequential antigenic change that *Schistosoma mansoni* or *Plasmodium falciparum* (malaria) engage in.

Once inside a host, bacteria and viruses are subject to selective pressures from antibodies and T-cells. These pressures would drive them towards camouflage and indistinguishability from host cells if they would remain in the same host long enough or if the pressures in different hosts would be the same. The well studied shifts of influenza virus strains are due to the selective pressures from host antibodies (Kendal et al., 1984). Thus in a clone where individuals have identical 'self' defining antigens, viruses, bacteria and fungi would evolve that would be indistinguishable from self and which could not be attacked by antibodies and by T-cells.

Self-defining antigens thus must change from generation to generation and from individual to individual. The polymorphism of the MHC complex and the numerous alleles at the MHC loci are effectively randomized by sexual recombination. Proof of the effectiveness is the rejection of tissue transplants when recipient and donor are not homozygous twins. Even when donor and recipient are matched for MHC compatibility, rejection still takes place and immune suppressive drugs must be given. MHC alleles are approximately selectively neutral. Small, isolated populations would in the course of time lose selectively neutral alleles through drift. Bremermann (1980) has proposed a mechanism of sperm selection in vertebrates that would enhance MHC diversity and which would counteract genetic drift. This mechanism would be analogous to mechanisms of pollen selection in flowering plants (DeNettancourt, 1977). *Existence of the conjectured mechanism still remains to be confirmed* [but see the last chapter by Stearns – Ed.]. (The ill effects of genetic drift in small isolated human populations were apparent in mountain villages in Europe prior to the onset of contemporary mobility and mixing. They can also be observed in the Nile valley where frequent first cousin marriages aggravate the effect. Genetic drift must have been a problem during much of hominid evolution when populations were small. Genetic drift seems to be a problem in rare species, like the cheetah, that survive in small numbers.)

## Split genes, introns, exons, and enzyme variety

For a long time the generation of antibody diversity (more than $10^7$ lympho-cyte clones of different specificity) from a few germ line genes (Watson, 1976) remained a mystery. Then exons and introns and mRNA splicing were discovered and it was realized that variety is being generated by soma-tic recombination and sliding splice sites (Alberts et al., 1983).

Split genes are the rule rather than the exception in eukaryotic cells and their function has remained somewhat of a puzzle. They have been recog-nized as sources of rapid generation of polymorphism. Craik et al. (1983, p. 1125) write: 'A comparison between eukaryotic gene sequences and pro-tein sequences from homologous enzymes from bacterial and mammalian organisms shows that intron-exon junctions frequently coincide with vari-able surface loops of the protein structures' . . . 'since intron-exon junctions map to protein surfaces, the alterations mediated by sliding of these junc-tions can be effected without disrupting the stability of the protein core' . . . 'junctional sliding provides a means for diversification of genes. This important function may be one evolutionary reason for the existence of seg-mented genes.'

Just as in sexual recombination, the benefit for evolution in the long run cannot be the selective force that maintains the intron-exon mechanism. Instead the appearance of the splicing sites on the core of enzymes that leaves the cores of enzymes intact fits perfectly with our hypothesis that variety is a defense against the molecular intrusions of microparasites. Toxin binding sites would not be the same as the active sites of enzymes. Hence variation of the binding sites can throw off the parasites without interfering with enzyme function.

## Conclusion

We have presented a theory of evolutionary stability of sexuality which identifies three different circumstances that favor sexual reproduction over asexuality.

1) Recombination is a repair mechanism for microorganisms that live in temporally or spatially inhomogeneous environments.

2) Recombination in a polymorphic population is a source of diversity which confounds pathogen adaptation in gene-on-gene interactions.

3) Recombination in a polymorphic host population is a source of indi-viduality with respect to 'self'-defining MHC allele combinations (diplo-types) that distinguish self from nonself in systemic immune defenses.

Proposal 1 could be tested in experiments with bacterial cultures, for example by shifting *E. coli* from a growth medium in which genes for lactose metabolism have been repressed to a medium rich in lactose. Proposal 2 also could be tested in cultures of bacteria and bacterial viruses (phages).

Both *E. coli* and phages have been studied extensively since the early days of molecular biology and experiments could build upon this knowledge.

Diseases of crops and forests provide a natural laboratory for our theory. Assays of better discriminating power than electrophoretic methods should be developed for the study of polymorphism (or the lack of it) in natural and cultivated crops and forests. The genomes of pathogen strains could be sequenced, the conformations of their gene products determined, and differences between strains characterized, as has been done for some human viruses, notably the influenza virus. The gene-on-gene interaction of host proteins with corresponding parasite proteins could be studied. A better understanding of these interactions, their polymorphism, and their population dynamics could be important for the genetic management of crops and forests.

Proposal 3 could be tested in naturally occurring populations of parthenogenetic lizards and fish. Tissue typing could determine whether such populations are clones. Estimates of the time elapsed since becoming parthenogenetic would be important. Sexually reproducing but highly inbred populations of mice are equivalent to parthenogenetic clones. Susceptability of such populations to disease could be studied in the laboratory and compared with susceptability of wildtype mice. Here the theory of commensalism between host and parasite would have to be taken into account. Even if the hosts were to be defenseless and at the mercy of their parasites, the latter still would have to be commensal.

Until the late 1970s genes, except for rare mutations, were assumed to be quite stable. The purported stability was reflected in the models of mathematical genetics. Then transposable elements, introns and exons, mRNA splicing, sliding splice junctions, and gene conversion were discovered. New types of mathematical models are required.

## New mathematical models

The newly discovered richness of processes causing genetic variability is too complex for realistic mathematical models of the molecular changes involved. Instead one can characterize alleles through their phenotypic effect on model parameters. (This approach is analogous to the game models of Maynard Smith, 1982, where the model parameters are values that determine game strategies. Mutations are treated as perturbations of these values.) Molecular processes generate breakthroughs from time to time: alleles that confer host resistance, a parasite allele that overcomes the resistance conferred by a specific host allele, or a new parasite allele that changes the trade-off between transmission rate and damage to the host.

We are thus led to models of host-parasite population dynamics that allow subpopulations of host and parasite strains carrying different alleles, alleles which cause different values of the phenotypic parameters: birth and death rates, excess death rates, and strain specific transmission/infection rates. New mutant strains cause perturbations in the system. The effect of perturbations can be studied: whether the mutant strain spreads or dies out, whether it comes to dominate, or coexists in a polymorphic equilibrium.

For an asexually reproducing host population Bremermann and Fiedler (1985) have shown that a stable polymorphic equilibrium between hosts and parasites is possible, where each parasite strain regulates a corresponding host strain to a level below the carrying capacity of the environment and where the total population nevertheless approaches the carrying capacity. If a host strain appears that is resistant to all parasite strains it will drive susceptible hosts to low population levels until the corresponding parasite infections can no longer be sustained and die out. After this the host population remains polymorphic indefinitely, with the resistant strain dominating, unless resistance involves a cost in the form of a lower reproductive rate. Bremermann (unpublished) has recently extended these results to sexually reproducing host populations for the case of gene-on-gene resistance and recessive susceptibility.

The fate of parasite mutants that differ in their trade-off between transmission rate and damage to the host have been modeled by Bremermann and Thieme (1987). If hosts are susceptible to all strains but infection by one strain precludes a simultaneous infection by another, then asymptotically all strains die out except those that maximize the basic reproductive rate of the parasite (as defined by Anderson and May, 1982). This corroborates a static optimization argument by Bremermann and Pickering (1983) and is a special case of a game-theoretical optimization model.

Game-theoretical models of populations where mutants affect strategy parameters have provided many insights (Maynard Smith, 1982). Host-parasite interactions can be viewed as games in which each player optimizes (through mutants) its own fitness. Game theory then yields criteria for optimal strategies such as the Nash equilibrium in noncooperative games (Bremermann and Pickering, 1983).

Optimal strategies, however, do not always correspond to dynamic population equilibria. The pay-off in these population games is in terms of increased or decreased subpopulations whose alleles determine the strategies which they play. The game models thus are imbedded in dynamic population models. In other words, game models are simplifications of the type of models described above, where mutant strains may invade the population. Optimal game strategies may be easier to compute than dynamic equilibria. Sufficient conditions for optimal game strategies to be locally stable dynamic equilibria have been given by Maynard Smith (1982) (Evolutionarily Stable Strategies' – ESS for short).

In the models of Bremermann and Fiedler and Bremermann and Thieme the game approximation was not necessary. In Bremermann and Pickering, however, the full dynamic description would have been so complex as to be intractable. Here the game-theoretical formulation is used as a simplification. It predicts the breakdown of host-parasite commensalism under certain conditions. In a forthcoming paper (Bremermann, unpublished) it will be shown how such a breakdown may be responsible for the high mortality that is associated with AIDS.

## Summary

A theory of sexuality and polymorphism is proposed in which diversity at the molecular level is the adaptive response of multicellular organisms to the challenge of microparasites that have smaller genomes, shorter generation times and that can evolve more quickly than their hosts. The theory has implications for genetically homogenized crops and other cultivated plants as well as for immunology. A different function of sexuality is proposed for microorganisms that reproduce both asexually and sexually. Several possible experimental tests are discussed. Mathematical modelling techniques are outlined qualitatively and compared with game-theoretical methods which may be interpreted as simplifications of population dynamic and genetic equilibria. Some results about equilibria, stability and extinction in the population dynamics of polymorphic host-parasite populations are referenced.

*Note added in proof.* In his review of Margulis and Sagan's recent book "Origins of Sex: Three Billion Years of Genetic Recombination", Maynard Smith (Nature, Jan. 22, 1987), finds "attractive" the idea that cycles of mixis and meiosis in protozoans are a response to a cyclical environment. This hypothesis has previously appeared in Bremermann, 1979, Appendix B, together with suggestions how to test it experimentally. – In the same review Maynard Smith proposes that viruses (like bacteriophage lambda), that insert into a bacterial genome, and which abandon the host when it has been damaged (for example by UV), do so as part of a strategy to save their own genes, rather than those of the host. The same suggestion was also made in Bremermann, 1980b and 1983b, and analyzed by means of a host-parasite population model. The advantages of lytic versus lysogenic infection depend upon the density of the host population and the genetic survivability of the host. This question is important for the evolutionary origins of sex.

## Addendum: How can the theory be tested?

*Direct tests*

The theory predicts that genetically homogeneous host populations provide a target of opportunity for microparasites to evolve 'molecular technology' that enables them to overcome host defenses. Crop failures can be seen as uncontrolled experiments confirming this prediction. Controlled laboratory experiments would be desirable, but are inherently difficult: the time until a pathogen evolves to which the host population is susceptible may be too long for the researcher's patience, grant support, or life-span. Even if in a laboratory a host population gets sick, the experiment may not be very convincing: laboratory populations get sick all the time – for other reasons. One would have to show that the microparasite has actually evolved and has broken through host defenses.

Actually, in a different context, an experiment with genetically homogeneous populations of mammals has been going on for decades. There now exist many strains of mice that have been inbred (by full sib matings) for over 100 generations. The common ancestor of these strains is estimated to have lived about 150 years ago (Fitch and Atchley, 1985). At least 16 strains are well documented since 1920. Inbred mice have been studied extensively in immunology and cancer research. With little additional effort they could serve to test our theory.

### Tumor viruses and cancer in inbred mice

Inbred strains of laboratory mice have a high incidence of cancer. In the extreme case, 100 % of C3H virgin females developed mammary tumors in their first 8.8 months (Murphy, 1966, p. 521).

Since the mouse strains are genetically homogeneous, our theory predicts that pathogens evolve which escape immune recognition. Now laboratory populations of inbred mice are kept under highly sanitary conditions, and diseased stock is destroyed and prevented from infecting healthy populations. Tumor viruses, however, are another matter. In contrast to human cancer, murine cancer has a strong viral component (T.-W. Fiennes, 1982; Gross, 1983). Tumor viruses are slow viruses, and they are commensal (in the sense of Anderson and May): their damage to the host is a compromise between their need to optimize their own reproduction and shedding, and their need not to kill the host in which they reproduce too quickly (Bremermann and Pickering, 1983).

The propensity of inbred strains of mice to develop tumors may thus constitute a confirmation of our theory. It would be desirable to test the antigenicity of the tumor viruses in coevolved hosts, as well as wildtype populations, and to test their ability to induce tumors in wildtype populations. It would also be interesting to compare the tumor viruses of different inbred mouse strains, and to cross test their tumorigenicity.

Since tumor viruses can be transmitted vertically as well as horizontally, this question connects with the issue of infection of offspring by their own parents (Rice, 1983a; Stearns, this volume).

### Plants: Seeds of destruction

Clonal propagation is popular in horticulture. We have pointed already to the long list of crop failures as uncontrolled experiments that confirm our theory, which predicts that history will repeat itself as long as genetically homogeneous cultivar populations are propagated without preventive genetic management.

The latest addition to clonal propagation are orchids, which otherwise show great variability and are difficult to grow. In the Netherlands clonal propagation of orchids has become a major industry. It remains to be seen if varieties will become susceptible to disease.

Experiments are also underway with clonal propagation of trees. Depending upon the time constants, genetically homogeneous forests could be an invitation for commercial disaster. It has been argued that the widespread damage to forests in Europe is in part due to the lack of genetic variability after centuries of tree farming. However, experts do not agree on the causes.

*Recency of parthenogenetic species*

Other things being equal, asexually reproducing species have a reproductive advantage over their sexually reproducing relatives (Cuellar, 1977; Maynard Smith, 1979; Cherfas and Gribbin, 1984). Our theory predicts that sooner or later the balance tips when a clone-specific pathogen appears. The time-span during which an asexual species will flourish before facing extinction cannot be predicted with precision, but it is bound to be short in terms of evolutionary time-spans.

Asexual reproduction is not uncommon among plants (cf. the chapter by Bierzychudek). In conformity with our theory, as a rule, such species bear the mark of recency: they seem to have evolved from sexually reproducing ancestors not very long ago (on an evolutionary time-scale). The dandelion, *Taraxacum,* is an example: the asexual species still produce bright yellow flowers, as if to attract pollinators, but pollination is no longer required, for seeds are produced apomictically (Bierzychudek, 1985). Parthenogenesis is less common among animals. For a survey (especially lizards), see Cuellar (1977). As a rule, parthenogens have close relatives that reproduce sexually (Williams, 1975; Maynard Smith, 1986). The *hypothesis of recency* can now be tested independently of morphology, and with greater precision, by means of molecular clocks. Other models (sex as insurance against environmental heterogeneity and environmental shifts) also predict recency. However, if a molecular test of recency were to fail, it would falsify all these theories, including ours. In order to distinguish between them, one should look for asexual species in environments that remain stable over long stretches of time, such as the deep sea.

Since the timescales are short by geological standards, and plants rarely make good fossils, recency up to now had to be judged by taxonomic criteria. Now, however, a more precise documentation of phylogeny has become possible. Protein sequencing and DNA sequencing allow comparison of species at the molecular level. In all organisms there are many amino acid and basepair substitutions which are selectively neutral (Kimura, 1983a). Upon speciation, neutral mutations accumulate, but the process is random, and the genomes of the evolving species drift apart, becoming more and more dissimilar. The discrepancy can be measured and related to the time since divergence. Mammals, for example, diverge roughly at the rate of 0.2% of DNA basepair substitutions per million years.

The methodology is quite novel. It was first suggested by Zuckerkandl and Pauling in 1962. Initially proteins were compared. DNA sequencing has been possible only since 1977. The method has been applied mainly to mammals and birds, especially to the question of human-primate phylogeny. Plant molecular biology is still in its infancy.

Individual proteins evolve at different rates. It has been postulated that, averaged over the entire genome, DNA substitutions accumulate at a uniform rate (Sibley and Ahlquist, 1984) and thus can serve as a 'molecular clock'. There has been discussion whether the rate of the clock is uniform across a wide range of species, or whether rates differ (Lee et al., 1985; Britten, 1986). The differences, however, amount at most to a factor of 5. Calibrations of the clocks are made by dating selected phylogenetic branch-points by means of the fossil record, by the separation of the Old World and New World through continental drift (80 million years ago), and by the colonization of volcanic islands, such as the Hawaiian islands, which emerged from the sea (Beverly and Wilson, 1985).

The literature contains mainly data about the molecular clocks of vertebrates (especially mammals and birds), fruitflies, and sea urchins. For shorter timespans, of the order of thousands of years, restriction site polymorphisms of globin genes in populations have been used as markers. In this way the origin of contemporary human populations has been traced. It has been estimated, for example, that anatomically modern man evolved in Africa some 50,000 to 100,000 years ago (Smith and Spencer, 1984) and that all populations outside Africa arose from a very small group that migrated from Africa some 50,000 years ago (Wainscoat et al., 1986). So far, only questions of human phylogeny have been studied by this method. Still in its infancy, molecular clocks promise to become as useful for phylogeny as radiometric dating for geology.

It should be interesting to see the molecular clock methodology applied to plants. For the question of sexual versus asexual reproduction, plants offer more opportunities than animals since parthenogenetic animal species are rare. It should be of interest to assess whether molecular evolution proceeds at the same rate in sexually and asexually reproducing species. According to Kimura (1983a, p. 42), the rate of substitution in an asexually reproducing population is about the same as in a sexually reproducing population, provided it consists of a large number of lines which are subject to frequent extinction and subsequent replacement. (This is important; all the data in the literature on molecular clocks refer to sexually reproducing species.)

*Faster rates of microparasite evolution*

An important feature of our theory is the notion that microparasites are capable of rapid adaptive evolution by which they could take advantage of host clones. There are few data on pathogen evolution, with the exception of the influenza virus and the AIDS virus. (Various names have been suggested for this virus: HTLV–III, LAV, ARV, and HIV.) The AIDS virus, since its discovery in 1984, has received extensive scrutiny. For several isolates the entire genome has been sequenced, for others the viral proteins have been sequenced; differences have also been assayed with restriction maps. The results are extraordinary: the substitution rate is of the order of 10 million times faster than the rate for mammals and birds!

Even though the first known cases of AIDS occurred less than 10 years ago, there are now considerable sequence differences between different isolates. Ratner, Gallo and Wong-Staal (1985) report a heterogeneity as high as 17% in the extracellular portion of the envelope *(env)* protein (between clones from New York and California) with an overall difference of 9.2%. They report greatest nucleotide conservation in the *gag, pol* and short open reading frame *(sor)* genes. Clones from the same area (New York) differed typically at 1.5% of the nucleotides and 2.2% of the amino acids. (For primates this would correspond to about 10 million years since separation of the ancestral lines!) Sanchez-Pescador et al. (1985) point out that many other viruses show sequence variation during passage. Montelaro et al. (1984) have demonstrated this phenomenon for equine infectious anemia virus (EIAV), which leads to differences in the *env* protein of progeny virus, probably as a consequence of immunological selective pressures in the host.

How can this discrepancy in substitution rates be explained? Note that retroviruses have the benefit of the host's 'molecular technology': the same fidelity of DNA replication, transcription and translation. There is one difference, however: The viral genome is reverse transcribed from RNA into DNA by reverse transcriptase, then transcribed into RNA again. The transcription from DNA to RNA has a much lower fidelity than DNA replication (by a factor of $10^4$ to $10^5$, Kirkwood et al., 1986). The mutation rate therefore is this much higher for the virus. The second difference is that in the host the virus is under incessant immune pressure. The immune system can respond to any new antigenetic determinant within days. It is indeed the envelope protein which evolves most rapidly, while the *env* and *pol* sequences of the genome are much more conserved.

The rapid evolution of the AIDS and similar viruses confirm the view that microparasites can evolve orders of magnitude faster than their hosts.

*Variety through recombination of exons*

According to our theory the function of recombination is to generate molecular novelty in membrane receptors and other target molecules of pathogen attack. The problem is that amino acid substitutions that are buried in the interior of an enzyme or channel may disrupt the function while doing little about the 'antigenicity' with respect to the pathogen. The solution to the problem seems to be genes segmented into exons and introns and processing of messenger RNA. Introns are ubiquitous in eukaryotes (while absent in prokaryotes – Darnell et al., 1986). We propose that the generation of immunoglobulin variety and 'defensive variety' (as emphasized by our theory) are both based on recombination of exons with introns serving as spacers and sites of crossing over.

The generation of diversity for immunoglobulins has been studied extensively for many years. (For a review see Tonegawa, 1983; Hood et al., 1985). The generation of variety involves somatic recombination, a process by which any one of a family of V gene segments is combined with a J and C gene segment. Randomly selected pairs of V and J genes create a variety of V-J coding pairs, which are then combined with one or several C gene segments (Lewin, 1985, Chap. 37). In this way the immune system in the mouse generates between $10^6$ and $10^9$ different types of antibody molecules. This repertoire is able to bind any invading antigen, and the binding clone of lymphocytes is then stimulated to proliferate till the antigen is eliminated (Alberts et al., 1983).

It has recently been shown that in the MHC (major histocompatibility) complex of the mouse there exist introns that are 'hot-spots' of recombination (Kobori et al., 1986). The recombination site studied by Kobori is associated with a tetramer tandem repeat sequence. Assembly from gene segments which are separated by introns is not limited to the immunoglobulins. Most eukaryotic genes are assembled from segmented coding sequences (exons) which are separated by noncoding introns. Studies of the human betaglobin locus and human insulin locus also indicate hotspots of recombination, each associated with tandem repeat sequences which may promote crossing over. Kobori conjectures that recombination, generally, is not random and may be concentrated in hotspots, scattered throughout the genome.

The principle of splicing segments of precursor RNA into mRNA makes sense if molecular variety is to be generated. The splice junctions of exons tend to translate into loops on the surface of proteins (Craik et al., 1983). There is no consensus about the function of the exons and introns. It is clear, however, that recombination is not a molecular accident but a controlled process that reassorts subunits of genes and thus creates molecular variety. We propose that exons are like interchangeable parts that can be exchanged without disrupting protein function while altering structural details that are exploited by parasites. Thus variety is generated systematically, and without loss of function of the proteins in question. Random mutations, in contrast, would be disruptive, and they are much rarer than recombination.

Without the need for variety the complicated processing of hnRNA and the degradation of the larger part of it into monomers would seem to be wasteful. The crux of the matter is what the variety is for. In their monumental text, 'The Molecular Biology of the Cell', Alberts et al. (1983, p. 471) state that recombination events create new types of proteins by joining coding domains, and that without introns there would be fewer opportunities. In contrast to this author they see the generation of variety in the context of long-term evolution, not anti-parasite defense. While the process undoubtedly is beneficial, evolution occurs on long timescales, and their argument is equivalent to the old group selectionist argument that recombination occurs because it speeds up evolution. We propose that the highly complex molecular machinery is primarily a defense against microparasites. Prokaryotes do not have this machinery, and they are parasitized by numerous bacterial viruses. Yeast, in contrast, is parasite free.

*The ultimate sexually transmitted parasite?*

Another puzzle, that has only recently come into focus, is the prevalence of noncoding DNA. According to the latest estimates only 5 % to 10 % of the DNA in the human genome is coding, possibly even less (Cavalier-Smith, 1985). The noncoding DNA has been called 'selfish' (Orgel and Crick, 1980; Doolittle and Sapienza, 1980). It has also been referred to as 'junk DNA' (Alberts et al., 1983). Cavalier-Smith (1985) has compiled data on the percentages of non-transcribed DNA in different species, which are rarely below 50 % and as high as 90 %. Also, closely related species, which presumably would need similar genes, can have vastly different amounts of DNA ('C-values'). This is known as the 'C-value paradox' (Lewin, 1985, p. 294–295).

Successful male gametes could spread selfish DNA rapidly throughout a population. Thus, selfish DNA would be the ultimate sexually transmitted disease, with the genome itself as vector (Hickey, 1982). Asexually reproducing species would be in a better position to suppress selfish DNA and thus be at an advantage, unless noncoding DNA has a function after all. While the papers in Cavalier-Smith (1985) present data on many species, the authors have not investigated a possible correlation between sexual and asexual reproduction. Of course, if noncoding DNA has the same function in either case, then no correlation can be expected. In the following we propose that the puzzle of noncoding DNA, again, may have a solution in that the phenomenon is part of the defenses against microparasites.

*Absorber theory of noncoding DNA*

We would like to propose that noncoding DNA could serve as a 'trap' for retroviruses. Retroviruses insert their genome into the host genome where it gets replicated along with the host DNA. For viral reproduction transcription of the virus is necessary, but transcription is regulated and depends upon the site. A virus that inserts at random thus may not be transcribed at all. The host can defend itself by having many 'dummy sites' which are not transcribed. In fact transcription signals are located many basepairs upstream from the start of transcription. In either case, whether insertion is random or site specific, a host can protect itself from virus attack by providing as many sites as possible where viruses are not transcribed and therefore not multiplied.

The function of the noncoding DNA would be analogous to that of the control rods in a nuclear reactor that trap enough neutrons to prevent a runaway chain reaction producing neutrons which split atoms which produce more neutrons which split more atoms, etc. Both a chain reaction and virus replication are 'renewal processes' which can be prevented from becoming a runaway reaction by the insertion of absorbers. No sexual recombination is required for having 'absorber sequences'. However, unequal crossing over, especially in tandem repeat sequences, can quickly increase or decrease the amount of satellite DNA and would allow a species to adjust its 'absorber sites' to the level necessary to prevent viral chain reactions within its particular niche. Cavalier-Smith has documented the rapid evolution of the C-value in related species.

*Mate choice, sperm selection*

For inbred strains of mice, female choice of mates with different MHC alleles over mates with identical MHC alleles has been reported (Yamazaki et al., 1976). Subsequent experiments have confirmed that congenic mice can make this discrimination on the basis of scent and the scent of urine (Yamazaki et al., 1979). This demonstrates that behavioral mechanisms exist which promote MHC polymorphism. In mammals, including man, there exist other mechanisms, notably spontaneous abortion, that promote MHC polymorphism. For a brief review see Jones and Partridge (1983) [and references cited in Stearns' final chapter – Ed.].

This author (Bremermann, 1980) has suggested that there should exist mechanisms in mammalian fertilization that favor heterozygosity at the MHC loci (and possibly at other loci). The mechanisms envisioned were: interaction with lymphocytes as sperm traverse the uterus toward the fallopian tubes, and secondly, interaction with the zona pellucida as sperm compete to enter the ovum. Analogous selection processes take place at stigma, style, and ovule of flowering plants and are governed by one or several self-

160

incompatibility loci (DeNettancourt, 1977). Experimental work (elsewhere) to test this hypotheses is in progress. The phenomenon, if it exists, could be important for otherwise unexplained cases of infertility.

Fitch and Atchley (1985) studied genetic variation at 97 loci in ten commonly used inbred strains of mice. They studied 10 strains of mice whose history is well documented and which descend from a common ancestor 150 years ago. They studied these strains to determine whether the observed genetic drift could be in accordance with the amount of difference predicted by the theory of molecular clocks (which were discussed above). To their surprise they found that the variation was greatly in excess of what could have been expected statistically if present strains had emerged by random sampling of the originally present alleles. After considering various hypotheses they conclude that the data can best be explained if there has been a mechanism selecting for heterozygosity. They suggest that there was such a selection early on when breeders selected the healthiest animals for further breeding: the healthiest animals being the most heterozygous (due to the inbreeding depression that occurs when an inbred population is being established). It is not clear, however, whether the genetic load of mutations that are detrimental in homozygotes, and which has a transitory effect when an inbred population is first established, is great enough to account for the heterozygosity reported by Fitch and Atchley. Our postulated sperm selection in favor of heterozygotes could be a contributing effect, or possibly even the dominant effect.

### Summary

The theory predicts that inbred strains of mice are susceptible to tumor viruses that have escaped immune recognition. The theory implies routine evolutionary extinction of asexual species and that existing parthenogenetic species would have to have evolved recently from sexually reproducing precursors. This could be tested by determining the amount of DNA divergence between parthenogenetic and related sexually reproducing species. In mammals and birds genetic substitutions occur at comparable rates, but the AIDS virus evolves about 10 million times more rapidly. This confirms the basic imbalance of rates of evolution between microparasites and metazoan hosts, which is fundamental to the theory. The organization of most eukaryotic genes into exons and introns facilitates the generation of variety of gene products, and the molecular mechanism is similar to the mechanism that generates antibody diversity in the immune response and antigenic variation in trypanosomes. It has been proposed that noncoding 'selfish' DNA is the ultimate sexually transmitted disease. If this were the case, then asexually reproducing species would have an added advantage. An alternative hypothesis is proposed: noncoding DNA could provide insertion sites for retroviruses that would prevent them from being transcribed and replicated and thus moderate their proliferation much as absorber rods moderate proliferation of neutrons in a nuclear reactor. Flowering plants have pollen selection mechanisms that enforce heterozygosity at one or more loci. It has been proposed that analogous sperm selection mechanisms exist in mammals. Such a process would account for observation of a mysterious excess molecular divergence between different strains of inbred mice.

# Resolving the paradox of sexual reproduction: A review of experimental tests

P. Bierzychudek

Theoretical biologists have responded enthusiastically to the challenge of resolving the paradox of sex: why is sexual reproduction so widespread even though its practitioners suffer a twofold numerical disadvantage? At least five books (Ghiselin, 1974; Williams, 1975; Maynard Smith, 1978; Bell, 1982; Shields, 1982) and a myriad of papers have sought to identify conditions that could mitigate the twofold disadvantage of sex. Curiously, however, even though over a decade has elapsed since the publication of the earliest of these books (time enough for two generations of doctoral theses), this flurry of theoretical work has not been followed by anything near an equivalent amount of experimental activity.

Here I review the few experiments that either address directly any of the various hypotheses that have been proposed or that examine the assumptions underlying those hypotheses. Since there are not many such experiments, this review will be brief, and one of its aims will be to persuade laboratory and field workers of the need for additional experimental work. Before examining the data, I will briefly review the essential elements of the different hypotheses that have been proposed. Because everyone who writes about these hypotheses describes them differently and gives them different names (Williams, 1975; Bell, 1982; Felsenstein, 1985; Antonovics et al., in press), and because the hypotheses themselves have evolved over time, it is important to clarify what I mean by, for example, 'lottery models' or 'minority-advantage models'.

## Hypotheses for/models of the evolution of sex

*Hypotheses focusing on genetic diversity among progeny*

Most theoretical studies of the advantages of sexual reproduction focus on the advantages associated with the production of genetically-variable progeny, the most obvious consequence of sexual reproduction. These can be divided into two classes, depending on whether their salient feature is 1) the operation of frequency-dependent selection, of 2) the existence of environmental heterogeneity, either in space or in time. (However, at least one model – the first – possesses both of these features.)

*Natural enemies and minority advantage.* Levin (1975) was the first to point out that predators, pathogens, and parasites would depress the fitness of common phenotypes to a greater extent than that of rare ones. This idea, sometimes termed the Red Queen hypothesis, has been expanded upon by many other workers (Jaenike, 1978; Glesener, 1979; Hamilton, 1980; Hamilton et al., 1981; Hutson and Law, 1981; Bell, 1982; Hamilton, 1982; Rice, 1983a; Bremermann, 1985), who have described the advantage of rarity either as reduced apparency to predators or as escape from exploitation by means of the possession of novel defenses. This hypothesis at first glance does not seem to require the existence of environmental heterogeneity for its operation. However, it is based on the expectation that the organisms comprising a species' biotic environment are changing over time, at least behaviorally, and possibly genetically as well. As predators exploit common prey types, novel types will enjoy an advantage, but only until their frequency increases and predators respond either by a shift in behavior or by a coevolutionary genetic change. This change in the biotic environment over time generates negative heritability in fitness, a condition originally described by Maynard Smith (1978) as a powerful selective force favoring sex. Maynard Smith described this condition as an implausible one, but he was focusing on change in the physical environment rather than change in the biotic environment; later workers (Bell, 1985b; Bremermann, 1985; Stearns, 1985) have emphasized that predators and disease could generate negative fitness heritability. In this model, the advantage to sex is independent of population density.

This hypothesis is currently enjoying considerable favor, though May and Anderson (1983a) caution that unless the pathogens in question are lethal, the frequency-dependent selective forces they generate will be too small to overcome the twofold numerical advantage of asexual genotypes. They concede, however, that this conclusion depends on the specific details of the parasite-host model used.

*Sib competition.* A second cause of frequency-dependent selection is sibling competition. If a genotype's success in a patch of competing sibs is inversely related to the number of individuals competing for the same resources, then a rare phenotype (with different resource requirements) will have higher fitness (Williams, 1975; Maynard Smith, 1978; Young, 1981; Price and Waser, 1982). This hypothesis does not depend on the existence of patches that are physically different from one another, but it does rely on there being a spectrum of resources available within a patch that can be partitioned among siblings. Under this hypothesis, the advantage to sex is density-dependent: the denser the progeny, the more severe the competition between them, the greater the advantage of novelty. This hypothesis has been termed by Young (1981) the Elbow-Room model to distinguish it from other hypotheses that also depend on sib competition.

This Elbow-Room model is extremely similar to Bell's Tangled Bank model (1982, 1985; see also Ghiselin, 1974), in which the environment is viewed as consisting of multiple niches. It is assumed that genotype x environment interactions exist, such that no one genotype can be 'best' in all of the niches. A sexual parent is expected to have greater success in such an environment because it can fill more of the niches.

*Lottery models and habitat heterogeneity.* A related group of hypotheses are known as lottery models (Williams and Mitton, 1973; Williams, 1975; Maynard Smith, 1978; Taylor, 1979; Bulmer, 1980a) or the Best Man model (Bell, 1982). Here again the environment is considered as a series of patches, but the heterogeneity exists between patches rather than within them. Each patch is colonized by progeny from several parents, but only one individual 'wins' (survives to reproduce). As was true of the Tangled Bank model, these models require the existence of a genotype x environment interaction. However, selection is frequency-independent.

Lottery models have two variants. In one version, each parent need contribute only one of its progeny to each patch. When this is the case, it is necessary that the environment be heterogeneous in a way that Bell (1982) has termed 'capricious', i. e. it is necessary that there be negative heritability of fitness from one generation to the next (Maynard Smith, 1978). In a different version, several progeny from each parent are dispersed to each patch. This removes the stricture of negative fitness heritability, and only requires that the heritability of fitness be zero (i. e. that environments be temporally unpredictable). In other words, it is expected that the fitness of asexual offspring will decrease as they are dispersed to increasing distances from the parent. Under these conditions, sexual parents are considered to have a greater chance of producing a winner, because of the greater variance in fitness expected among the progeny of a sexual parent (thus the appellation 'best man'). The more siblings dispersed to each patch per parent, the greater the chance that a sexual parent will produce the winner.

These models only predict an advantage for sex when there is extreme truncation selection (i. e. only one survivor per patch). When selection is relaxed, the advantage to sex disappears.

### DNA repair hypothesis

Not all of the theories that have been advanced to explain what maintains sexual reproduction focus on the importance of producing genetically-variable progeny. Bernstein and his colleagues at the University of Arizona (Bernstein et al., 1981; Bernstein, 1983; Bernstein et al., 1985) argue that recombination serves other functions besides the generation of genetic variation among progeny, and that these other functions could be the primary selective forces maintaining sexual reproduction. They point out that double-stranded DNA damage, such as that caused by the oxygen radicals generated during cellular respiration, can only be repaired if a suitable template exists. The process of meiosis, during which chromosomes synapse with homologues, provides that template. Therefore, they propose that the sexual reproduction evolved and is maintained for its ability to provide a means of repair of DNA damage.

## Tests of the models' assumptions

The assumption upon which all the above models are predicated is that there is a twofold numerical disadvantage associated with the production of sexual progeny. Several workers have compared the fecundity of sexual parents with that of closely-related parthenogens or apomicts. Although not very many such studies have been done, preliminary indications are that the 'twofold disadvantage' is an over-estimate of the 'cost of sex' in insects (the only animal group in which parthenogenesis is common), but an under-estimate of its cost in plants. Lamb and Willey (1979) review studies on hatching success in laboratory populations of insects, and conclude that, on average, parthenogenetic insects have a hatching success only 67% that of sexual insects.

Templeton (1982) describes evidence that the fecundity of newly-risen parthenogenetic lines of *Drosophila* may be much less than that of sexual lines, and, even after such lines have become well-established, their rate of production of females may only equal that of sexual lines, rather than exceeding it. By contrast, Michaels and Bazzaz (1986) and Bierzychudek (in press) have demonstrated that apomictic femals of species of *Antennaria* (Asteraceae) produce significantly more seed than sexual females growing in the same habitat. This implies that apomictic females are producing even more than twice as many female seeds as their sexual relatives (if we assume that one-half of a sexual female's seeds are female, whereas all of an apomictic females' seeds are female). In both of these species of *Antennaria,* sexual females require visitation by insects in order to produce seed, and Bierzychudek (in press) suggests that it is insufficient pollination that limits the fecundity of sexual females. Such pollinator limitation is probably not a rare phenomenon (Bierzychudek, 1981).

The lottery model assumes that the progeny of a sexual parent will display greater variance in fitness than will the progeny of an asexual parent. Since asexual progeny are genetically invariant, this assumption seems like a reasonable one, but it is possible that the genetic differences among sexual progeny are irrelevant to important phenotypic characters like fitness. Bergelson and Bierzychudek tested this assumption by growing seed progeny from both sexual and asexual females of *Antennaria parvifolia* under controlled conditions. After one month (well before plants had attained sexual maturity), plants were harvested and their sizes measured. Contrary to expectation, sexual progeny failed to exhibit significantly more variance in plant size or in germination time than apomictic progeny did. The largest individuals at the end of the experiment were progeny of asexual females (Bergelson and Bierzychudek, manuscript). If seedling size is a good predictor of lifetime fitness, then these results suggest that sexual progeny may not necessarily be more likely to produce individuals with highest fitness.

The major portion of the experimental work testing these hypotheses or their underlying assumptions has been conducted by Janis Antonovics and his students and colleagues at Duke University. All of the members of this group have used the same experimental system: a short-lived perennial grass, *Anthoxanthum odoratum* L., sweet vernal grass. *Anthoxanthum* is a sexual species, reproducing by means of outcrossed seed (it is self-incompatible). However, individuals are also capable of tillering, i. e. of spreading vegetatively. The experiments of Antonovics and his colleagues have taken advantage of *Anthoxanthum's* ability to produce offspring that are genetically identical to their parent (by tillering) as well as offspring that are genetically different from their parent and siblings (by seed production). To assure that morphological differences between sexual and asexual progeny do not confound experimental results, the seed progeny are reared until they are equal in size to the vegetatively-produced tillers. Thus every individual *Anthoxanthum* genotype can be made to produce both sexual and asexual progeny that are 1) morphologically similar, but 2) genetically different. These two progeny types are then transplanted to either greenhouse or (more generally) field situations, in experimental designs tailored to test particular assumptions or predictions.

Experiments performed by Kelley (1985), working in association with Antonovics, to test the lottery model automatically provided a test of another of this model's assumptions, that of extreme truncation selection. The patches Kelley created (see 'Tests of the Models' Predictions – Lottery Models') rarely produced only a single survivor, and Kelley was forced to redefine the criteria for judging the winner of these contests. Browne (1980), who tested the lottery model in 'patches' of brine shrimp, also failed to observe patches being taken over by a single individual.

Obviously, all of the models that focus on the advantage of producing genetically-diverse progeny contain the implicit assumption that the probability of progeny success is a function of individual genotype. But in the case of non-motile organisms, the characteristics of an individual's microenvironment may overwhelm any genotype effect. Schmitt and Antonovics (1986a) tested the assumption that genotypes matter by growing seeds from six families of *Anthoxanthum odoratum* in the field. Their results suggest that parental genotype does indeed affect the performance of offspring, and that even half-sibs from the same maternal family may vary in performance. However, in a related experiment, the susceptibility of seedlings to aphid infestation did not depend on parental genotype (Schmitt and Antonovics, 1986b).

The results of Schmitt and Antonovics (1986a) also provide good evidence for the existence of genotype x environment interactions on a very local spatial scale, another assumption critical to many of the hypotheses for the advantage of sex. In another experiment with *Anthoxanthum,* however, genotype x environment effects, though present, were quite small (Antonovics et al., in press).

## Tests of the models' predictions

*Natural enemies and minority advantage.* Antonovics and Ellstrand (1984) tested the prediction that rare genotypes would enjoy higher fitness by planting asexual progeny from different genotypes of *Anthoxanthum* in two experiments of complex design; the essential feature of both experiments was that each genotype served alternately as a majority and as a minority type. The fitness of each genotype in each situation was assessed as the product of yearly survival and fecundity (number of inflorescences) over a 3-year period. Skewed fitness distributions prevented the data from being analyzed with parametric tests, and the complexity of the experimental designs created small sample sizes, further reduced statistical power and prevented a comparison of each genotype's performance when a minority with its performance when in the majority. However, the overall result was that, for one experiment, the mean fitness of majority genotypes was only 66 % of the mean fitness of minority genotypes, and, for the second experiment, the mean fitness of genotypes with one or two alien neighbors was nearly two times larger than the mean fitness of genotypes with no alien neighbors. Antonovics and Ellstrand (1984) were unable to identify the mechanism(s) responsible for the minority advantage they observed.

Additional evidence in support of this hypothesis comes from work done by Schmitt and Antonovics (1986b), again using *Anthoxanthum.* Seeds from reciprocal diallel crosses were planted into the field 1) alone, 2) surrounded by sibling neighbors, or 3) surrounded by unrelated neighbors. Unexpectedly, many of the seedlings were attacked by aphids, and the survivorship of plants surrounded by siblings (i. e. genetically similar individuals) was 77 % that of plants whose neighbors were unrelated (and thus genetically more dissimilar). Since this differential survivorship was not observed for non-infested plants, the advantage was assumed to be in some way related to the presence of the pests, but the mechanism is not understood.

Using as an experimental system the black pineleaf scale that infests Ponderosa pine, Rice (1983b) tested whether one generation of sexual reproduction could provide the host's offspring with protection against its parent's parasites. Rice transplanted scales from donor trees to a variety of new hosts with different degrees of genetic relatedness to the donor: hosts unrelated to the donor, outbred progeny of the donor (both egg and pollen progeny), inbred progeny of the donor, hosts cloned from the donor, and the donor itself. If the ability of the scale to establish itself on a new host is a function of the new host's genetic similarity to the previous host, then we would predict that scale survivorship would decline with increasing genetic distance. The results that Rice observed were quite ambiguous. While scale survivorship on inbred progeny was indeed significantly higher than scale survivorship on unrelated hosts, other pairwise comparisons were either not significantly different or gave results that were opposite to those predicted. The fact that Rice observed no significant difference in the survival of scales transferred to unrelated trees when compared to their survival on cloned trees or on the original host itself indicated that genetic similarity between donor and recipient was of little importance (Rice, 1983b). One detail of this experiment's design makes these results particularly difficult to interpret: all the trees used in this experiment, both donors and recipients, had been grafted onto unrelated rootstocks. How much of the tree-scale interaction is dependent on the genotype of the branch and how much on the genotype of the rootstock is unknown, but it seems likely that this grafting introduced an additional source of phenotypic variation.

Taking advantage of a mixed population of sexual and asexual (apomictic) *Antennaria parvifolia,* Bierzychudek (unpub. data) measured rates of predation by a specialized lepidopteran seed predator. The amount of loss sustained by sexual females was not significantly different from the losses suffered by asexual females. Differential rates of seed predation, then, appear not to be an adequate explanation for the persistence of sexual females in these populations.

*Sib competition.* Using the same experimental system as in their other experiment, Ellstrand and Antonovics (1985) grew genetically uniform (cloned) tillers and genetically variable (sexually-derived half-sib) tillers in a density gradient under natural conditions. Sexual and asexual progeny were not competing directly; each plot consisted of only a single progeny type. While the sexual progeny had significantly greater survivorship, fecundity, and growth than the asexual progeny (the overall relative fitness of the asexual progeny was 0.88), their advantage did not increase with increasing density, and so was unlikely to be a competitive advantage. In fact, the advantage enjoyed by the sexually-produced progeny was most marked at the lower densities. Ellstrand and Antonovics (1985) speculate that the sexual progeny's advantage might have been due to their escape through rarity from pathogens and parasites (as in the earlier experiment), but the idea that pathogens and parasites would be most destructive when their hosts are at low densities is counterintuitive. Another possible explanation for their results is that the progeny produced by seed would have had the chance to shed any pathogens their female parents may have had, an opportunity not available to the cloned progeny. If this latter explanation is correct, then the advantage of sex in this case rests not on the production of genetically-variable progeny, but on the production of propagules that can escape parental contamination. In such a situation, apomictically-produced seed, despite being genetically uniform, would be similarly protected. A similar experiment that avoids this complication of parental contamination is currently being conducted by Bierzychudek (unpub. results). This experiment is designed to compare the success of sexual and apomictic progeny of *Antennaria parvifolia* growing at two different densities. Both types of progeny are derived from seed, and thus differ only in their degree of genetic variation.

Another test of the sib competition model was performed by Weeks (1986) using laboratory populations of sexual and asexual lines of tadpole shrimp, *Triops longicaudatus*. Shrimp were grown in aquaria at a range of densities. Each aquarium contained either asexual or sexual shrimp, so the two types were not competing directly. The results were equivocal, with sexual lines outperforming asexual ones with respect to only some of the fitness measures used (survivorship, individual growth, individual biomass, but not fecundity), and then only at the lowest densities. Contrary to the model's predictions, at higher densities, the advantage to sexual lines decreased rather than increased.

Weeks used only 2 sexual lines for his experiments, one of which was completely homozygous and non-polymorphic at all of the 14 loci that were sampled electrophoretically. The only evidence that this line was sexual is that it produced both male and female progeny. The other sexual line showed rather low levels of variability as well; only 2 of 14 loci sampled were polymorphic (Weeks, 1986). These electrophoretic data make the results of the experiment difficult to evaluate; perhaps the experiments failed to confirm the model's predictions because the sexual lines Weeks chose did not have typical levels of genetic variability (Weeks, pers. comm.).

*Lottery models.* To determine whether the kind of sibling competition invoked by the lottery model operates in nature, Kelley (1985) performed additional field and greenhouse experiments with cloned and seed-derived tillers of *Anthoxanthum*. In Kelley's first experiment, the progeny of 'sexual' and 'asexual' parents (i. e. seed-derived and clonally-derived progeny) were planted together in patches. Each patch contained progeny from at least one 'sexual' and one 'asexual' parent (some treatments used 2 parents of each type); the number of progeny per patch from each parent, depending on treatment, was either 2 or 4. An additional treatment consisted of single tillers from each parental type. The sib competition version of the lottery model predicts that the relative advantage of sexual offspring should increase with numbers of progeny per patch. While Kelley did observe a significant overall advantage for sexual progeny (with total reproductive rates of 1.37:1), this advantage did not vary with patch size.

The lottery model is based on the assumption that only one individual survives per patch, and predicts that this 'winner' will be a sexual individual.

Kelley's patches rarely produced only a single survivor, so he defined the 'winner' as the individual with the highest lifetime fecundity, and assessed 'success' by the number of patches 'won' by a sexual offspring. By this criterion, sexual offspring enjoyed a greater than twofold advantage overall, and the magnitude of their advantage did increase with patch size. Though Kelley was unable to pinpoint a mechanism for the advantage he observed, he speculates that pathogens or parasites are likely to constitute the important selective force. However, Kelley's results can also be explained as demonstrating the 'escape' of seed progeny from their parents' pathogen loads, as was the case in the previous experiment.

Kelley (1985) performed a variant of this study in the greenhouse, with fitness measured as total fecundity over two years. The results were strikingly different from the results of the field experiment. Not only were there no significant differences in fitness between the two progeny types, but the 'winners' (as measured in the field experiment) were more often asexual than sexual, and the 'success' of the sexual progeny did not increase with patch size. Kelley offers two hypotheses to explain these different results: 1) greenhouse plantings were denser than field plantings, and Ellstrand and Antonovics (1985) have provided evidence that in *Anthoxanthum* there is an inverse density-dependent advantage for sexual progeny; 2) the selective factor that favored sex in the field (e. g. a pathogen or pest) may have been absent from the greenhouse.

A final experiment performed by Antonovics' group that is relevant to the lottery models is a study by Antonovics and Ellstrand (1985). Sexually-derived and clonally-derived tillers of *Anthoxanthum* were planted in the field around parent plants; the planting pattern was designed to simulate the natural pattern of seed dispersal around parents. The lottery model predicts that the fitness of asexual offspring will decrease as they are dispersed to increasing distances from the parent. However, the advantage enjoyed by the sexual tillers was independent of their distance from the parent. At very long distances from the parents (greater than 40 m), only sexual tillers had high fitness, but because few *Anthoxanthum* seeds disperse beyond 5 m, the significance of these results is questionable (Antonovics and Ellstrand, 1985).

Kelley believes that these results, taken together, fail to provide support for the sib competition version of the lottery model. Instead, Kelley believes that the influence of predators, pathogens, and parasites is a much more likely explanation for the maintenance of sexual reproduction. However, these experiments of Kelley and colleagues may give unrealistic importance to pathogens; because the 'asexual' progeny have been propagated vegetatively, instead of by seed, any pathogens a parent might possess can be transmitted directly to its asexual offspring, but not to its sexual ones. Pathogen transmission may be far less important for an organism that produces asexual propagules by other means (e. g. apomictic seed production, parthenogenetic eggs).

Browne (1980) conducted an experiment with sexual and parthenogenetic strains of brine shrimp, *Artemia salina,* very similar to Kelley's greenhouse experiment. Though Browne did not intend his experiment to be an explicit test of the lottery model, its design allows it to serve as one. Within individual culture jars, Browne placed either 40 sexual shrimp (20 males and 20 females) and 20 asexual female shrimp (high density treatment) or 12 sexual shrimp (6 males and 6 females) and 6 asexual females (low density treatment). The 2 strains were allowed to compete for 150 days, by which time all jars consisted solely of members of one of the two strains. In the low-density treatment, sexual individuals prevailed in 6 out of the 8 replicates; in the high-density treatment, sexual individuals prevailed in all 5 replicates. Thus, these results, unlike Kelley's, are in accord with the predictions of the sib competition version of the lottery model: the advantage of sex is more pronounced with larger numbers of individuals/patch. Browne speculates that asexual strains manage to survive, despite their poorer competitive abilities, as 'fugitives', continually dispersing to new habitats (Browne, 1980).

*DNA repair hypothesis.* Experimental testing of the DNA repair hypothesis has just begun, and no results have been published yet. Michod and coworkers at the University of Arizona are currently engaged in such tests using *Bacillus subtilis,* a soil bacterium. Some *Bacillus* cell lines are competent at transformation (the bacterial analog of sexual reproduction), while others are not. Michod and his colleagues have been comparing the survivorship of these two kinds of cell lines after exposure to agents known to cause DNA damage, and are finding that 'sexual' lines show higher survivorship under these conditions, as predicted by the DNA repair hypothesis (R. Michod, pers. comm.).

**Critique of empirical approaches**

So far, none of the experimental work that has been conducted has produced particularly satisfying results. Some experiments have been unable to demonstrate that sexual progeny really do enjoy an advantage over asexual progeny; others have succeeded in making this demonstration, but have not led to an understanding of a mechanism for the observed advantage. No one hypothesis stands out as having received an overwhelming amount of support. Why has so little progress been made toward solving this problem?

One of the most serious impediments to our understanding has been the confusion of models for the evolution of sex with hypotheses about what actually maintains sexual reproduction in nature. The models I described in the first part of this paper are logical syllogisms, descriptions of the necessary outcomes of particular sets of assumptions. If constructed with care, such models are flawless; if their assumptions hold, then the predictions will be obtained. To understand why sex occurs in nature, one has to evaluate the extent to which different suites of assumptions hold, or the manner in which opposing forces interact. Let me illustrate my point with a simplistic example.

Suppose I wanted to understand why the sidewalk in front of my house was wet, or why I frequently observed wet sidewalks. I can construct 3 'models' for the 'evolution' of wet sidewalks: model A states that when it rains, sidewalks get wet; model B states that sidewalks get wet when one waters a lawn with a poorly positioned sprinkler; model C states that rivers overflowing their banks wet nearby sidewalks. To understand why sidewalks get wet, it is of little avail to turn on a sprinkler and observe that it indeed waters the sidewalk. Rather, one would want to gather information about how often it rains in the area of interest, or how close sidewalks usually are to the banks of a river.

In the case of the evolution of sex, it does little good to set up an experiment that re-creates a model's assumptions, no matter how carefully, and then to observe the (inevitable?) outcome. The most useful studies are those that explore the models' assumptions: are they biologically realistic or not, likely or unlikely? Are genetically-variable progeny really different enough to enjoy reduced sibling competition? How common are genotype x environment interactions? Are fitnesses of asexual progeny dependent on their distance from the parent? What are the differences in fitness spread between sexually-produced and asexually-produced progeny? A few of the studies reviewed here have tried to answer such questions; many more are needed.

By focusing on the assumptions rather than on the outcome, on the confusing and strange results rather than on the expected ones, we will be better able to understand the nature of the selective forces that are at work, and will be able to gain some real understanding of the ways that predators, pathogens, sibling competition, environmental patchiness, and mutational repair have contributed to the maintenance of sexual reproduction. I think it is unlikely that such studies will ever result in our pinpointing 'the' advantage of sexual reproduction; instead, I expect that we will learn that a wide variety of selective forces have contributed to its success. We should be searching not for a simple answer, but for an appreciation of the factors that determine the relative importance of these different forces.

**Summary**

Though theoretical biologists have responded enthusiastically to the challenge of resolving the paradox of sex, there has not been an equivalent amount of experimental testing of the hypotheses they have put forward. This paper briefly reviews the essential elements of the models proposed to explain the predominance of sexual reproduction, and describes the existing experimental evidence. Some of the experimental approaches have involved tests of the models' assumptions; work done to date, for example, suggests that the theoretically twofold 'cost of sex' may be an over-estimate for animals, but an under-estimate for plants [see also the chapter by Lewis – Ed.]. The assumption that truncation selection occurs in nature has found no experimental support. Other experiments have tested the models' predictions. There is little evidence that sibling competition is mitigated by genetic diversity among progeny. Several studies have provided evidence that minority genotypes have higher fitness, but generally fail to reveal a mechanism for that advantage. Overall, no one model emerges as a likely single 'cause' for the maintenance of sexual reproduction. Future progress is likely to come, not from further tests of the models' predictions (which are necessary logical outcomes of particular sets of assumptions), but rather from examining the frequency with which their assumptions are met in nature.

Acknowledgment. I am indebted, as usual, to Peter Kareiva, my favorite gadfly, for his cantankerous intolerance of fuzzy thinking, my own not excepted. I am grateful to Janis Antonovics and Norm Ellstrand for providing me with unpublished theses and manuscripts. Field work on *Antennaria* has been supported by NSF grant BSR-8407468; field assistance has been provided by J. Banks, J. Bergelson, V. Eckhart, G. Hammond, M. Nuttall, B. Roy, J. Smith, R. Smith, and J. Winterer.

# Genotypic characteristics of cyclic parthenogens and their obligately asexual derivatives

P. D. N. Hebert

## Introduction

Cyclic parthenogenesis is a mode of reproduction which involves the more or less regular alternation of asexual and sexual reproduction. During the asexual phase of the life cycle individuals develop from unfertilized eggs as opposed to other forms of asexuality in which new individuals develop from somatic tissues. In organisms employing this latter breeding system the timing of asexual reproduction varies from immediately after zygote formation (polyembryony) until late in adult life (fission). Many somatic asexuals also reproduce sexually and hence employ a breeding system which is very similar to cyclic parthenogenesis. While the distinction of somatic and cyclic parthenogens may be hard to justify on genetic criteria, their separation has a long historical basis (von Siebold, 1857; Weismann 1893) founded on the unique ability of cyclic parthenogens to produce both sexual and asexual eggs. Bell (1982) has advocated extending the use of the term cyclic parthenogenesis to include organisms which alternate between obligate self-fertilization and outcrossing. This seems unwise, for such a breeding system has genetic consequences very different from those of conventional cyclic parthenogenesis. Selfing leads to a rapid loss of heterozygosity (Wright, 1969), while during apomictic parthenogenesis heterozygosity is maintained or even amplified (Berger, 1976). On this basis the present review considers only traditional cyclic parthenogens.

The 1000 or so animal species which reproduce by obligate parthenogenesis are scattered, at low incidence, through a broad range of taxa (Bell, 1982). By contrast the 15,000 species reproducing by cyclic parthenogenesis belong to just seven taxonomic groups (Table 1). This species richness provides clear evidence for both the adaptive radiation and long-term persistence of cyclically parthenogenetic groups and indicates that this breeding system carries with it none of the disadvantages associated with obligate asexuality. While theoretical work has failed to provide a simple basis for the superiority of sexual reproduction, it is generally agreed that cyclic parthenogens provide important information on the likely nature of this advantage. Specifically it has been argued that the co-occurrence of both modes of reproduction implies short-term selection for the sexual process (Bell, 1982; Maynard Smith, 1978; Williams, 1975). Otherwise, it has been argued, cyclic parthenogens would make a rapid and frequent transition to obligate asexuality.

Table 1. Taxonomic diversity of groups reproducing by cyclic parthenogenesis. * Other members of this taxonomic rank employ a primary mode of reproduction which differs from cyclic parthenogenesis.

| Class | Order | Families | Genera | Species | Reference |
|-------|-------|----------|--------|---------|-----------|
| Rotifera* | Monogononta | 23 | 98 | 2000+ | Dumont, 1983; Koste, 1978 |
| Crustacea* | Cladocera | 8 | 52 | 450 | Bowman and Abele, 1982 |
| Trematoda* | Digenea | 137 | | 5000 | LaRue, 1957; Schmidt and Roberts, 1985 |
| Insecta* | Homoptera* | 3 | 500 | 4000 | Carter, 1971; Eastop and Hille Ris Lambers, 1976 |
| Insecta* | Hymenoptera* | 1* | 60 | 2000+ | Weld, 1952; Menke, pers. comm. |
| Insecta* | Diptera* | 1* | 13 | 20 | Pritchard, 1960; Wyatt, 1961, 1967 |
| Insecta* | Coleoptera* | 1 | 1 | 1 | Barber, 1913; Borror et al., 1976 |

This chapter aims to review information available on genotypic diversity in groups reproducing by cyclic parthenogenesis, after first providing summary information on the life history and genetic system employed by each group. Work on the genetics of natural populations has focused largely on the cladocerans and aphids (Young, 1983), but undoubtedly many of these results apply to cyclic parthenogens in general. The review additionally aims to consider the genetic characteristics of obligate asexuals derived from cyclic parthenogens and the forces which promoted this shift in breeding system. The latter analysis makes it unlikely that cyclic parthenogens play the pivotal role, which has been envisaged, in elucidating the nature of the selective forces maintaining sex. Specifically the conclusion that short-term selection is required to sustain the sexual phase of the life cycle is dubious. The data additionally raise doubts about the applicability of optimality arguments to the evolution of breeding systems.

## Taxonomic diversity

The rotifers, cladocerans and trematodes are each comprised of a number of higher taxonomic categories and represent the most taxonomically diversified groups of cyclic parthenogens (Table 1). All three groups have a world-wide distribution and likely originated in the Paleozoic, although confirmatory fossils are known only for the Cladocera (McKenzie, 1983). The insect groups reproducing by cyclic parthenogenesis are likely more recent in origin. On the basis of their taxonomic diversity the Aphidoidea should be the most ancient of these groups, and the oldest aphid fossils date from the Permian (Dixon, 1985). The cynipid wasps are extremely diverse at the species level, but the cyclic parthenogens all belong to one subfamily. This group is primarily restricted to the Holarctic; its distribution mirrors that of its predominant host, the plant genus *Quercus*. Paedogenetic cecidomyiids are currently also known only from the Holarctic, but studies of adult morphology suggest that they occur as well in Africa (Gagne, 1979) and Australia (Wyatt, 1967). Fossil records suggest that paedogenesis evolved in the group more than 30 million years ago (Gagne, 1973). It is noteworthy that the cyclically parthenogenetic cecidomyiids are known not only from the primitive subfamily Lestriminae, but also from the subfamily Porricondylinae, suggesting that cyclic parthenogenesis evolved at least twice (Wyatt, 1967). Nicklas (1960) has suggested that bisexual forms have instead originated from an ancestral cyclical parthenogen, but this seems unlikely as the most primitive Cecidomyiidae all reproduce sexually. The single species in the Micromalthidae is apparently native to the southeastern USA, although it is now broadly distributed because of its predilection for creosote-soaked timber. Its position as a taxonomically isolated and primitive coleopteran requires further verification (Borror et al., 1976).

Although the taxonomic composition of groups recognized as cyclic parthenogens has remained stable over the past half century, there is a need for further work on the digenetic trematodes and the micromalthid beetles. Several studies have suggested that adults of *Micromalthus debilis* are either sterile or reproduce parthenogenetically (Kuhne, 1972; Scott, 1936). If either conclusion is correct, the species would reproduce by obligate rather than cyclic parthenogenesis. The retention of functionless males is known in other obligately asexual taxa, but the production of sterile females would be without precedent. Most of the digenetic trematodes are known to alternate between sexual and asexual reproduction, but there has been a long controversy concerning the origins of the asexual offspring. The most recent information (Clark, 1974; Whitfield and Evans, 1983) suggests that these offspring arise from somatic tissues rather than eggs, implying that the group reproduces by somatic rather than cyclic parthenogenesis. Because of the continuing uncertainty over their breeding systems, the present review has taken the conservative position of including both taxa.

While they may be excluded in future treatments, other taxa probably merit inclusion. There is evidence that many rhabdisoid nematodes reproduce by cyclic parthenogenesis (Noble and Noble, 1982; Schmidt and Roberts, 1985). For instance, members of the rhabdisoid genus *Strongyloides* reproduce parthenogenetically in their vertebrate hosts. The asexual eggs which they produce may develop into infective larvae, continuing the phase of asexual reproduction, but under favourable conditions they develop into adult males and females. *S. ratti* employs an XO sex determination mechanism (Bolla and Roberts, 1968), but in other species sex appears to be environmentally controlled (Moncol and Triantophyllou, 1978; Triantophyllou and Moncol, 1977). Eggs from bisexual matings invariably develop into infective larvae which re-establish the asexual phase of the life cycle (Moncol and Triantophyllou, 1978). Interpretation of the rhabdisoid breeding system is complicated by the fact that, in some bisexual matings, the sperm seems to degenerate before fusion with the egg nucleus (Bolla and Roberts, 1968; Triantophyllou and Moncol, 1977). Such species are evidently obligate parthenogens that alternate between apomixis and gynogenesis. The retention of males by so many species in the group makes it unlikely that all bisexual matings involve gynogenesis. More work on this group is needed, but it seems likely that at least some rhabdisoids reproduce by cyclic parthenogenesis.

Similarly some members of the hymenopteran family Aphelinidae may be cyclic parthenogens. One fourth of the species in the genus *Aphytis* are capable of producing female offspring parthenogenetically, and these species continue to produce a low incidence of males that appear functional (Rosen and DeBach, 1979; Rossler and DeBach, 1972; Viggiani, 1984).

## Life cycles

While all cyclic parthenogens alternate between parthenogenetic and sexual reproduction, this is only enforced on a generation to generation basis in the cynipid wasps (Table 2). Members of the asexual generation emerge from galls in late fall or early spring and lay their eggs solitarily or in small groups. The larvae from these eggs develop in galls and emerge as sexually reproducing adults in early summer. In *Neuroterus lenticularis* (Doncaster, 1914, 1916) there are two types of asexual females – those which produce only male and those which produce strictly female offspring. The haploid eggs of the male-producers arise via a single maturation division, while the diploid eggs of female-producers lack this division. The males and females arising from different asexual lines then mate and produce the fertilized diploid embryos which give rise to the asexual generation. Unfertilized eggs produced by sexual females are either non-viable or produce sterile males (Doncaster, 1916). The daughters of a single sexual female consist entirely of female- or male-producers, so there are two genetically different types of sexual females. The genetic basis of the difference is unclear, but other Cynipinae deviate from this simple pattern, as the asexual females of some species are able to produce both male and female offspring (White 1973). In these latter species it is unclear whether all eggs undergo a maturation division followed by the restoration of diploidy in some eggs by polar body fusion, or whether only a fraction of the eggs undergo a maturation division. Although there is a clear need for more information on the mechanism employed to accomplish it, there is no question that cynipids show a regular alternation of sexual and asexual reproduction.

Table 2. Sex determination system, parthenogenetic stage and duration of parthenogenesis in groups reproducing by cyclic parthenogenesis.

| Taxon | Parthenogenetic stage | Sex determination | Duration of parthenogenesis |
|---|---|---|---|
| Rotifera | Adult | Haplodiploid | Unlimited |
| Cladocera | Adult | Environmental | Unlimited |
| Digenea | Larva | Hermaphroditic, ZW | 2–5 generations |
| Aphidoidea | Adult | XO | Unlimited |
| Cynipinae | Adult | Haplodiploid | 1 generation |
| Cecidomyiidae | Larva, Pupa | Chromosome exclusion | Unlimited |
| Micromalthidae | Larva | Haplodiploid | Unlimited |

The digenetic trematodes are the only other group of cyclic parthenogens which have a strict limit to the number of generations of parthenogenetic reproduction. Digenetic trematodes employ two hosts – a vertebrate in which the sexual phase of the life cycle occurs and a mollusc (rarely an annelid) in which asexual reproduction occurs. The eggs produced by the sexual generation are released into water and hatch into miracidia which enter the intermediate host and metamorphose into sporocysts. Each sporocyst produces a new generation of larvae called redia, which themselves subsequently give rise to motile larvae termed cercariae. The cercariae leave the intermediate host and may either actively penetrate the definitive host or await passive ingestion. Thus the typical digenetic trematode passes through just two asexual generations – the sexually produced miracidium, after transforming into a sporocyst, produces redia asexually, and the redia reproduce asexually to produce cercariae. Variation in the life cycle does exist among species, and often includes an additional sporocyst or redial generation, but few if any species have more than four generations of asexual reproduction between bouts of sexuality.

In contrast to trematodes and cynipids, the remaining groups of cyclic parthenogens are able to continue asexual reproduction for an indefinite number of generations. The life cycles of the aphids, rotifers and cladocerans share many similarities. All-female populations are typically initiated each spring from sexually produced resting eggs. These females produce parthenogenetic eggs that undergo embryogenesis before being released. The resulting offspring continue to reproduce asexually until a change in environmental conditions provokes a switch to sexual reproduction. In rotifers these cues elicit the production of haploid eggs, which develop into males if unfertilized and into resting eggs if fertilized. In cladocerans, sex is determined by environmental conditions and male offspring are genetically identical to females. Environmental cues subsequently cause females to switch to the production of haploid eggs which require fertilization. In aphids the switch to sexuality requires the production of both males and sexual females.

The remaining two groups of cyclic parthenogens are paedogenetic. The cecidomyiids produce parthenogenetic eggs while in the larval or pupal stage and sexual eggs as adults. Single individuals do not produce both egg types – rather the reproductive behaviour of neonates is determined by environmental conditions. The timing of asexual reproduction varies among taxa. For example, under favourable conditions *Heteropeza pygmaea* completes its paedogenetic life cycle as a first instar larva, while *Mycophila speyeri* never reproduces before the second instar (Wyatt, 1961). Poor conditions elicit sexual forms, but third instar larvae of both species reproduce paedogenetically if transferred from such an environment to favourable conditions (Harris, 1925; Wyatt, 1961). The two cecidomyiid genera *(Henria, Tekomyia)* which show pupal paedogenesis belong to different subfamilies. However, *Henria* is closely allied to species in the genus *Leptosyna* showing larval paedogenesis, while *Tekomyia* is closely related to larval paedogens in the genus *Mycophila*. The evolution of pupal paedogenesis is understandable in light of the interspecific variation in the timing of larval paedogenesis. By delaying paedogenesis to the end of larval life, pupal paedogens increase their generation time, but this is likely counterbalanced by increased fecundity. Pupae of *Henria* produce up to 96 offspring (Wyatt, 1961), while paedogenetic larvae of *Mycophila speyria* produce only 10–20 offspring (Nicklas, 1960).

The sole species in the coleopteran family Micromalthidae possesses a life history which is similar to that of the cecidomyiids. Paedogenetic larvae produce either diploid females by thelytoky or haploid males by arrhenotoky (Barber, 1913; Scott, 1936, 1938). The cues provoking the transition from female to male production are unclear. Female offspring are released in groups of 10–20 first instar larvae, but only a single male egg develops. This male is cannibalistic, and feeds, until pupation, upon the female which produced it. Occasionally, female larvae produce a male egg which fails to develop; such individuals resume the production of female offspring. Under unfavourable conditions, such as dryness (Scott, 1938), female larvae fail to reproduce paedogenetically and instead develop into adults. The adults appear to be sterile (Kuhne, 1972; Scott, 1938) and if so dispersal of the species rests upon the movements of first instar larvae.

## Genetic systems

From a genetic standpoint the cytological nature of the parthenogenetic phase of the life cycle is extremely important in deciding the genotypic structure of the population. If parthenogenesis were automictic, segregation would occur at heterozygous loci, and clonal lineages would become increasingly homozygous. Cytological work on aphids and cladocerans suggested the possibility of recombination (Bacci and Vaccari, 1961; Cognetti, 1961), but breeding studies on individuals heterozygous at allozyme loci have failed to reveal segregation (Blackman, 1979; Hebert, 1986; Young, 1983).

With the exception of hermaphroditic groups, all cyclic parthenogens require a sex determination mechanism which permits the recovery of males after a series of parthenogenetic generations (table 2). Clearly the XY sex determination mechanism is forbidden, but a remarkable array of other sex determination systems are employed. Males of rotifers, cynipid wasps, and the beetle *Micromalthus* are haploid, while those of aphids are XO. A reduction division leading to the loss of either an entire chromosome set or a sex chromosome permits male production in these groups. Cladocerans employ environmental cues to alter sex, while most trematodes are hermaphroditic and avoid the problem entirely (Benazzi and Benazzi Lentati, 1976).

However, species in the family Schistosomatidae possess two sexes which are often strikingly dimorphic. Although early cytological work failed to identify chromosomal differences (Short, 1957), recent C-banding studies have shown that males are homogametic and females heterogametic (Grossman et al., 1981). Sex is determined during fertilization and all progeny of one miracidium have the same sex (Cort, 1921). Allozyme studies have now identified sex-linked markers (Jelnes, 1983), providing a basis for sexing larval forms (sporocysts, redia, cercaria) which otherwise lack dimorphism. The bisexual schistosomes are generally thought to have evolved from hermaphroditic ancestors, but the selective forces favouring this transition are unclear.

The Cecidomyiidae employ the most complicated mechanism of sex determination (White, 1973). Immediately after fertilization all cecidomyiid zygotes contain a large number of chromosomes, but most are excluded from somatic cell primordia during early embryogenesis with the complete set being retained only in the gametic tissue. The chromosomes can be divided into two components – the E chromosomes which are destined for exclusion and the S chromosomes which are retained in the somatic cells. Superimposed on this exclusion of E chromosomes is the loss of one half the S chromosomes from the somatic cells of incipient males. For example, in *Miastor metraloas* the total chromosome number is 48, but only 12 and 6 chromosomes are present in the somatic cells of females and males, respectively (White, 1973). Similarly the total chromosome count for the strain of *Heteropeza pygmaea* studied by Panelius (Panelius, 1971) was 55, with female somatic cells containing 10 chromosomes and male somatic cells just 5 chromosomes. Thus the cecidomyiids employ chromosome exclusion as a sex determination system which is similar in many respects to the haplodiploid system employed by many other cyclic parthenogens.

## Evolution of cyclic parthenogenesis

The rarity of cyclic parthenogenesis has generally been attributed to the difficulty of evolving a mechanism permitting the co-ordinated occurrence of sexual and asexual reproduction (Lynch and Gabriel, 1983; White, 1973). Each of the seven groups reproducing by cyclic parthenogenesis is clearly derived from ancestors which employed only sexual reproduction. However, there has been little examination of how this transition from sexuality may have occurred. Hebert (1987) has argued that two preadaptations were critical to the evolution of cyclic parthenogenesis in the Cladocera – the production of two egg types and the adoption of environmental sex determination. The production of two egg types provided two intracellular environments, so that it was possible for a gene to suppress meiosis in only one segment of the life cycle. The adoption of environmental sex determination permitted the recovery of males after a period of female parthenogenesis. Other cyclic parthenogens possess analogous life history attributes. All employ a sex determination system which permits male production after parthenogenesis and, except for the cynipids, all produce two egg types. The gene(s) suppressing meiosis in this latter group may be temperature-sensitive in their expression, as opposed to being expressed only in certain egg types.

The argument that the evolution of cyclic parthenogenesis requires specific preadaptations hinges on the supposition that the attributes shared by cyclic parthenogens represent characters evolved prior to the adoption of parthenogenesis. Establishing the sequence of acquisition of traits rests largely on circumstantial evidence, especially in the more ancient groups. For instance, branchiopod crustaceans produce two egg types so frequently that it is reasonable to assume that the trait was present in the ancestral cladoceran lineage. The timing of acquisition of environmental sex determination in the Cladocera is less clear, as all cladocerans employ it, but it is unknown in other branchiopods. In ancient groups such as the Cladocera, the loss of ancestral groups presents a major obstacle in elucidating the sequence of acquisition of significant aspects of the genetic system. In some of the more recently evolved cyclic parthenogens the evidence is more direct. The haplo-diploid sex determination system employed by the Cynipinae is characteristic of all hymenopterans. Similarly the complex reductional sex determination mechanism found in the cyclically asexual genera of cecidomyiids is also found among bisexual species in the same family and the related family Sciariidae (Matuszewski, 1982; White, 1973). Cases such as these provide clear evidence for the preadaptation hypothesis and suggest that parthenogenesis may well have been the final attribute embedded into an already atypical genetic system. If valid, this conclusion has the corollary that only a small proportion of all breeding systems are subject to the transition from bisexuality to cyclic parthenogenesis.

## Genetic effects of cyclic parthenogenesis

Parthenogenetic organisms are capable of both initiating populations from a single individual and of a subsequent rapid increase in population size. As a consequence one might expect cyclic parthenogens to show a greater level of inbreeding than is typical of bisexual outcrossers. Prior work has shown that self-fertilizing organisms possess much lower levels of allozyme variation than bisexuals (Nevo et al., 1984) and one might expect the same trend in cyclic parthenogens. In addition, if the sexual phase of the life cycle occurs infrequently, the genotypic composition of local populations should deviate from that typical of randomly mating sexual populations. Specifically, one expects genotypic frequencies in natural populations of cyclic parthenogens to deviate from Hardy-Weinberg equilibrium, and one expects linkage disequilibrium between variants at different gene loci (Hebert, 1974b; Young, 1979a). Competitive interactions among genotypes might also lead to reduced genotypic diversity as highly fit genotypes increased their relative abundance.

These deviant genotypic characteristics are only likely to be important in taxa where parthenogenesis may continue for an extended period of time, so they are unlikely in the cynipids and trematodes. The remaining groups have the capability of extended parthenogenesis, but environmental conditions may limit the length of their asexual bouts. Thus the cladoceran and rotifer faunas of ephemeral ponds are regularly re-established from sexually produced resting eggs. However, in permanent habitats, such as lakes, one may anticipate a long sequence of parthenogenetic generations. Similarly, aphid populations in mesic areas and the paedogenetic cyclic parthenogens in favourable habitats can continue asexual reproduction for long periods of time. It is reasonable to expect an association between habitat stability and the genotypic structure of populations, insofar as stability is inversely correlated with the amount of sexual recruitment.

## Allozyme variation in cyclic parthenogens

Allozyme surveys have not yet been carried out on the cyclically parthenogenetic cynipids, cecidomyiids, or the sole micromalthid. Work on rotifers has also been too limited (King and Snell, 1977; Snell and Winkler, 1984) to indicate the level of variation in this group with the exception of King's (1980) survey of a single population of *Asplanchna girodi* which was polymorphic at 1 of 10 loci, with individual heterozygosities averaging 4.8%.

Most of the work on allozyme variation in digenetic trematodes is also limited in scope, with the exception of Fletcher et al.'s (1981) study of genetic variation in 22 laboratory strains of *Schistosoma mansoni*. Their analysis showed little genetic divergence between New and Old World strains, in accordance with evidence that the parasite first reached the western hemisphere via the slave trade. Strains of *S. mansoni* were, on average, polymorphic at 13 % of their loci, while individual heterozygosities averaged 4 %. Strains which had been maintained in the laboratory for a long period were much less variable; a strain isolated in 1950 was heterozygous at only 0.2 % of its loci (Fletcher et al., 1981; LoVerde et al., 1985). Considerably more variation resided in the species gene pool, than in that of any single population, as 37 % of the loci were polymorphic when all strains were considered. Less detailed allozyme surveys have been carried out on other schistosomes (Rollinson and Gebert, 1982) and confirm the general presence of genetic variation in this group. Because they possess distinct sexes, the results on the schistosomes may be atypical; other trematodes may have less heterozygosity because of the opportunity for self-fertilization. The only survey of genetic variation in a non-schistosome trematode revealed esterase variation among individuals of *Fasciola hepatica* (Alcaino et al., 1976), but genetic interpretation of the data was uncertain.

In contrast to the limited work on other cyclic parthenogens, a substantial amount of information is available on allozyme variation in natural populations of cladocerans and aphids. The results of these surveys suggest that local populations of both groups are relatively invariant and that individuals are heterozygous at a small percentage of their loci.

Levels of genetic variation have been studied among members of three cladoceran genera. Initial work on *Simocephalus serrulatus* suggested that this cladoceran was highly polymorphic (Smith and Fraser, 1976), but reanalysis indicated that several species had been confused (Hann and Hebert, 1982). More recent studies on four species in the genus (Table 3) indicate that individual heterozygosities average just 4.9 %, while populations are polymorphic at 10 % of their loci. Work on species in the genus *Daphnia* suggests that low levels of variation are typical of cladocerans. Interestingly, the sole North American member of the genus *Daphniopsis* appears to possess as much variation as species belonging to the more speciose cladoceran genera, suggesting that variation in the latter groups has not been augmented by introgressive hybridization. Populations of *D. magna* do show geographic variation in their levels of polymorphism, as populations from both western Europe and North America are much more variable than those at the eastern boundary of the species range (Table 3).

Calculations of individual heterozygosity and the proportion of loci polymorphic in single populations conceal the extent of genetic variation which exists among conspecific populations. In contrast to more highly vagile groups, cladocerans show large gene frequency differences among local populations (Hebert, 1974a; Hebert and Moran, 1980; Korpelainen, 1984) and macrogeographic differentiation is striking (Hebert, 1986). As a consequence, each local population contains only a small proportion of the variation present in the species gene pool. For instance, individual populations of *D. magna* from East Anglia were polymorphic at just 11 % of their loci,

Table 3. Levels of genetic variation in aphid and cladoceran species reproducing by cyclic parthenogenesis. He = individual heterozygosity (%); H = % polymorphic loci; * when a variable number of individuals was analyzed for each locus, the # of isolates is the mean of three numbers; ** Est-2 deleted.

| Taxon | Site | #Isolates* | #Loci | He | H | Reference |
|---|---|---|---|---|---|---|
| **Aphididae** | | | | | | |
| Acyrtosiphon pisum | Finland | 8 | 18 | 1.4 | 5.5 | Suomalainen et al., 1980 |
| Aphis fabae | Germany | 40 | 7 | – | 43.0 | Tomiuk and Wohrmann, 1980 |
| Aphis pomi | Germany | 227 | 25 | | 8.0 | Tomiuk and Wohrmann, 1980 |
| | Germany | 1512 | 12** | 0.0 | 0.0 | Tomiuk and Wohrmann, 1980 |
| Sp A | Canada | 290 | 18** | 1.1 | 5.6 | Singh and Rhomberg, 1984 |
| Sp B | Canada | 125 | 19 | 0.0 | 0.0 | Singh and Rhomberg, 1984 |
| Aphis sambuci | Germany | 47 | 9 | 3.7 | 11.0 | Tomiuk and Wohrmann, 1980 |
| Macrosiphum euphorbiae | USA | 47 | 18 | 7.0 | 27.7 | May and Holbrook, 1978 |
| Macrosiphum funestrum | Germany | 432 | 8 | 0.0 | 0.0 | Tomiuk and Wohrmann, 1980 |
| Macrosiphum rosae | Germany | 20 | 7 | 0.0 | 0.0 | Tomiuk and Wohrmann, 1980 |
| | Ontario | 851 | 30 | 2.9 | 16.6 | Rhomberg et al., 1985 |
| Myzus persicae | Germany | 1918 | 20 | 4.4 | 10.0 | Tomiuk and Wohrmann, 1980 |
| | England | 35 | 11 | – | 9.0 | Wool et al., 1978 |
| | USA | 260 | 19 | 0.0 | 0.0 | May and Holbrook, 1978 |
| Rhodobium porosum | Germany | 45 | 13 | 0.0 | 0.0 | Tomiuk and Wohrmann, 1980 |
| Sitobion avenae | England | 92 | 26 | 2.3 | 61.5 | Loxdale et al., 1985 |
| Wahlgreniella nervata | Germany | 214 | 10 | 3.7 | 20.0 | Tomiuk and Wohrmann, 1980 |
| **Cladocera** | | | | | | |
| Daphnia carinata | Australia | 720 | 16 | 2.1 | 6.7 | Hebert and Moran, 1980 |
| Daphnia cucullata | Finland | 356 | 7 | 8.8 | 43.3 | Korpelainen, 1985; pers. comm. |
| Daphnia longispina | Finland | 1572 | 7 | 4.3 | 48.9 | Korpelainen, 1985; pers. comm. |
| Daphnia magna | England | 1512 | 15 | 6.6 | 11.1 | Hebert 1974a |
| | Finland | 2890 | 7 | 7.7 | 49.7 | Korpelainen, 1985; pers. comm. |
| | N.W.T. | 400 | 10 | 8.7 | 20.0 | Loaring, 1982 |
| | Manitoba | 576 | 16 | 0.06 | 1.2 | Crease, 1986 |
| Daphnia pulex | Europe | 374 | 7 | 3.7 | 60.0 | Korpelainen, 1985; pers. comm. |
| Daphniopsis ephemeralis | Ontario | 96 | 19 | 3.5 | 21.0 | Schwartz and Hebert, 1986 |
| Simocephalus congener | Ontario | 24 | 12 | 9.2 | 16.7 | Hann and Hebert, 1986 |
| Simocephalus exspinosus | Ontario | 64 | 12 | 1.9 | 2.8 | Hann and Hebet, 1986 |
| Simocephalus serrulatus | Ontario | 263 | 12 | 5.0 | 11.9 | Hann and Hebert, 1986 |
| Simocephalus vetulus | Ontario | 338 | 12 | 3.6 | 9.7 | Hann and Hebert, 1986 |

but 30% of the loci were polymorphic among populations in this region. Broader geographic surveys have shown that variation exists at 50% of the loci. Such evidence indicates that the gene pools of cladoceran taxa are not invariant. Rather the gene pool of each species is fractured into a number of weakly polymorphic lines as a consequence of inbreeding.

The lack of variation evident within local cladoceran populations is even more strikingly apparent in the aphids. Among the 11 aphid species which have so far been analyzed (Table 3) individual heterozygosities average just 1.9%, while the proportion of loci polymorphic in local populations is 10.1%. Again there is much evidence of geographic variation in gene frequencies, suggesting that local population surveys provide a misleading impression of the extent of variation in the gene pool.

When all available data involving the survey of 10 or more loci are pooled they suggest that local populations of cyclic parthenogens are polymorphic at 12% of their loci and that individual heterozygosities average 3.2%. These are much lower than values (40%, 10%) typical of bisexual invertebrates (Nevo et al., 1984). When the lack of variation in local populations is coupled with the evidence for pronounced inter-populational differentiation, the results suggest that cyclic parthenogenesis promotes inbreeding and local gene pool homozygosity in a fashion similar to self-fertilization.


## Genotypic frequencies in natural populations of cyclic parthenogens

While basic surveys of allozyme diversity have now been carried out on 5 of the 7 groups reproducing by cyclic parthenogenesis, spatial and temporal shifts in genotypic frequencies have been examined only in natural populations of aphids and cladocerans and to a lesser extent of rotifers. Recent reviews (Blackman, 1985; Hebert, 1986) on the two former groups discuss the results of this work in more detail than is possible in this chapter.

The only detailed study of the genetics of a natural rotifer population revealed striking temporal shifts in its genotypic composition. The most dramatic of these involved the replacement of one clonal group by a second which possessed different alleles at 6 of the 10 loci which were surveyed (King, 1977a). As later analysis revealed both major life history differences (Snell, 1979) and complete reproductive isolation between the two groups, a shift in the frequency of sibling species rather than conspecific clones appears to have occurred. This re-interpretation is significant as this study provided the primary support for King's (1977a) complete genetic discontinuity model which proposed both that clonal variants within rotifer species exhibited large fitness differences and that inbreeding was intense (King 1977a, 1977b). In fact the little data available suggest that local rotifer populations are panmictic and that fitness differences are modest. Thus, the more common of the two clonal groups of *A. girodi* was polymorphic at one locus and showed only minor shifts in genotypic frequencies.

When first surveyed, the population was fixed for heterozygotes, but after re-establishment from sexual eggs genotypic frequencies were in Hardy-Weinberg equilibrium. After a two-month period of stability, genotypic frequencies shifted (King, 1980; Snell, 1980), but the changes were no more dramatic than those observed in cladoceran and aphid populations. The more spectacular examples of life history variation and shifts in genotypic composition in rotifer populations are attributable to taxonomic problems.

Genetic studies on natural populations of cladocerans have focused largely on members of the genus *Daphnia*. The first surveys showed that the genotypic characteristics of English populations of *D. magna* were effected by environmental conditions (Hebert, 1974b; 1974c).

Populations in intermittent habitats possessed genotypic frequencies that were both relatively stable and in good congruence with Hardy-Weinberg expectations. By contrast populations in permanent habitats showed striking Hardy-Weinberg deviations, often as a consequence of heterozygote excess. Rapid shifts in genotypic composition were observed in these populations as well as non-random associations between variants at different loci. This last observation indicated that selection was operating on gene complexes as opposed to single loci. More detailed studies by Young (1979a, 1979b) and Carvalho (1985) have confirmed the generality of these observations, but have additionally shown that a large number of genotypes co-occur in permanent populations. There is no evidence that selection eliminates all but the fittest clone, and indeed such exclusion would be expected only if one clone had the highest fitness under all conditions and if sexual recruitment was absent. Clonal diversity in most permanent populations is likely in a continuing state of disequilibrium resulting from the interplay between the release of new diversity via sexual recruitment and its erosion by selective processes (i. e. competition, predation).

It has proven difficult to identify other cladocerans with a population structure similar to that of permanent populations of *D. magna*. Habitat permanency is clearly an insufficient condition for the emergence of these genotypic attributes. Thus Australian populations of *D. carinata* occupy permanent habitats, but diapause each summer as sexual eggs. Their populations are effectively intermittent and not unexpectedly they show the genotypic attributes of such populations (Hebert, 1980). Many of the cladocerans in lake habitats are known to overwinter as parthenogenetic adults and thus seemed likely to possess a population structure similar to that of permanent populations of *D. magna*. However, surveys of *D. galeata* populations in several German lakes revealed that genotypic frequencies were in Hardy-Weinberg equilibrium and that temporal shifts were small (Mort and Wolf, 1985, 1986). These results are not universal – surveys of the same species in Czechoslovakian reservoirs (Hebert, 1987) revealed marked Hardy-Weinberg disturbances. The genotypic composition of *Daphnia* in the German lakes may indicate either that adults fail to overwinter or that sexual recruitment is high. Apparently the emergence of a genotypic structure similar to that seen in permanent populations of *D. magna* demands both the opportunity for continued parthenogenetic reproduction and a low level of sexual recruitment.

Further studies on intermittent populations of cladocerans have shown a tendency for genotypic frequencies to deviate from Hardy-Weinberg proportions as a consequence of heterozygote deficiency (Korpelainen, 1986; Lynch, 1983; Schwartz and Hebert, 1986). In addition evidence has accumulated for significant shifts in genotypic frequencies both within and between individual annual cycles. The deviations from Hardy-Weinberg equilibrium and the shifts in genotypic frequencies are, however, much smaller in magnitude than those in permanent populations. The heterozygote deficiencies are likely a result of the gene frequency shifts, as such deficiency is the expected result of admixture of offspring from a series of temporally isolated populations with different gene frequencies. The genotypic frequency shifts may arise as a consequence of the smaller number of sexual recruits which refound populations each year. The shifts are not solely a consequence of sampling effects on genotypic frequencies at single loci. If this were the case, genotypic frequencies would remain stable after refounding, which is not always the case. It appears that the frequencies of allozyme genotypes are often also driven by chance linkage associations which arise because of the small number of founders. Similar associations arise in sexual populations, but have a much stronger effect on asexuals, because favoured genotypes are not disrupted by recombination and are able to replicate for many generations. Studies on intermittent populations have failed to reveal evidence of non-random associations between genotypic variants at different loci (Hebert, 1974c; Korpelainen, 1986). These analyses undoubtedly underestimate the extent of linkage disequilibrium, as most if not all of the loci which have been examined are unlinked.

The most detailed studies on the genotypic structure of aphid populations have been carried out on *Macrosiphum rosae*. Work on European populations of this species has focussed on patterns of variation at the MDH and PGM loci (Tomiuk and Wohrmann, 1981, 1984; Wohrmann, 1984). Temporal analysis of local populations showed that genotypic frequencies at the MDH locus were unstable and often deviated from Hardy-Weinberg expectations as a result of heterozygote deficiency. Work on *M. rosae* in Ontario showed a similar pattern of heterozygote deficiency at polymorphic loci (Rhomberg et al., 1985). Temporal surveys revealed rapid shifts in the frequencies of multilocus genotypes in the Ontario populations. There was a tendency for a reduction in clonal diversity over the year, supporting Tomiuk and Wohrmann's (1984) observation of lower clonal diversity in populations which reproduced asexually for long periods of time. The rapid shifts in genotypic frequencies suggest that each population is dominated by a few clones. The absence of concordant trends in genotypic frequencies among different populations suggests that there is a unique clonal array at each site. The frequency changes at the allozyme loci are undoubtedly caused largely by non-random associations with other selectively important variation. Surveys on a broader scale revealed clonal variation in gene frequencies among European populations (Wohrmann, 1984), but in Canada microspatial variation in genotypic frequencies was as great as that observed between populations several hundreds of kilometres apart (Rhomberg et al., 1985).

The genotypic structure of the other aphid species which have been studied resembles that of *M. rosae*. Thus, a survey of anholocyclic populations of *Myzus persicae* revealed strong disequilibrium between allelic variants at different loci and microspatial variation in genotypic frequencies associated with differing host plant preferences (Baker 1979). Populations of *Sitobion avenae* showed a higher incidence of heterozygote deficiency than *M. rosae*. Indeed heterozygotes were rare or absent at nine of ten polymorphic loci in this species. Moreover, it was uncertain whether the sole exception represented true heterozygosity or gene duplication (Loxdale et al., 1985).

The work on aphid populations has revealed a population structure reminiscent, in some respects, to permanent populations of cladocerans. Both groups show microgeographic heterogeneity and rapid shifts in genotypic frequencies coupled with non-random associations between allelic variants at different loci. The most striking difference between the groups is the lower level of heterozygosity in aphid populations. The heterozygote deficiencies suggest either strong inbreeding during the sexual phase, or the presence of a mechanism promoting homozygosity during parthenogenesis (Cognetti, 1961; Pagliai, 1983). The occurrence of segregation during parthenogenesis seems to have been ruled out (Blackman, 1972; Tomiuk and Wohrmann, 1982), but it also seems unlikely that sexual matings regularly involve close relatives. The heterozygote deficiencies instead may indicate the selective superiority of the few inbred clones which do arise. In this regard it may be important that past surveys have examined crop-infesting aphids, whose genotypic structure has undoubtedly been affected by biological and chemical control efforts. Resistance to insecticides has, for instance, been linked to shifts in esterase activity (Beranek, 1974; Lewis and Madge, 1984; Weber, 1985). If outcrossing disrupts the gene complexes necessary for insecticide resistance then the heterozygote deficiencies in aphid populations are simply an indirect consequence of such chemical challenges. It seems desirable, therefore, to study aphid species which live under less disturbed conditions.

**Transitions to obligate parthenogenesis**

There are reports of obligately asexual taxa among all of the major cyclically parthenogenetic groups. The term has, however, been used to describe a rather diverse array of shifts in the breeding system, ranging from cases in which the sexual phase of the lifecycle has been truncated to those in which the lifecycle is complete, but reproduction is always asexual. Evidence for the adoption of asexuality is generally circumstantial, relying upon the absence of either males or sexual females.

Recognition of obligate asexuals is difficult, especially in the cynipids and rotifers. Galls produced by the bisexual generation of cynipids are less conspicuous than those of the asexual generation and adults of the two generations are often morphologically divergent enough to have originally been placed in different genera. Sexual forms were, for instance, thought to be absent in the cynipid genus *Disholocaspis* (Muesbeck et al., 1951) which contains more than 30 species and seemed to represent a rare case of taxonomic diversification in an agamic group. However, recent studies, (D. C. Dailey, pers. comm.) have shown that at least one species in the genus is a cyclic parthenogen. Although the occurrence of obligate asexuality in the cynipids is widely acknowledged (Bell, 1982; White, 1973), there appear to be only a few species in which it has been firmly established (Stille, 1980). Obligate asexuality also appears rare in the best-studied genera of rotifers such as *Asplanchna* and *Keratella,* although Ruttner-Kolisko (1946) reported a population of *K. quadrata* that produced diploid resting eggs. Dumont (1983) points out that males and resting eggs have been observed in fewer than 1 % of all rotifer species, but, because of their small body size and short lifespan, detection is difficult. Dumont has argued that the restricted ranges of certain rotifer species may be due to their inability to produce resting eggs. Such forms might represent obligate asexuals arising as a consequence of lifecycle truncation.

Evidence of a less anecdotal nature is available on the incidence of obligate asexuals in certain genera of cladocerans, aphids and digenetic trematodes. Three of the 50 or so recognized species in the cladoceran genus *Daphnia (D. cephalata, middendorffiana, pulex)* reproduce by obligate parthenogenesis. Only one of these species *(middendorffiana)* seems to consist entirely of asexuals. The other species include obligately and cyclically parthenogenetic populations, which rarely co-occur. Obligately asexual populations of *D. pulex* predominate over much of temperate and arctic Canada (Hebert, 1983; Hebert et al., 1987), but cyclically asexual populations are common in the USA (Innes et al., 1986; Lynch, 1983). In all three *Daphnia* species, life cycle complexity has been retained – both subitaneous and resting eggs are produced ameiotically (Hebert, 1981; Schrader, 1926; Zaffagnini and Sabelli, 1972).

In contrast, cases of obligate parthenogenesis in the Aphididae appear invariably to involve a truncation of the life cycle. Anholocyclic aphids consist of clones which lack the ability to produce males and sexual females and as a consequence also resting eggs. Such clones are restricted to areas where the climate is sufficiently mesic to allow the long term survival of parthenogenetic individuals. The loss of sexual forms is, in some cases, simply due to the absence of inductive cues. Anholocyclic strains of the introduced aphid *Therioaphis trifolii* produced sexual forms once their range extended into northern sections of the USA (Dixon, 1985). The genetic control of life cycle truncation has been studied in *Myzus persicae*. Some clones in this species produce both sexual females and males (holocyclic) while others produce only males (androcyclic) and some produce neither (anholocyclic). The life cycle differences are inherited monofactorially with androcycly recessive to holocycly (Blackman, 1972). The simple genetic control, with holocycly as the dominant trait, ensures that the three breeding systems can be regenerated each year. Other aphids are more permanently linked to obligate asexuality. Species such as *Myzus dianthicola* and *M. ascalonicus*, as well as genera such as *Forda* and *Trama* appear to consist exclusively of anholocyclic forms (Blackman, 1980; 1981).

Many species in the closely related homopteran family, the Adelgidae, show a similar truncation of the life cycle (Carter, 1971). These forms reproduce by continued parthenogenesis on a single host, while cyclically parthenogenetic adelgids alternate between two hosts. The incidence of anholocyclic forms appears to be much greater in the adelgids than aphids and all anholocyclic adelgids are capable of producing overwintering eggs. As a consequence, parthenogenetic lineages can persist indefinitely and many anholocyclic adelgids are recognized as distinct species. The success of adelgids in evolving asexual forms capable of producing overwintering eggs is likely a consequence of the fact that both parthenogenetic and sexual females in cyclically parthenogenetic adelgids produce eggs, while parthenogenetic aphids produce only subitaneous young. If this difference has been a significant preadaptation in the transition to obligate asexuality, the phylloxerans, which share oviparity with the adelgids, should also show a high incidence of asexuals. Information on phylloxeran breeding systems is scant, but several obligately asexual species are known.

Cytological work has verified the occurrence of obligate parthenogenesis in several genera of digenetic trematodes including *Bunodera* (Cannon, 1971), *Fasciola* (Sakaguchi, 1980) and *Paragonimus* (Shimazu, 1981). In all cases the asexual forms have retained the life cycle complexity typical of digenetic trematodes. All of the asexuals are triploids, but have closely related diploids that reproduce by cyclic parthenogenesis. In some cases diploid and triploid conspecifics co-occur in a single host. Several other trematodes ordinarily reproduce by cyclic parthenogenesis, but have the facultative ability to complete their life cycle parthenogenetically. For example, in the absence of a mate, *Schistosomatium douthitti* produces miracidia via automictic parthenogenesis (Short and Menzel, 1959), more than 90% of which are haploid. The haploids have very low fitness and rarely colonize an intermediate host, but the parthenogenetically produced diploids appear to have normal viability (Whitfield and Evans, 1983). At least two species in the genus *Schistosoma* are able to produce miracidia gynogenetically (Taylor et al., 1969; Vogel, 1942); they employ the sperm of a related species to initiate the development of parthenogenetically produced eggs. In the

remaining groups of cyclic parthenogens – cecidomyiids, and micromal-
thids, there are no reports of secondarily derived obligately asexual stocks.

## Genetic diversity in obligate asexuals

Although many cyclic parthenogens have made the transition to obligate
asexuality, very little work has been done on their genetics. Cytogenetic
studies have established that all asexual trematodes are triploid (Whitfield
and Evans, 1983), but DNA quantification studies have shown that asexual
cladocerans include both diploid and polyploid clones (Hebert, 1987). A
clear geographic pattern is evident in *Daphnia pulex* with polyploids most
prevalent at high latitudes. Work on obligately asexual aphids has not
revealed polyploidy, but has shown that their karyotypes do not consist of
homologous chromosome pairs, suggesting that these asexual taxa may
have arisen via hybridization (Blackman, 1981).

Studies of genotypic diversity in asexuals derived from cyclic parthenogens have so far only been caried out on two cladocerans *(Daphnia pulex, D.
middendorffiana)* and on two cynipids *(Diplolepis rosae, D. mayri)*. These
two groups possess a fundamental difference in their mode of reproduction.
Cytological and breeding studies have shown that both *Daphnia* species
reproduce by apomictic parthenogenesis (Hebert and Crease, 1983), while
the *Diplolepis* species are automictic parthenogens (Stille, 1980, 1985).
These cytological observations imply that each *Diplolepis* species should
consist of homozygous clone(s), while the clones of the *Daphnia* species
could potentially be highly heterozygous.

Allozyme surveys indicate that local populations of both *Daphnia* species
contain few multi-locus genotypes and that frequencies of these genotypes
deviate strongly from Hardy-Weinberg expectations. Individual genotypes
show average heterozygosities in excess of 15 %, far higher than the values
typical of cyclically parthenogenetic *Daphnia*. The clonal arrays of single
populations remain stable over several years, in contrast to permanent
populations of cyclic parthenogens, in which new genotypes are regularly
introduced by sexual recruitment. Moreover, in regions where polyploids
occur, obligate asexuals frequently show unbalanced heterozygote
phenotypes at polymorphic loci. While local populations of *D. pulex* and *D.
middendorffiana* contain an average of only 3 clones, regional diversity is
far greater. For example, a survey of 156 asexual populations of *D. pulex* in
the Great Lakes watershed revealed 145 clones, and analysis indicated that
these were only a small proportion of the clones present in this area (Hebert
et al., 1987).

The clones are not ecological analogues; there is direct evidence of diversification in important life-history characters (Loaring and Hebert, 1981)
and indirect evidence, such as striking shifts in clone frequencies across
physico-chemical gradients (Weider and Hebert, 1987). Ascertaining the

source of this variation is important. Have the differences originated after the monophyletic loss of sex or do they represent variation captured from the cyclic parthenogenetic ancestor via the polyphyletic loss of sex? Recent work (Crease, 1986; Stanton, 1987) combining both allozyme and mitochondrial DNA analysis has shown that clones with divergent allozyme phenotypes often have identical mitochondrial phenotypes. This lack of concordance between divergence in the nuclear and mitochondrial segments of the genome can only be explained if sex has been lost polyphyletically.

The mechanism producing this repeated loss of sexuality appears to be understood (Hebert, 1981). Approximately 50 % of the clones whose females reproduce by obligate asexuality continue to produce males. Since the gene that suppresses meiosis during oogenesis is not completely penetrant during spermatogenesis, these males produce some normal sperm. Breeding studies (Innes and Hebert, 1987) have shown that these males can transmit the gene(s) which suppresses meiosis during oogenesis, and thus obligate asexuality is 'contagious'. This mechanism provides a simple explanation for the tremendous genetic diversity found in obligately asexual lineages of *Daphnia,* and a similar mechanism could account for the adoption of asexuality in hermaphroditic lineages, such as the digenetic trematodes (Igenike and Selander, 1979). It may be less likely in cyclic parthenogens which rely on meiosis to produce males, but scenarios permitting the spread of sex-limited meiosis suppressors can be developed even in these groups.

Allozyme studies on the automictic parthenogen *Diplolepis mayri* (Stille, 1985) revealed three closely related clones in northern Sweden and a fourth very genetically divergent clone in the south. A more detailed survey of genetic diversity in *D. rosae* revealed eight clones, each homozygous, as expected, at all 27 loci (Stille, 1985). Only 4 clones were present among the 284 individuals from Sweden, but clonal diversity appeared higher in Germany as 4 clones were represented in 16 individuals. A single clone made up more than 80 % of the Swedish isolates and the other three clones differed from it by a substitution at just one locus. This dominant clone was also found in Germany, together with two clones differing from it by single-locus substitutions. A fourth German clone was more distinctive, and possessed substitutions at several loci. Similarly two isolates from Greece possessed a genotype which was very distinct from that seen at the other sites. The results suggest that *D. rosae* includes a number of distinct asexual lineages, the most common of which have given rise to numerous mutational derivatives. Ecological differences among clones, even those which have arisen as mutational derivatives, are suggested by the largely allopatric distributions of the dominant Swedish clones.

The genetic surveys on both obligately asexual cladocerans and cynipids make it clear that there is likely to be high genotypic diversity in other obligate asexuals. The extent of this diversity is undoubtedly important in explaining the broad geographic distribution and clear success, at least in the short-term, of many obligate asexuals. The work on cladocerans indicates that most of this diversity reflects the polyphyletic origin of asexuality rather than diversification after the loss of sex. The work on *Daphnia* has also suggested that the road to asexuality may not be unidirectional. The persistence of cyclic parthenogens and their ability to interbreed with males from obligately asexual lineages raises the possibility that genetic variation from asexuals may, on occasion, be reintegrated into a sexually reproducing population.

## Summary

Among extant taxa it appears that cyclic parthenogenesis has evolved on fewer than ten occasions. Evidence suggests that these groups are all derived from bisexual ancestors with genetic systems possessing specific preadaptations for the breeding system shift. Its apparently independent origin in two cecidomyiid lineages suggests that, when such preadaptations exist, the adoption of cyclic parthenogenesis may be a highly probable event.

The genetic consequences of this transition are not dramatic; the genotypic characteristics of most cyclic parthenogens are similar to those of bisexuals. Specifically, genotypic frequencies at polymorphic loci are generally in Hardy-Weinberg equilibrium and linkage disequilibrium between variants at different loci is rare. Levels of variation in local populations of cyclic parthenogens are lower than those typical of sexual outcrossers, but the lack of variation appears to be a consequence of founder effects, as regional gene pool diversity is not similarly impoverished. An unusual genotypic structure occasionally develops in populations from habitats in which sustained parthenogenetic reproduction is possible and sexual recruitment is low. Under these circumstances low fitness genotypes are winnowed from the population by natural selection. The truncation of genotypic diversity often produces Hardy-Weinberg deviations and linkage disequilibrium. The depletion of clonal diversity rarely leads to the domination of a habitat by a single clone; an array of ecologically diversified clonal variants remains. The level of clonal variation in such populations is occasionally augmented by sexual recruitment, but the persistence of clonal diversity in local populations of obligate parthenogens makes it clear that clonal coexistence is not a consequence of such recruitment.

Cyclic parthenogens have made the transition to obligate asexuality with high frequency, but there is little evidence to support the argument (Williams, 1975) that such shifts result from the relaxation of the short-term selection pressures supposedly necessary to sustain the sexual phase of the life cycle. Maynard Smith (1978) has recognized that in groups such as the aphids, which have never evolved the ability to produce their diapausing eggs asexually, sexuality may simply be retained as a consequence of its role in the formation of these eggs. However, he accepted that short-term selection was required to maintain sexuality in groups such as the cladocerans in which asexual forms have demonstrated their ability to retain life cycle complexity.

The argument that cyclic parthenogens are poised on the brink of relinquishing asexuality assumes that the transition to obligate parthenogenesis is easily accomplished. In fact, the transition to an obligately asexual life cycle that produces asexual eggs and lacks males likely requires gene substitutions at several loci. Among the cladocerans, for example, it is clear that the genes terminating meiosis do not also control male production, leading to the retention of males by lines adopting obligate asexuality. The sexual phase of the life cycle can be dislodged, but only by mutations with the specific effect of suppressing meiosis during oogenesis but not spermatogenesis. When such mutations occur, they can spread through populations with scant regard for their effect on short-term fitness.

While experimental studies have so far only shown that the loss of sex in cladocerans is the product of sex-limited meiosis suppressors, there is no case in which the loss of sex has been linked to environmental parameters. When the data on cladocerans are coupled with those showing that hybridization is responsible for asexuals derived from bisexual taxa (Moore, 1984), there is general support for the conclusion that asexuality originates as a product of internal genetic factors rather than as a response to selection pressures imposed by the environment. It can be argued that selection intervenes after their origin – the longer persistence of asexuals in certain environments might account for geographic patterns in the prevalence of asexuality (Glesener and Tilman, 1978). However, demonstrating even this subsidiary role for selection is complicated by the fact that most asexuals are polyploid (Bell, 1982). Regions, such as the arctic, which are rich in asexuals may simply be habitats where polyploidy is favoured.

The need for more information on the factors which have promoted the abandonment of sexuality is obvious. Asexuals derived from cyclic parthenogens represent a significant fraction of all asexuals and should figure prominently in these studies, as should bisexual groups such as the ostracods (Bell, 1982) which show an unusually high frequency of transitions to asexuality. If the limited information now available proves representative, there is little support for arguments which assume breeding system optimization.

Acknowledgments. Drs J. M. Campbell (Micromalthidae), A. Borkent, R. J. Gagné (Cecidomyiidae), D. C. Dailey, A. S. Menke (Cynipidae), J. C. Gilbert (Rotifera), P. Mackay, A. G. Robinson (Aphididae), and R. Footit (Adelgidae) generously shared information on the groups upon which their own research has focussed, while Dr L. Weider provided helpful comments on the manuscript. A grant from the Natural Sciences and Engineering Research Council of Canada aided in the preparation of this chapter.

# Patterns in plant parthenogenesis

P. Bierzychudek

## Introduction

Several recent reviews have described the ecological and geographical patterns of parthenogenesis in animals (Bell, 1982; Glesener and Tilman, 1978) in an effort to determine which selective forces best explain the maintenance of sexual reproduction. These reviews conclude that parthenogenetic lineages tend to be found at higher latitudes, at higher elevations, in more xeric conditions, in more disturbed habitats, and on more island-like habitats than their sexual relatives. In general, it is argued that asexual taxa are more widely distributed than are their sexual relatives, with sexuals limited to habitats in which the important selective forces are principally biotic rather than physical.

These patterns have been explained in four distinct ways. First, Glesener and Tilman (1978) have interpreted these patterns as evidence that frequency-dependent selection, in the form of interactions with predators, parasites, pathogens, and competitors, is instrumental in maintaining sexual reproduction. Bell (1982), in an exhaustive review of animal distributions, came to a similar conclusion, as did Levin (1975), who claimed that there existed a positive correlation between the degree of recombination engaged in by plants and the intensity of pest and pathogen pressure to which they were subjected. Secondly, other authors, especially botanists (e. g. Stebbins, 1950), cite the superior colonization abilities of asexual taxa as the reason for their success in marginal habitats. The third explanation was offered by Lynch, who has asserted that disjunct distribution patterns provide a prezygotic isolating mechanism that prevents parthenogens and sexuals from interfering with one another's genetic integrity, and thus producing offspring of inferior fitness. However, even when some other isolating mechanism has evolved, permitting coexistence, the range of sexual individuals is generally less extensive than that of parthenogens. Lynch's second hypothesis explained this difference as a consequence of selection on parthenogens for the acquisition of 'general-purpose' genotypes (Baker, 1974), genotypes that are able to survive under a wide range of extreme conditions. Thus, the first hypothesis asserts that competition between sexual and parthenogenetic genotypes will cause sexuals to be displaced except where sex provides an 'escape' from biotic interactions. The other three do not assume that such competition will necessarily occur, and attribute the observed distribution patterns to other causes: parthenogens' superior colonization abilities, avoidance of destabilizing hybridization, or parthenogens' enhanced abilities to respond to selection for general-purpose genotypes.

This paper is an attempt to evaluate these hypotheses in the light of information on the distribution patterns of asexually-reproducing plants. I have restricted myself to a consideration of plants that are apomictic in the strict sense, i. e. that produce seed mitotically, without benefit of fertilization. I hope to answer the following questions:

1) Are there any distinct patterns to the ecological or geographical distributions of apomictic plants?

2) If such patterns can be identified, is there any evidence suggesting what forces might cause those patterns?

3) Are there any differences, either in the patterns or in their causes, between plants and animals?

I also plan to address a potentially important issue that has received very little attention: the fact that polyploidy is an important correlate of parthenogenesis. Most asexual taxa are also polyploid taxa, and the possible influence of this factor must be considered in any explanation of distribution patterns.

## Biology of apomixis

Many plant species are able to produce seeds asexually. While specific developmental pathways vary widely between and even within particular species, the endpoint is always the same: the embryo within these seeds is genetically identical to its maternal parent. There are two major categories of apomixis (Gustafsson, 1946a). In one, termed 'adventitious embryony', the embryo (sporophyte) arises directly from the ovular tissue of the parent, and the process of alternation of generations is completely bypassed. Rather little is known about the biology of these species, and so this review will focus on the other type of apomixis, often called 'gametophytic apomixis'. Gametophytic apomixis involves a distinct gametophytic generation, which arises either by mitosis or by a very much altered version of meiosis in which chromosome pairing and reduction in chromosome number do not occur. The embryo develops autonomously, without fertilization. Some species of gametophytic apomicts (termed 'pseudogamous') require pollination for successful reproduction, because the endosperm must be fertilized in order to develop. In these pseudogamous species, the paternal parent makes no genetic contribution to the embryo itself. Gametophytic apomixis may be either facultative or obligate; facultative apomixis is perhaps more common (Clausen, 1954).

Gametophytic apomixis occurs most commonly among three families of angiosperms: the Compositae, the Rosaceae, and the Gramineae (Gustafsson, 1947), but has been reported from at least 17 other families as well (Nygren, 1954). Khokhlov (1976) claims that apomixis has been reported in over 300 plant genera in 80 families, but these numbers are unsubstantiated; the definition of apomixis he uses is not made clear. Gametophytic apomixis is confined nearly exclusively to herbaceous and woody groups that are long-lived and cross-fertilized (Gustafsson, 1946b); annual apomicts are extremely rare. Undoubtedly this is because most gametophytic apomicts arose as allopolyploids involving the hybridization of two sexual parents, and since most annuals self-fertilize to a large extent (Baker, 1959), such hybridization is not possible.

It is almost always the case that gametophytic apomicts are polyploid while their sexual relatives are diploid (Stebbins, 1950). Undoubtedly the hybrid origin of most apomicts is a major determinant of this correlation. The same correlation between breeding system and ploidy level exists among animals (Lokki and Saura, 1980; Lynch, 1984; White, 1973). However, one difference between plants and animals is that the converse is not so often true for plants as it is for animals. Nearly all known polyploid animals reproduce parthenogenetically, while only a small fraction of the polyploid plant species are apomictic (Lewis, 1980; Stebbins, 1980).

### Distributional patterns of apomicts and sexuals

*Range size*

Support for the claim that apomicts have ranges that are greater than the ranges of their sexual relatives can be found among many groups of angiosperms. The quality of this support varies. Sometimes descriptions of species' ranges are very anecdotal; other times extensive collections have been made, many herbarium specimens examined, and reliable range maps constructed. Sometimes a group is well-enough known that phylogenetic relationships are reasonably clear, but in many cases this information is not available. Table 1 lists those taxa for which good distribution maps have been published, and for which an attempt has been made to determine the sexual progenitor of an apomictic line.

It is often the case that sexual species are sympatric with their apomictic relatives, and occupy an area in the center of the apomictic range; this pattern characterizes *Crepis* (Babcock and Stebbins, 1938), *Parthenium* (Rollins, 1949), *Eupatorium* (Sullivan, 1976), and *Townsendia* (Beaman, 1957). Examples of the opposite trend, however, are not uncommon. Some of these are listed in Table 1. More anecdotal cases include *Potentilla,* European *Antennaria,* and the genus *Cotoneaster. Potentilla glandulosa,* which is sexual, is 'geographically widespread'; *P. gracilis* (an apomict) is more restricted (Babcock and Stebbins, 1938). The sexual species of *Antennaria (A. dioica, A. carpatica)* are 'widespread' (Porsild, 1965), while the

Table 1. Taxa containing both sexual and apomictic elements for which range maps or precise distribution descriptions are available. For each taxon I have noted 1) whether or not the range of the apomictic element is appreciably greater than that of the sexual, and 2) whether or not the apomictic element ranges further north than the sexual. All species are from the northern hemisphere.

| Taxon | Family | Apomictic range greater? | Apomictic range further north? | Reference |
|---|---|---|---|---|
| *Arnica alpina* | Compositae | yes | yes | Barker, 1966 |
| *A. amplexicaulis* | Compositae | yes | yes | Barker, 1966 |
| *A. chamissonis* | Compositae | yes | yes | Barker, 1966 |
| *A. lessingii* | Compositae | yes | no | Barker, 1966 |
| *A. lonchophylla* | Compositae | yes | yes | Barker, 1966 |
| *A. longifolia* | Compositae | yes | yes | Barker, 1966 |
| *A. louiseana* | Compositae | yes | yes | Barker, 1966 |
| *Bouteloua curtipendula* | Gramineae | no | no | Gould, 1959 |
| *Calamagrostis lapponica/ C. stricta* | Gramineae | no | no | Greene, 1980 |
| *C. stricta* ssp. *inexpansa* | Gramineae | no | no | Greene, 1980 |
| *Crepis acuminata* | Compositae | no | no | Babcock and Stebbins, 1938 |
| *C. bakeri* | Compositae | yes | yes | Babcock and Stebbins, 1938 |
| *C. exilis* | Compositae | yes | yes | Babcock and Stebbins, 1938 |
| *C. modocensis* | Compositae | yes | yes | Babcock and Stebbins, 1938 |
| *C. monticola* | Compositae | yes | yes | Babcock and Stebbins, 1938 |
| *C. occidentalis* | Compositae | yes | yes | Babcock and Stebbins, 1938 |
| *C. pleurocarpa* | Compositae | yes | yes | Babcock and Stebbins, 1938 |
| *Eupatorium altissimum* | Compositae | yes | yes | Sullivan, 1976 |
| *E. cuneifolium* | Compositae | yes | yes | Sullivan, 1976 |
| *E. lechaefolium* | Compositae | yes | yes | Sullivan, 1976 |
| *E. leucopis* | Compositae | yes | yes | Sullivan, 1976 |
| *E. pilosum* | Compositae | yes | yes | Sullivan, 1976 |
| *E. rotundifolium* | Compositae | yes | yes | Sullivan, 1976 |
| *E. sessilifolium* | Compositae | yes | yes | Sullivan, 1976 |
| *Hieracium pilosella* | Compositae | ? | yes | Turesson and Turesson, 1980 |
| *Parthenium argentatum* | Compositae | yes | yes | Rollins, 1949 |
| *Poa cusickii* ssp. *cusickii* | Gramineae | yes | yes | Soreng, 1984 |
| *P. fendleriana* var. *fendleriana* | Gramineae | yes | yes | Soreng, 1984 |
| *P. fendleriana* var. *longiligula* | Gramineae | yes | yes | Soreng, 1984 |

| Taxon | Family | Apomictic range greater? | Apomictic range further north? | Reference |
|---|---|---|---|---|
| *Taraxacum* ssp. | Compositae | ? | yes | den Nijs and Sterk, 1980; Sterk et al., 1982 |
| *Townsendia condensata* | Compositae | yes | yes | Beaman, 1957 |
| *T. exscapa* | Compositae | yes | yes | Beaman, 1957 |
| *T. grandiflora* | Compositae | no | no | Beaman, 1957 |
| *T. hookeri* | Compositae | yes | yes | Beaman, 1957 |
| *T. incana* | Compositae | yes | yes | Beaman, 1957 |
| *T. leptotes* | Compositae | yes | yes | Beaman, 1957 |
| *T. montana* | Compositae | yes | yes | Beaman, 1957 |
| *T. Parryi* | Compositae | yes | yes | Beaman, 1957 |
| *T. Rothrockii* | Compositae | no | no | Beaman, 1957 |
| *T. scapigera* | Compositae | no | no | Beaman, 1957 |
| *T. spathulata* | Compositae | no | no | Beaman, 1957 |
| *T. strigosa* | Compositae | no | yes | Beaman, 1957 |

apomicts *(A. alpina, A. Porsildii, A. Nordhageniana)* are endemic in Scandinavian mountains. Sax (1954) maintains that diploid species of *Cotoneaster* (which are mostly sexual) are geographically limited relative to apomictic species, but her data provide no support for this claim: among triploids (almost certainly apomictic), 86 % of 37 species have a relatively limited geographic distribution, whereas among diploids and tetraploids (most of which are sexual), 93 % of 14 species have limited ranges. *Dichanthium* (Celarier et al., 1958), *Calamagrostis* (Gustafsson, 1947), and *Sorbus* ((Fahraeus, 1980) also appear to be exceptions to this trend. Stebbins (1971) argues that conformity with the trend is a function of the age of the species complex; only in older, 'mature' complexes will the apomicts be widespread and the sexuals rare and scattered. To evaluate whether, as a general rule, the geographic ranges of apomicts tend to exceed those of sexuals, I performed a sign test on the data in Table 1. Of a total of 41 species pairs (representing 8 genera), the range of the apomictic member of the pair was greater 76 % of the time. This value is significantly greater than the 50 % expected by chance ($p < 0.01$).

*Latitudinal trends*

There is also good support for the notion that apomicts range into higher latitudes than do their sexual relatives. Again, Table 1 lists only those taxa for which good distributional information is available. There are, in addition to these, more anecdotal reports. For example, Porsild (1965) reports that in North America, the percentage of *Antennaria* species (Compositae) that are sexual declines with increasing latitude. The species of blackberry, *Rubus* (Rosaceae), in Sweden and Great Britain are predominantly apomictic; each country is host to only one sexual species. But in more southerly, continental Europe, there are many more sexual species (Haskell, 1966). Muntzing (1966) concludes that *Poa alpina* (Gramineae) in Sweden is predominantly apomictic, and points out that purely sexual strains do exist, but only in more southern countries.

A sign test comparing the ranges of the apomictic taxa in Table 1 with those of their closest sexual relatives revealed that the apomictic member of the pair had the more northerly range significantly more often (76% of the time) than the 50% expected by chance (p < 0.01). This test included 43 species, representing 10 genera.

*Elevational trends*

The trend for apomicts to range further north than their sexual relatives is paralleled by an elevational trend: apomicts are often found at higher elevations than their sexual relatives. This trend is striking in *Townsendia,* the 'alpine daisy'. In *T. grandiflora,* sexual individuals are much more common than apomictic ones, but the single apomictic collection represents the highest elevational record for the species. In *T. scapigera,* apomicts are also uncommon; they occur only at or near mountain summits, despite the fact that the sexual individuals can be found at elevations anywhere from 4600 to 10,000 feet. In *T. rothrockii,* a species in which sexuals and apomictics are equally wide-ranging geographically, the apomictic individuals are the only ones that have been observed above 10,000 feet. Apomictic *T. leptotes* are not only found over a wider geographical range than sexuals, but they tend to reach higher elevations as well. *T. condensata,* a primarily high-altitude species, is almost completely apomictic; the only sexual population known occurs on a south-facing slope, at the lowest elevation recorded for the species (Beaman, 1957). Another example of this tendency is provided by a pair of species in the genus *Draba* (Cruciferae). *Draba streptobrachia,* the apomictic form of the sexual *D. spectabilis,* is found only above treeline in the Colorado Rockies, while its sexual counterpart ranges from 8000 to 13,000 feet (Price, 1980).

*Occupancy of glaciated areas*

The elevational trend just described is related to the fact that apomicts have a tendency to colonize once-glaciated areas, leaving their sexual relatives behind in the process. For example, Haskell (1966) argues that the distribution of *Rubus* (blackberry) species in Europe (apomictic species predominate in the more northerly areas) is a consequence of glaciation, the apomicts being better able to colonize areas opened by glacial retreat. Similarly, only apomictic *Antennaria* have been collected from northeastern Canada and Greenland, which were largely glaciated (Porsild, 1965).

Table 2. Members of groups containing both sexual and apomictic members for which information about occupancy of formerly-glaciated areas is available. For each taxon I have listed its breeding system as sexual or apomictic, and noted whether it has been collected from 1) unglaciated areas only, 2) glaciated areas only, or 3) both.

| Taxon | Breeding system | Family | Unglaciated only | Glaciated only | Both | Ref. |
|---|---|---|---|---|---|---|
| *Antennaria parlinnii* | S | Compositae | | | X | Bayer and Stebbins, 1983 |
| *A. parlinnii* | A | Compositae | | X | | Bayer and Stebbins, 1983 |
| | | | | | | |
| *Arnica acaulis* | S | Compositae | X | | | Barker, 1966; Wolf, 1980 |
| *A. alpina* | S | Compositae | | X | | Barker, 1966 |
| *A. alpina* | A | Compositae | | X | | Barker, 1966 |
| *A. amplexicaulis* | S | Compositae | | X | | Barker, 1966 |
| *A. amplexicaulis* | A | Compositae | | | X | Barker, 1966 |
| *A. angustifolia* | S | Compositae | X | | | Wolf, 1980 |
| *A. angustifolia* | A | Compositae | | | X | Wolf, 1980 |
| *A. chamissoni* | S | Compositae | | | X | Barker, 1966 |
| *A. chamissoni* | A | Compositae | | | X | Barker, 1966 |
| *A. cordifolia* | A | Compositae | | | X | Wolf, 1980 |
| *A. diversifolia* | A | Compositae | | | X | Barker, 1966 |
| *A. fulgens* | S | Compositae | | | X | Barker, 1966; Wolf, 1980 |
| *A. latifolia* | S | Compositae | | | X | Wolf, 1980 |
| *A. lessingii* | S | Compositae | | X | | Barker, 1966 |
| *A. lessingii* | A | Compositae | | X | | Barker, 1966 |
| *A. lonchophylla* | S | Compositae | X | | | Barker, 1966 |
| *A. lonchophylla* | A | Compositae | | X | | Barker, 1966 |
| *A. longiflora* | S | Compositae | | | X | Barker, 1966 |
| *A. longiflora* | A | Compositae | | | X | Barker, 1966 |
| *A. louiseana* | S | Compositae | | X | | Barker, 1966 |
| *A. louiseana* | A | Compositae | | | X | Barker, 1966; Wolf, 1980 |
| *A. mollis* | A | Compositae | | | X | Barker, 1966 |
| *A. montana* | S | Compositae | X | | | Wolf, 1980 |
| *A. nevadensis* | A | Compositae | | | X | Wolf, 1980 |
| *A. Parryi* | A | Compositae | | | X | Barker, 1966 |
| *A. Rydbergi* | A | Compositae | | X | | Barker, 1966 |
| *A. sororia* | S | Compositae | X | | | Barker, 1966; Wolf, 1980 |
| *A. unalaschensis* | S | Compositae | | | X | Barker, 1966; Wolf, 1980 |
| | | | | | | |
| *Bouteloua curtipendula* | S | Gramineae | | | X | Gould, 1959 |
| | | | | | | |
| *Calamagrostis cairnii* | S | Gramineae | X | | | Greene, 1980 |
| *C. cinnoides* | S | Gramineae | | | X | Greene, 1980 |
| *C. deschampoides* | S | Gramineae | | X | | Greene, 1980 |
| *C. epigejos* | S | Gramineae | | | X | Greene, 1980 |
| *C. lapponica* | A | Gramineae | | X | | Greene, 1980 |
| *C. pickeringii* | S | Gramineae | | X | | Greene, 1980 |
| *C. porteri* | S | Gramineae | X | | | Greene, 1980 |
| *C. purpurescens* | S | Gramineae | | | X | Greene, 1980 |
| *C. purpurescens* | A | Gramineae | | | X | Greene, 1980 |
| *C. stricta* | S | Gramineae | | | X | Greene, 1980 |
| *C. stricta* | A | Gramineae | | | X | Greene, 1980 |
| | | | | | | |
| *Crepis acuminata* | S | Compositae | | | X | Babcock and Stebbins, 1938 |
| *C. acuminata* | A | Compositae | | | X | Babcock and Stebbins, 1938 |
| *C. bakeri* | S | Compositae | X | | | Babcock and Stebbins, 1938 |

204

| Taxon | Breeding system | Family | Unglaciated only | Glaciated only | Both | Ref. |
|---|---|---|---|---|---|---|
| C. bakeri | A | Compositae | X | | | Babcock and Stebbins, 1938 |
| C. barbigera | A | Compositae | X | | | Babcock and Stebbins, 1938 |
| C. bursifolia | S | Compositae | X | | | Babcock and Stebbins, 1938 |
| C. elegans | S | Compositae | | | X | Babcock and Stebbins, 1938 |
| C. exilis | S | Compositae | | | X | Babcock and Stebbins, 1938 |
| C. exilis | A | Compositae | | | X | Babcock and Stebbins, 1938 |
| C. intermedia | A | Compositae | | | X | Babcock and Stebbins, 1938 |
| C. modocensis | S | Compositae | X | | | Babcock and Stebbins, 1938 |
| C. modocensis | A | Compositae | | | X | Babcock and Stebbins, 1938 |
| C. monticola | S | Compositae | X | | | Babcock and Stebbins, 1938 |
| C. monticola | A | Compositae | X | | | Babcock and Stebbins, 1938 |
| C. nana | S | Compositae | | X | | Babcock and Stebbins, 1938 |
| C. nicaensis | S | Compositae | X | | | Babcock and Stebbins, 1938 |
| C. occidentalis | S | Compositae | X | | | Babcock and Stebbins, 1938 |
| C. occidentalis | A | Compositae | | | X | Babcock and Stebbins, 1938 |
| C. pleurocarpa | S | Compositae | X | | | Babcock and Stebbins, 1938 |
| C. pleurocarpa | A | Compositae | X | | | Babcock and Stebbins, 1938 |
| C. rubra | S | Compositae | X | | | Babcock and Stebbins, 1938 |
| C. runcinata | S | Compositae | | | X | Babcock and Stebbins, 1938 |
| C. setosa | S | Compositae | X | | | Babcock and Stebbins, 1938 |
| C. tectorum | S | Compositae | | | X | Babcock and Stebbins, 1938 |
| Draba spectabilis | S | Cruciferae | | | X | Price, 1980 |
| D. streptobrachia | A | Cruciferae | | X | | Price, 1980 |
| Eupatorium altissimum | S | Compositae | X | | | Sullivan, 1976 |
| E. altissimum | A | Compositae | | | X | Sullivan, 1976 |
| E. cuneifolium | S | Compositae | X | | | Sullivan, 1976 |
| E. cuneifolium | A | Compositae | X | | | Sullivan, 1976 |
| E. lechaefolium | S | Compositae | X | | | Sullivan, 1976 |
| C. lechaefolium | A | Compositae | X | | | Sullivan, 1976 |
| E. leucopis | S | Compositae | X | | | Sullivan, 1976 |
| E. leucopis | A | Compositae | X | | | Sullivan, 1976 |
| E. pilosum | S | Compositae | X | | | Sullivan, 1976 |
| E. pilosum | A | Compositae | X | | | Sullivan, 1976 |
| E. rotundifolium | S | Compositae | X | | | Sullivan, 1976 |
| E. rotundifolium | A | Compositae | X | | | Sullivan, 1976 |
| E. sessilifolium | S | Compositae | X | | | Sullivan, 1976 |
| E. sessilifolium | A | Compositae | | | X | Sullivan, 1976 |
| Hieracium pilosella | S | Compositae | | | X | Turesson a. Turesson, 1960 |
| H. pilosella | A | Compositae | | | X | Turesson a. Turesson, 1960 |
| Parthenium argentatum | S | Compositae | X | | | Rollins, 1949 |
| P. argentatum | A | Compositae | X | | | Rollins, 1949 |
| Poa alpina | S | Gramineae | X | | | Muntzing, 1966 |
| P. alpina | A | Gramineae | | | X | Muntzing, 1966 |
| Spiranthes casei var. casei | A | Orchidaceae | | X | | Catling, 1982 |
| S. casei var. novaescotiae | A | Orchidaceae | | X | | Catling, 1982 |
| S. cernua var. cernua | A | Orchidaceae | | | X | Catling, 1982 |

| Taxon | Breeding system | Family | Unglaciated only | Glaciated only | Both | Ref. |
|---|---|---|---|---|---|---|
| *S. cernua* var. *odorata* | A | Orchidaceae | X | | | Catling, 1982 |
| *S. magni-camporum* | A | Orchidaceae | | X | | Catling, 1982 |
| *S. magni-camporum* | S | Orchidaceae | | | X | Catling, 1982 |
| *S. ochroleuca* | A | Orchidaceae | X | | | Catling, 1982 |
| *S. ochroleuca* | S | Orchidaceae | | | X | Catling, 1982 |
| *Taraxacum* spp. | S | Compositae | | | X | Richards, 1973 |
| *Taraxacum* spp. | A | Compositae | | | X | Richards, 1973 |
| *Townsendia annua* | S | Compositae | X | | | Beaman, 1957 |
| *T. condensata* | S | Compositae | X | | | Beaman, 1957 |
| *T. condensata* | A | Compositae | | | X | Beaman, 1957 |
| *T. excinia* | S | Compositae | X | | | Beaman, 1957 |
| *T. exscapa* | S | Compositae | X | | | Beaman, 1957 |
| *T. exscapa* | A | Compositae | | | X | Beaman, 1957 |
| *F. fendleri* | S | Compositae | X | | | Beaman, 1957 |
| *T. florifer* | S | Compositae | X | | | Beaman, 1957 |
| *T. formosa* | S | Compositae | X | | | Beaman, 1957 |
| *T. glabella* | S | Compositae | X | | | Beaman, 1957 |
| *T. grandiflora* | S | Compositae | X | | | Beaman, 1957 |
| *T. hookeri* | S | Compositae | X | | | Beaman, 1957 |
| *T. hookeri* | A | Compositae | | | X | Beaman, 1957 |
| *T. incana* | S | Compositae | X | | | Beaman, 1957 |
| *T. incana* | A | Compositae | X | | | Beaman, 1957 |
| *T. leptotes* | S | Compositae | X | | | Beaman, 1957 |
| *T. leptotes* | A | Compositae | | | X | Beaman, 1957 |
| *T. mensana* | S | Compositae | X | | | Beaman, 1957 |
| *T. mexicana* | S | Compositae | X | | | Beaman, 1957 |
| *T. montana* | S | Compositae | | | X | Beaman, 1957 |
| *T. montana* | A | Compositae | | | X | Beaman, 1957 |
| *T. Parryi* | S | Compositae | | | X | Beaman, 1957 |
| *T. Parryi* | A | Compositae | | | X | Beaman, 1957 |
| *T. Rothrockii* | S | Compositae | X | | | Beaman, 1957 |
| *T. Rothrockii* | A | Compositae | | X | | Beaman, 1957 |
| *T. scapigera* | S | Compositae | | X | | Beaman, 1957 |
| *T. scapigera* | A | Compositae | | X | | Beaman, 1957 |
| *T. spathulata* | S | Compositae | X | | | Beaman, 1957 |
| *T. spathulata* | A | Compositae | X | | | Beaman, 1957 |
| *T. strigosa* | S | Compositae | X | | | Beaman, 1957 |
| *T. strigosa* | A | Compositae | X | | | Beaman, 1957 |
| *T. texensis* | S | Compositae | X | | | Beaman, 1957 |

Those taxa for which detailed distributional data are available are listed in Table 2. I performed a $\chi^2$-squared test on these data to establish whether a significant association existed between breeding system and occupancy of formerly-glaciated areas. I scored each taxon for 1) breeding system (sexual or apomictic) and 2) whether or not it ranged into areas that were once glaciated. The limits of glaciation used were as described by Flint (1957). This survey included 130 species (or races) in 13 genera. 75 % of 57 apomictic taxa ranged into areas opened by glacial retreat, whereas only 41 % of 73 sexual taxa did so; this positive association between apomixis and colonization of glaciated areas is significant ($x^2 = 15.35$, $p < 0.005$).

*Disturbance*

Several authors have tried to make the case that apomicts are found in sites that are more disturbed than those of their sexual relatives. Stebbins (1958) claims that the apomictic species of *Hieracium, Ixeris,* and *Taraxacum* are found in sites he characterizes as 'unstable' (cultivated or abandoned fields, or roadsides), while the habitats of their sexual relatives are more likely 'intermediate' (forest clearings, dry slopes, meadows, steppes) or even 'stable' (forests, swamps). Of the British Columbia species of *Hieracium* (Guppy, 1978), the sole apomict, *H. umbellatum,* occurs in successional or disturbed habitats. Only one of the six sexual species grows in similar places; the others occur in stable or slowly-changing plant communities. The apomictic *Draba streptobrachia* is found on more disturbed sites (scree slopes and talus edges, where plant cover is low) than is its sexual relative *D. spectabilis,* which grows near streambanks and in moist meadows (Price, 1980). The eastern North American species of *Calamagrostis* that are sexual occupy 'relatively stable, late-successional habitats', while the apomicts 'colonize disturbed, open habitats' (Greene, 1984). Den Nijs and Sterk (1980), however, found no support for this generalization in their study of *Taraxacum* in Europe. In a test of a 1966 contention by Furnkranz that diploid (sexual) *Taraxacum* prefer relatively undisturbed sites, while polyploids (apomicts), with their higher levels of seed production, are better adapted than diploids to disturbed habitats, den Nijs and Sterk assayed the ploidy levels of plants at the same sites used by Furnkranz. They saw no evidence of obvious differences in habitats occupied by the two types of individuals. The most intensively disturbed sites showed no increase in frequency of polyploid individuals in the intervening 15 years.

Another contradiction to this generalization is provided by species of *Eupatorium,* both in northeastern North America and in the Rokko Mountains of Japan. In North America (Sullivan, 1976), the habitats of the sexual and apomictic races of each species tend to be very similar and not differentiated by degree of disturbance. For *E. altissimum,* both apomicts and sexuals are found in open woods, clearings, and alkaline prairies (though only the apomicts also grow on the ballast of railroad tracks). Both sexual and apomictic *E. lechaefolium* occupy woods and woods edges as well as old fields, fire lanes, and highway medians. In this species, the sexual individuals are more common in the disturbed sites than in the relatively undisturbed ones, the opposite of the expected pattern. *E. rotundifolium* is found in both stable and disturbed sites, regardless of breeding system. In the Japanese species *E. chinense,* which includes both diploid sexuals and polyploid apomicts, the sexuals (which are more slender and shorter than the apomicts) cannot compete with the tall grasses and forbs found in grasslands, roadsides, and forest edges. They are restricted to scree slopes and rocky areas, which are species-poor and lack tall competitors. Thus in this case it is the sexual species, not the apomicts, that occupy the habitats that are most frequently disturbed (Watanabe, 1982).

Of seven sexual-apomictic pairs representing four genera, in only two (29 %) was the apomictic member found in more disturbed sites. In all other cases there was either no difference, or else the sexual member occupied the more disturbed areas.

*Other trends*

I was able to find no information at all about the relative frequencies of occurrence of apomicts and sexuals on islands. However, there are a few indications that the habitat of apomicts may sometimes be more arid than that of sexuals. Sexual *Draba spectabilis* grows near streambanks and in moist meadows, while apomictic *D. streptobrachia* is found on dry scree and talus slopes (Price, 1980). Sexual species of *Antennaria* (Compositae) in the Elk Mountains of Colorado are usually found in wet meadows and on the edges of ponds and streams; their apomictic counterparts occur in dry rocky meadows and on talus slopes as well (Bierzychudek, personal observation). Of the four species of beggar's tick, *Bidens* (Compositae), in northwestern Ontario, one, *B. cernua,* is sexual, while the others *(B. frondosa, B. connata,* and *B. vulgata)* are apomictic. All four species are weedy annuals that grow in disturbed sites, from cracks in sidewalks to riverbanks, but the sexual species is restricted to the wettest sites: beaver dams, riverbanks, and wet ditches (Crowe and Parker, 1981).

**Possible explanations for observed patterns of distribution**

Most of the trends that have been identified for animal parthenogens are apparently equally valid for plants. The ranges of apomicts are larger, and often extend to higher latitudes and to higher elevations than do the ranges of their sexual counterparts; apomicts seem more tolerant of arid habitats than their sexual counterparts are. The response of apomicts to disturbance, though, is not so clear-cut; while in many groups, apomicts seem to show weedier tendencies, this behavior is not statistically significant. And little information exists on the relative frequency of apomicts and sexuals in island or island-like habitats.

Glesener and Tilman (1978) and Bell (1982) maintain that high latitudes, high elevations, and arid habitats all represent situations in which biological interactions are relatively unimportant. Under such circumstances, they argue, and only under such circumstances, does the greater reproductive capacity of parthenogenetic animal species confer an advantage. But in situations where competitive ability or escape from predators, parasites, and pathogens is an important prerequisite for persistence, the genetic variability that characterizes sexual progeny provides them with an advantage great enough to override their numerical disadvantage.

Other authors (e. g. Stebbins, 1950) cite the superior colonization ability of apomicts, their ability to found populations with a single individual, as the reason for their success in these marginal conditions, or in newly-opened habitats such as those made available by the retreat of the Pleistocene ice sheets. This explanation assumes that sexual populations could succeed in these areas once they became established, but that their establishment is extremely unlikely.

Lynch (1984) finds neither of these explanations sufficiently general. He finds the evidence for competitive inferiority on the part of parthenogens scanty and unconvincing, and rejects the colonization ability argument on the grounds that in animals, parthenogens generally have a lower reproductive rate than sexuals, and are often quite immobile. For example, most of the parthenogenetic insects are flightless (Lokki and Saura, 1980).

The hypothesis offered by Lynch (1984) in place of these is the 'general-purpose genotype' hypothesis. Lynch suggests that obligate parthenogens are under intense selection to produce highly generalized genotypes that are able to survive under a wide variety of conditions. It is only such genotypes, he argues, that can survive extinction by environmental fluctuations. He further maintains that selection for such 'general-purpose genotypes' can be more effective in parthenogenetic populations than in corresponding sexual ones. As pointed out by Templeton (1982), selection acts on the composite properties of parthenogenetic genotypes, whereas in sexual populations, selection operates only on additive genetic variance (which apparently contributes little to tolerance of extreme environments – Jinks et al., 1973). Thus, as a result of the effectiveness of this intense clonal selection, extant parthenogens have acquired the ability to be extremely generalized, which permits them to occupy a wider variety of conditions than their sexual relatives.

In the sections that follow, I will evaluate each of these hypotheses in light of the information available about the biology and distribution patterns of sexual and apomictic plant taxa.

*Biotic interactions*

What is the evidence that, in areas where biotic interactions are especially important, sexuals enjoy advantages over apomicts? If the genetic variability among sexual progeny provides them with advantages in certain habitats, we expect not to see coexistence between sexual and apomictic taxa in those habitats, but rather dominance by sexual lines. And certainly there are some cases in which coexistence seems to be prohibited. Sullivan (1976) points out that in *Eupatorium altissimum, E. pilosum, E. rotundifolium,* and *E. sessilifolium,* all of which include both sexual and apomictic individuals, the two types never grow intermixed. Watanabe et al. (1982) mention that sexual and apomictic individuals of *E. chinense* are rarely seen growing together. And the apomictic *Draba streptobrachia* is never found intermixed with its sexual counterpart, *D. spectabilis* (Price, 1980). Lynch (1984) would argue that these cases can be explained equally well as the result of selection for prezygotic isolating mechanisms, to prevent destabilizing hybridization.

It is not uncommon, however, to find sexual and apomictic members of other genera growing side by side within the same 'population'. At high elevations in the Colorado Rockies, one frequently finds sites in which two or even three *Antennaria* species – some sexual, some apomictic – grow intermingled (Bierzychudek, personal observation). Bayer and Stebbins (1983) report finding sexual and apomictic individuals of *A. parlinii* in Ohio growing intermixed. In Europe, mixed populations of sexual and apomictic dandelions, *Taraxacum,* have been described by several authors (Van Loenhoud and Duyts, 1981; den Nijs and Sterk, 1980; Sterk et al., 1982). De Wet and Harlan (1970) report that the habitats of the apomictic grass *Dichanthium* are always simultaneously occupied by a sympatric sexual lineage.

Cases of coexistence may be more frequent than suggested by the literature, because it is often difficult to distinguish between sexual and apomictic females in the field, and mixed populations are more likely to be misinterpreted as uniform ones than vice versa. It seems quite difficult to evaluate the validity of the biotic interactions hypothesis on the basis of distributional data alone. The fact that sexuals and apomicts commonly coexist, while not sufficient to refute the hypothesis, does suggest that this explanation is perhaps too simplistic.

*Colonization ability*

Apomicts are seen to have two advantages over sexuals in terms of their potential to start new populations. First, they can found a breeding population with only a single colonist. Since their sexual relatives are almost always self-incompatible, the founding of a sexual population of plants requires at least two individuals that are in close proximity, and that are able to attract pollen vectors. Because of their inability to move and their requirement of an external agent of pollen transfer, plants are more severely handicapped than animals by low population numbers. Secondly, once established, apomictic populations are expected to increase faster than sexual ones, because all their reproductive energy can be devoted to the production of female progeny. Avoiding this 'cost of producing males' (Maynard Smith, 1978) theoretically confers a twofold advantage. That advantage may not be as great for plants as it is for bisexual animals. Since the majority of plants are hermaphroditic, they may already be devoting more than half their reproductive energy to ovule production. For pseudogamous plants in particular, which must continue to produce pollen, there may be no differential between sexual and apomictic rates of reproduction. But while the rate of population growth may be no greater for new populations of apomictic plants, ease of establishment will undoubtedly be greater.

Lynch (1984) cites many examples of parthenogenetic animals with lower fecundities than their sexual progenitors. I am aware of no similar handicap for plants. Indeed, quite the opposite is true. In very large populations of sexual *Antennaria parvifolia* in the Colorado Rockies, rates of seed production at best equal those observed in apomictic populations of the same species. Females in small sexual populations, however (which are more the rule), rarely set as much seed as apomictic females do; pollinator service is simply not adequate to fertilize more than a small proportion of the ovules produced (Bierzychudek, unpub. data). Soreng (1984) has observed a similar phenomenon in sexual and apomictic *Poa fendleriana*.

And, indeed, many observers of apomictic plants attribute their geographic distributions to their presumed greater colonizing ability: Bayer and Stebbins (1983) have done so for *Antennaria parlinii,* as have Haskell (1966) for *Rubus,* and Catling (1982) for *Spiranthes.* To accept this hypothesis, however, we must be willing to accept a non-equilibrium view of present-day species distributions, and believe that the 'marginal' habitat occupied by apomicts but not sexuals is not intrinsically unoccupiable by sexuals, but rather that populations in these areas are subject to such high rates of turnover that at any moment they seem to be dominated by the groups capable of faster re-establishment, the apomicts. Furthermore, we must be willing to believe that the time that has passed since the retreat of the Pleistocence ice sheets has not been sufficient to permit sexuals to successfully invade the areas made available by that retreat.

*General-purpose genotypes*

Apomictic lines have classically been regarded as evolutionary dead-ends that possess no genetic variation and are thus incapable of response to selection (Stebbins, 1950). In order to argue that selection has acted on apomicts to produce general-purpose genotypes, it is necessary to demonstrate that apomictic populations contain genetic variability on which selection can act. There is, in fact, considerable evidence that apomictic populations contain substantial amounts of genetic variability.

Most of the documentation of this variability has focused on morphological characteristics. For example, Babcock and Stebbins (1938) measured pappus length, number of florets, and length of outer bracts for 50 sexual individuals and 55 obligately apomictic individuals of the hawkweed *Crepis acuminata.* The range of variation they observed in these characters was nearly as great among apomicts as it was among sexuals. For example, sexual plants had a mean number of florets of $7.8 \pm 5.7$; for apomicts, this value was 7.4 ± 5.6. The difference between the two types of breeding systems was that, in the population of apomicts, most individuals were members of one of two easily recognized forms, whereas the sexual populations displayed more continuous variation.

Usberti and Jain (1978) grew individuals from both sexual and apomictic populations of the grass *Panicum maximum* in a randomized design under greenhouse conditions, and measured a variety of morphological and reproductive characteristics on 15 plants from each population, such as tiller number, plant height, biomass, panicle length, and days to flowering. Usberti and Jain were interested in comparing coefficients of variation in these characters among different types of populations, especially between sexual and apomictic populations. They detected insignificant differences in variability between the two breeding system types. Sexual populations did not show greater levels of variation; in fact, their values were exceeded by those of some of the apomictic populations. However, Usberti and Jain (1978) examined plants from only three sexual populations (vs 25 apomictic ones). And since they relied on agricultural institutions for their seed, it is not clear that all the collections represent seed from the same number of individuals. However, the apomictic populations were clearly quite variable.

There are several ways in which these high levels of variability can arise within apomictic populations. Since many plants that reproduce apomictically are not obligate apomicts, but instead produce both reduced and unreduced ovules in varying proportions, both types capable of being fertilized, the progeny of these plants can be extremely genetically variable, not only in allelic composition but also in ploidy level. And even among obligate apomicts, there are several ways to generate variability. Repeated hybridization between sexual parents continues to produce new apomictic genotypes. Abnormal mitosis, especially in these high polyploids, occasionally produces an ovule with one chromosome too many or too few (Sorensen and Gudjonsson, 1946). Point mutations occur and are perpetuated. Finally, autosegregation (bivalent formation) occasionally occurs, permitting crossing-over within the parental genome.

So while to my knowledge selection experiments have never been performed on apomictic plants, most populations appear to contain a pool of genetic variation sufficient to allow such experiments to be at least partially successful. Selection acting on this pool of variability could have at least two kinds of results:

1) Selection could produce an apomictic 'species' that is actually a collection of distinct genotypes, each specialized for a narrow habitat, but which in sum occupy a wide range of types of environment, or

2) selection could operate to produce a 'general-purpose genotype' that can occupy an extremely wide range of habitat types.

Which of these outcomes has occurred most frequently is difficult to determine. Circumstantial evidence in support of both scenarios exists. For example, most apomictic groups have been found to comprise a variety of genetically-distinct clones, an observation that is consistent with the first result. Turesson (1943) transplanted several clones from each of many apomictic *Alchemilla vulgaris* populations into a common garden, and observed considerable variation. Hull and Groves (1973) report at least three morphologically distinct types of skeletonweed, *Chondrilla juncea* (Compositae) from southern Australia, where this species is an obligate apomict. Morphological differences were conserved in a common environment. Similarly, apomictic *Poa pratensis* from the Sierra Nevada consists of many different genotypes, intermingled in the same meadow (Clausen and Hiesey, 1958). Twelve individuals sampled from near Mather and transplanted to a common environment all differed from one another with respect to one or more of these characters: habit, size, leaf color, susceptibility to disease, and flowering time. When eleven different individuals, sampled from different climatic areas, were cloned and grown under a variety of growth chamber conditions, they displayed significant differences in their growth responses to the various conditions (Hiesey, 1953).

*Taraxacum* is perhaps the best-studied example of clonal diversity; nearly 2000 'microspecies' (morphologically distinct forms) have been described (Richards, 1973). Den Nijs et al. (1978) sampled 50–100 individuals in each of many populations of apomictic *Taraxacum* in the coastal Netherlands. They found that most populations contained from 4 to 15 distinct morphotypes. Van Loenhoud and Duyts (1981) determined that within a single site, there can be up to 20 *Taraxacum* microspecies. Laboratory germination studies demonstrated that seeds of each microspecies responded differently to light and temperature regimes, and had different responses to storage, i e. that these morphological differences are associated with other differences that could have selective value. Ford (1981) showed that different *Taraxacum* clones responded in unique ways to competition with one another and with a variety of grass species.

Other studies have attempted to measure genetic variation by randomly sampling enzyme loci electrophoretically. Hancock and Wilson (1976) sampled 3 old-field populations of *Erigeron annuus* (one of the few annual apomicts) for genetic variation at 4 enzyme loci. They identified 17 distinct genotypes. Using 6 enzyme loci, Solbrig and Simpson (1974) demonstrated the presence of at least 4 distinct *Taraxacum officinale* genotypes within a $100 \, m^2$ area. (*T. officinale* in North America is obligately apomictic.) No one, however, has even done the sorts of studies that would be necessary to demonstrate that any of these clones is especially well-adapted for the particular microhabitat in which it occurs. Without such evidence, it is more reasonable to conclude that the clonal diversity we can observe within and between sites is simply a result of there being many independent sources of variability.

In support of the second proposed outcome of selection, there is evidence that some clones are in fact widespread. For example, Babcock and Stebbins (1938) reported that while *Crepis acuminata* populations near the center of distribution for the species contained considerable morphological diversity, variants were much less common in more remote areas, and the same type could be found in localities that were separated by hundreds of kilometers. Their explanation for this phenomenon was, first, that the source of new variants is repeated hybridization between sexual parents, so variants will be more common nearer sexual populations, and, second, that apomictic clones on the margins of the range will have been subjected to more intense selection, because their environments are more extreme.

Perhaps the best evidence for the existence of widespread genotypes comes from a study by Lyman and Ellstrand (1984), who surveyed 22 North American populations of *Taraxacum officinale* for variation at 5 enzyme loci. They used the phenotypic trait of seed color to further resolve differences. By these methods they demonstrated that their sample of 518 individuals included at least 47 distinct genotypes. The average number of genotypes per population was 5. Most clones (66%) were found in only one population, but a few were extremely widespread. One, for example, was found in 19 populations ranging from Vermont to California to Alaska. If this is indeed a single clone (and not a heterogeneous group, among which genetic variation existed but was not detected), it certainly seems to have the potential to be a general-purpose genotype. But a general-purpose genotype is widespread, in theory, because of a broad physiological tolerance, and we have no information on the fitness of these plants under different environmental conditions. So far, then, these few examples provide only very weak support for the idea of general-purpose genotypes as a general explanation for observed distribution patterns.

## Polyploidy and distribution patterns

When trying to construct an explanation for the distribution patterns of apomicts or parthenogens relative to sexuals, it is important to recognize that a striking correlate of asexual reproduction among both plants and animals is polyploidy. Recent observers of distributional trends among animals (Bell, 1982; Glesener and Tilman, 1978) have ignored this correlation, and have attributed the causes of those patterns to breeding system differences rather than to ploidy level differences. Because there are very few sexual polyploids among animals, deconfounding the two possible causal factors is quite difficult. However, polyploidy is common among sexually-reproducing plants, a fact that makes it possible to examine whether polyploidy alone might not provide an adequate explanation for the observed differences in distribution patterns between sexuals, which are generally diploid, and apomicts, which are nearly always polyploid. The evidence that polyploidy alone endows plants with characteristics that could influence their distribution patterns is in fact quite persuasive.

Long before evolutionary biologists began to notice the phenomenon of geographical parthenogenesis, botanists were drawing attention to the fact that, in flowering plants, the frequency of polyploids tends to increase with increasing latitudes, and often increases with increasing elevation (de Wet, 1980). Disclaimers have been made that such surveys in the past have failed to recognize that the frequency of plant growth-forms also varies with latitude (frequency of perennials increases at higher latitudes), and that since ploidy levels are correlated with growth form (perennials are more likely to be polyploid than annuals are), this trend is not as striking as it first appears (Ehrendorfer, 1980). However, an examination within taxonomic groups and within geographic areas still reveals the existence of these trends:

1) Diploids are often more limited in distribution than are their tetraploid relatives (Lewis, 1980; Roose and Gottlieb, 1976; Stebbins, 1971).

2) High polyploids often occur in upper alpine sites (Lewis, 1980).

3) Polyploids demonstrate a marked tendency to successfully colonize once-glaciated areas, while their diploid progenitors remain in unglaciated places, and are often limited to refugia (Ehrendorfer, 1980; Johnson and Packer, 1965; Lewis, 1980; Manton, 1934; Stebbins, 1950).

The explanation most often offered to account for these observations has been a genetic one, an argument that rests on the genetic variability presumed to characterize polyploids as a result of their hybrid orgin. Polyploids often arise from the fertilization of an accidentally unreduced diploid female gamete from one individual by a normal haploid male gamete from another. The triploid organism that results may produce unreduced triploid female gametes that are then fertilized by haploid male gametes to form tetraploid offspring (de Wet, 1980). Thus the genotype of many polyploids may contain complete or partial genomes from as many as three genetically distinct individuals. This condition can confer several advantages: 1) The multiple copies of genes so acquired can provide pathways for the development of new gene functions, and 2) by combining the genetic information of several individuals, offspring can be produced that have more physiological and ecological flexibility than any of their parents; polyploids so produced may be expected to be more heterozygous than any of their parents (Roose and Gottlieb, 1976). Indeed, since polyploids must originate sympatrically with their progenitors, their widespread success attests to the fact that at least some of the polyploids that are produced must be as fit or more fit than the genotypes that produced them (Schultz, 1980).

A variety of experiments has demonstrated that hybrids, and thus polyploids, can display a wider range of tolerance than their ancestors. For example, when Hiesey and Nobs (1970) produced hybrids between different ecological races of *Achillea millefolium,* some of the $F_1$ and $F_2$ progeny had ecological tolerances exceeding those of either parent. In a later study, Hiesey and Nobs (1982) crossed two different species of *Poa,* both of which were facultatively apomictic, then transplanted twelve lines of apomictic $F_2$ progeny to three different environments, along with their parents. One of the progeny lines outperformed both parents at all three transplant stations. Smith (1946) performed a comparison of the physiological and morphological characteristics of different chromosomal 'races' of *Sedum pulchellum*. She found, by growing plants in a common environment, that the hexaploids were more tolerant of extremes of soil moisture than were diploids, and that the competitive ability of seedlings increased with ploidy levels. This was true regardless of whether seedlings were competing with individuals of the same ploidy level or of different ploidy levels. Tal (1980) compared the response of diploid and tetraploid plants of cultivated tomato (these were autotetraploids, so they did not have the advantage of possessing two genomes) to salinity stress, and found the tetraploids to be more tolerant of salinity than the diploids. Tomkins and Grant (1978), who examined a variety of morphological and genetic factors for their ability to predict the response of 75 weed species to herbicides, found that the ploidy level of plants resistant to herbicides was significantly higher than the ploidy level of susceptible plants.

Finally, in a recent review of the role of polyploids in generating evolutionary novelty, Levin (1983) cites a number of examples of polyploid plants that are more resistant to pathogens and pests, or more tolerant of low nutrients, or more drought or cold resistant, than their diploid progenitors.

## Discussion

The differences in distribution patterns between sexuals and parthenogens have been recognized at least since 1928, when Vandel (1928) coined the term 'geographic parthenogenesis' for this phenomenon. The recent resurgence of interest in the evolution of sexual reproduction has seen this pattern cited as support for the hypothesis that sexuality is favored by biotic selection. However, proponents of this view have failed to note the correlation between breeding system and ploidy level that characterizes both plants and animals, and have ignored the potential confounding influence of polyploidy. Vandel (1940) himself drew attention to the correlation between polyploidy and parthenogenesis, and observed that it was possible in plants (though difficult in animals) to separate the two effects, so this observation is not original. However, in light of the attention being received by the 'biotic selection' explanation for the maintenance of sex, it seems important to point out that occupancy by parthenogens of wider ranges and more extreme conditions is just as likely to be the consequence of their high ploidy levels as of their breeding system; in fact, this appears to be the more parsimonious conclusion. It seems premature, then, to interpret the trends identified in this paper and by Glesener and Tilman (1978) and Bell (1982) as support for the biotic interaction hypothesis, or for any other hypothesis. Because of the confounding of breeding system and ploidy level, proper

216

testing of alternative hypotheses must be done by experimental studies in which ploidy level is held constant.

It also seems important to point out that the 'biotic interaction' explanation for these patterns rests on a completely untested assumption: that in areas of co-occurrence, parthenogens will experience intense competition with their sexual progenitors, who will not succeed unless their numerical disadvantage is countered by some corresponding advantage. It seems equally likely that the two forms will have quite different ecological properties, and are no more likely to be ecologically identical than any two congeners are. Indeed, it is hard to believe that the observed cases of coexistence between sexuals and parthenogens described in this paper are cases in which the competitive abilities of the two are exactly equal. Rather, it seems probable that the two forms are sufficiently ecologically distinct to coexist. If we do not assume that sexuals and parthenogens necessarily compete where they co-occur, then the phenomenon of geographical parthenogenesis is consistent with two other hypotheses: 1) that the greater range of parthenogens is a product of their greater potential for establishing new populations, and 2) that parthenogens are the product of selection for general-purpose genotypes. Descriptive studies alone will not permit us to distinguish among these hypotheses.

### Summary

Plant taxa that reproduce asexually display some distinct geographical and ecological patterns. A literature review reveals that such taxa 1) tend to have larger ranges, 2) tend to range into higher latitudes, and 3) tend to range to higher elevations than do their sexual relatives. Asexual taxa have a greater tendency than sexual taxa do to colonize once-glaciated areas. These trends have previously been identified as characteristic of parthenogenetic animals as well. While many authors have interpreted these trends as providing support for the 'biotic uncertainty' hypothesis for the maintenance of sex, these trends are consistent with several other interpretations as well. Furthermore, all of these interpretations have ignored the positive correlation that exists between ploidy level and breeding system: asexual plant and animal taxa are generally polyploid, while their sexual relatives are generally diploid. Evidence is presented for plants, and by extension for animals as well, that high ploidy levels alone – independent of breeding system – could endow individuals with the ability to tolerate these 'extreme' environments. For this reason, it appears premature to interpret observed distribution patterns as evidence to support hypotheses about what forces maintain sexual reproduction. Only experimental tests, using sexuals and asexuals of comparable ploidy levels, can permit us to discriminate among the alternatives.

Acknowledgments. This paper was written while I held a Visiting Professorship for Women in Science at the Botany Department of Duke University, supported by NSF grant R II-8310325. I would like to acknowledge the hospitality of Duke's Botany Department, and especially that of Janis Antonovics and his students. Results reported here were obtained in connection with field work funded by the Research Corporation, Pomona College's Schenck Fund, the Seaver Science Fund, and NSF grant BSR-8407468. N. C. Ellstrand, C. W. Greene, and M. Lynch generously provided me with unpublished manuscripts, I am grateful for the improvements suggested by J. Antonovics, H. G. Baker, J. Bergelson, N. C. Ellstrand, P. Kareiva, M. Lynch, P. Pack, and S. Stearns, and for discussions with G. L. Stebbins (who does not necessarily agree with the conclusions presented here).

# Sex allocation in animals

E. A. Herre, E. G. Leigh Jr and E. A. Fischer

## Introduction

There are numerous reasons why the study of sex allocation has attracted a great deal of both theoretical and empirical attention. To begin with, sex allocation is a ubiquitous phenomenon. All sexually reproducing animals exhibit some apportionment of reproductive effort between sexual functions. Sex ratios in organisms with separate sexes (gonochores), relative investment in sexual functions in simultaneous hermaphrodites, and age or size at which individuals change from one sexual function to another in sequential hermaphrodites are all forms that apportionment may take.

Empirically, the apportionment of effort between sexual functions and the relevant factors that are thought to affect that apportionment are often easily and unambiguously measured. Theoretically, sex allocation offers a tractable evolutionary problem. The character in question, the apportionment of effort between the sexual functions, is neatly circumscribed between 0 and 1, all female or all male. The effects of different allocation patterns on genetic representation in future generations are direct, large, and easily calculated. These factors allow the theory to yield precise quantitative predictions of both expected optima and, potentially, the dynamics of approach to those optima. The mechanisms and genetics of sex determination and whatever constraints that they may impose on adaptation are often clearly understood. Sex allocation therefore appears to provide an opportunity for determining the precision of an adaptation.

Sex allocation is also an interesting type of frequency dependent selection. For example, (at least in case of active parental control) a tendency to produce offspring of any particular sex ratio can only be subject to selection in subsequent generations through the differential mating success of those offspring (e. g. as realized in number of sets of grandprogeny). The success of those offspring in turn depends on the sexual composition of the population of available mates. The apparent intricacy of the problem led Darwin 'to leave its solution for the future'. Also, sex allocation patterns and the forces which select them have been the focus of a rekindled controversy over levels of selection. Finally, sex allocation has become one of the most successful branches of evolutionary theory. In no other field of evolutionary biology has it been possible to test and verify so many detailed quantitative predictions.

In this review, we will present derivations of major theoretical results for expected allocation patterns in outbreeding gonochores, simultaneous hermaphrodites, and sequential hermaphrodites, analyze the effect of violating the assumptions underlying that theory, and compare the predictions with empirical findings. Finally, we will identify problems confronting our present understanding of sex allocation and suggest some possible lines of future inquiry.

## 50:50, the basic prediction

With an insight that is fundamental to the study of sex allocation, Fisher (1930) showed that in large, outbreeding populations of gonochores the population is expected to apportion effort equally between the two sexual functions. This result follows directly from the fact that in most organisms, males, as a group, as well as females, as a group, each contribute half the genes for the subsequent generation. Therefore, if the sex ratio is $\frac{1}{2}$ (fraction of males), the average reproductive value of a male is equal to that of a female. If the sex ratio is not $\frac{1}{2}$, then the average reproductive value of members of the rare sex is higher and there is a selective advantage to individuals that produce offspring of that sex. Generally, in cases in which parental investment is equal for offspring of both sexes, this should lead to a sex ratio of $\frac{1}{2}$. However, if, for example, sons are inherently more costly to raise than daughters, then enough extra daughters should be reared so that the population expenditure on each sex remains the same. Therefore, the result is not necessarily that the sex ratio should be $\frac{1}{2}$ but rather that total populational investment in the two sexes should be equal. This most basic result has been reconfirmed many times by a variety of different methods (Charnov, 1982; Kolman, 1960; Leigh, 1970; MacArthur, 1965; Shaw and Mohler, 1953; Stubblefield, 1980; Uyenoyama and Bengtsson, 1979, 1981; Appendices A, B, C). In the following discussions we will treat equal investment as meaning equal ratios unless it is explicitly stated otherwise.

The prediction of equal investment in the sexes possesses many interesting properties. First, the prediction pertains to the sex ratios of populations, not of the offspring of individuals. As we have shown, if the sex ratio of a population is $\frac{1}{2}$, the average reproductive value of a son equals that of a daughter. Therefore, there is no selective advantage or disadvantage to an individual producing offspring in any particular sex ratio. It follows that selection on sex ratio in populations in equilibrium does not restrict genetic variation in sex ratio (see Appendix B). However, alleles that code for a sex ratio of $\frac{1}{2}$ always increase in frequency in a population if the sex ratio is not $\frac{1}{2}$. Therefore, $\frac{1}{2}$ generally will be the most efficient sex ratio for the spread of genes within a population, so that $\frac{1}{2}$ will also generally best represent the common interest of the genome of both parents or, for that matter, among any group of relatives (Fisher, 1930; Leigh, 1977; MacArthur 1961; Stubblefield, 1980). That is, in diploid organisms there is no conflict of interest with respect to sex ratio of offspring among sexual partners or their relatives.

So we have seen that $\frac{1}{2}$ is expected to come about due to the individual advantage that it gives within populations. On the other hand, populations would be capable of greater growth if they were female biased. Therefore, sex ratios of $\frac{1}{2}$ in outbreeding populations would be evidence of the predominance of selection within populations (individual selection) and female biased sex ratios would be evidence of the predominance of selection among populations (group selection). It is odd that there is so much evidence apparently supporting the predicted outcome of individual selection within sexually reproducing populations (Charnov, 1982; Clutton-Brock, 1986; Clutton-Brock and Iason, 1986; Hamilton, 1967; Leigh et al., 1985; Williams, 1979) while there is apparently so little supporting sexual reproduction as itself being an outcome of individual selection (Maynard Smith, 1978).

**Evidence for basic sex ratio predictions (water, water, everywhere . . .)**

Most mammals and birds show a nearly 1:1 populational sex ratio as predicted by theory (Clutton-Brock, 1986; Clutton-Brock and Iason, 1986). In those exceptions where there are statistically significant deviations from unity, the magnitude of the deviations are generally small. However, can the case studies of organisms in which the sex determining mechanisms essentially assure a 1:1 sex ratio (eg., XY) be considered as strong support for a theory which makes this prediction (Williams, 1979)? It may be that the mechanism itself was selected because it gives the ratio of $\frac{1}{2}$ (Bull, 1983). On the other hand, chromosomal sex determination may have assumed predominance for other reasons and 1:1 may not necessarily be selected for. For example, if the genes affecting sex ratio have pleiotropic effects on other characters, selection on those characters may skew the sex ratio to such an extent that with a few random deaths of the rare sex, the population could not reproduce itself. That is, the advantage of sex chromosomes may lie in the fact that this system provides stability (Rice, 1986) so that even in small populations, both sexes will generally be present. Whatever the reason for the predominance of chromosomal sex determination, such systems obscure the interpretation of observed sex ratios. Although there are cases of organisms with X−Y sex determination in which there is evidence that, at either the populational or the individual level, a sex ratio of 1:1 is not strictly adhered to (Blank and Nolan, 1980; Clark, 1978; Clutton-Brock, 1986; Clutton-Brock and Iason, 1986; Gutzke et al., 1985; Howe, 1977; Trivers and Willard, 1973), nonetheless, we might expect that the most useful tests of sex allocation theory should come from those species which lack the meiotic constraint of the X−Y system of sex determination.

In many species of turtles and crocodiles sex is determined by the incubation temperature of eggs (Bull, 1983). Those species that lack chromosomal sex determination, and therefore lack the meiotic constraint, frequently show quite striking female biases (Bull, 1986; Ferguson, 1982; Ferguson and Joanen, 1983; Mrosovsky et al., 1984; Schwarzkopf and Brooks, 1985; Webb et al., 1983). These female biases pose a serious problem for the theory which tells us to expect a 1 : 1 sex ratio (Bull, 1983). Are the observed biases simply transitory disequilibria? If, on the other hand, the observed biases do represent stable equilibria, what does that tell us about the theory? Is it simply wrong? Are there aspects of the natural population structure of these organisms which we are not taking into account? Further work is needed to clarify these issues.

Haplodiploid organisms also lack the constraints associated with heterogametic sex determination. While we will discuss other aspects of their sex allocation phenomena later in the review, some of the strongest evidence for equal investment in outbreeding populations come from these organisms. In general, solitary outbreeding Hymenoptera spend equal effort on each sex (Charnov, 1982; Frohlich and Tepedino, 1986; Hamilton, 1967). The parasitic wasp *Pachysomoides stupidus* puts mostly males or mostly females into any single host, but over all, half its young are male (Pickering, 1980). In an interesting six year study, Utah populations of the solitary nesting bee, *Osmia lignaria,* show dramatic between-year variation in population sex ratio (Tepedino and Torchio, 1982). A sample from any single year may show either significant female or male bias or no difference from 1 : 1. This result stresses the importance of caution in interpreting the findings of short-term studies. Do the biases represent the response to selection for overproduction of the rare sex in a population on its way to the expected equilibrium of 1 : 1? Do the biases or the 1 : 1 condition itself merely reflect random, non-selected swings?

Among humans and in various other well-studied vertebrate populations, heritable variation in the sex ratio is undetectably small (Clutton-Brock, 1986; Clutton-Brock and Iason, 1986; Williams, 1979). Sex ratio thus contrasts strikingly with most other studied characters which almost invariably show heritable genetic variation (Falconer, 1981). This contrast seems surprising; most characters are subject to stabilizing selection (Fisher, 1930) which restricts their heritable variation (Turelli, 1984), while, as we have seen, selection on the sex ratio does not restrict its genetic variance. The absence of genetic variation in sex ratio must therefore mean that mutation rarely affects sex ratio and that sex ratio is accordingly unresponsive to selection. Why, of all characters, should sex ratio be so particularly immune to the effects of mutation? This immunity may well reflect the common interest of the genome in honest meiosis (Leigh, 1971). However, it is important to distinguish the common interest of the genome with respect to honesty of meiosis from the common interest of the genome with respect to sex ratio. We will discuss circumstances in which the most efficient sex ratio for spreading genes (i. e. the sex ratio that best serves the common interest of the genome) may not be $\frac{1}{2}$ and, in fact, at the populational equilibrium sex ratio of $\frac{1}{2}$ all sex ratios are equally efficient. Nonetheless, in diploid organisms with a heterogametic sex determining mechanism (e. g. an X–Y system) a sex ratio of $\frac{1}{2}$ and the lack of heritable variation is likely to be an epiphenomenon resulting from the underlying common interest of the genome in honesty of meiosis.

Recent work points to exciting new possibilities for testing the prediction of equilibrium sex ratio of $\frac{1}{2}$. Cyclically parthenogenetic *Daphnia* (Barker and Hebert, 1986) show evidence for heritable variation in sex ratio when they switch to the sexual phase (D. Barker, pers. comm.). Many species of *Poeciliid* fish also show heritable variation in sex ratio, often as a result of unusual forms of chromosomal sex determination (Bull, 1983; Sullivan and Schultz, 1986; A. Basolo, pers. comm.). This variation could be exploited to determine experimentally whether the predicted population sex ratio of $\frac{1}{2}$ is achieved. Although there is highly suggestive evidence from fruit fly populations infected by sex ratio distorters that a sex ratio of $\frac{1}{2}$ will be reached (Lyttle, 1977, 1979, 1981), there is no demonstration of the evolution of the predicted sex ratio of $\frac{1}{2}$. Such a demonstration would be a powerful confirmation of the validity of sex allocation theory.


## Hermaphrodites

*Theory*

When might gonochores be replaced by a genotype where one individual performs both sexual functions? Let us suppose that a hermaphrodite which devotes a fraction $1 - k$ of its reproductive effort to producing eggs and the remainder to fertilizing the eggs of others, produces $nf(k)$ eggs and fertilizes $nm(k)$ eggs, while a female produces $nf(0)$ eggs and a male fertilizes $nm(1)$. If members of our population mate at random, and if we treat eggs and 'successful' sperm as haploid males and females, and their zygotes, whether gonochore or hermaphrodite, as matings between these haploids, then sex ratio theory tells us when selection favors hermaphrodites. In a population a fraction $k''$ of whose zygotes produce only male gametes and the rest, only female, selection favors a dominant mutant, all of whose zygotes devote a fraction $k$ of their effort to producing sperm (or ensuring their success), if $nf(k)/[(1 - k'')nf(0)] + nm(k)[k''nm(1)]$ exceeds 2. In a population composed exclusively of such hermaphrodites, selection favors a dominant mutant for gonochorism a fraction $k''$ of whose zygotes are 'male', if $(1 - k'')nf(0)/nf(k) + k''nm(0)/nm(k)$ exceeds 2.

If $nf(k) = nf(0)(1 - k)$, $nm(k) = knm(0)$, there is selective equipoise between such hermaphrodites and a gonochore genotype a fraction $k$ of whose zygotes are male. Usually, producing eggs and fertilizing them are sufficiently different occupations – at least for 'higher' animals and plants – that, other things being equal, it pays to specialize. Thus sexes should be separate unless there is some special advantage to combining the sexual functions (Charnov et al., 1976).

If the advantage to combining the sexual functions is great enough that, for some suitable value of $k$, a hermaphrodite produces more than half as many eggs as a female gonochore and fertilizes more than half as many eggs as a male, then selection favors hermaphroditism. In a population of gonochores, a mutant for such hermaphroditism would spread, while selection eliminates gonochore mutants from a population of such hermaphrodites. In this case, selection favors the value of $k$ for which the proportionate gain in number of eggs produced $[nm(k + d) - nm(k)]/nm(k)$ from a slight increase $d$ in effort fertilizing eggs balances the proportionate loss $[nf(k + d) - nf(k)]/nf(k)$ thereby imposed on eggs fertilized (Charnov, 1976; Leigh et al., 1976).

Finally, selection might favor a rare mutant for hermaphroditism in a population of gonochores, even though it would favor a mutant whose bearers were all gonochores of one sex in a population composed exclusively of such hermaphrodites. For example, an all-male mutant would be favored if $nm(0)/nm(k) > 2$. If so, selection would increase this mutant's frequency to the value $X$ where this mutant's bearers fertilized just twice as many eggs per zygote as the average for the population at large – where $nm(0) = 2Xnm(0) + 2(1 - X)nm(k)$. However, hermaphroditism cannot be maintained in a stable polymorphism with both male and female gonochores (Charnov et al., 1976).

## Simultaneous hermaphroditism

There are four categories of animals where one would most expect individuals to benefit from carrying on both sexual functions at once (Charnov et al., 1976; Fischer, 1980).

1) Simple animals, where different sex roles require little specialization.

2) Those sessile animals with planktonic larvae which can only fertilize near neighbors. By sacrificing less than half of a female's egg production, a sessile hermaphrodite can fertilize more than half as many eggs as a male.

3) Animals which, like most higher plants, carry out their male functions (such as making pollen) before bearing their young. Here, the energy devoted to male function does not divert equivalent energy from female function and a hermaphrodite can do more than half as well as a gonochore at either.

4) Animals whose simultaneous hermaphroditism permits a reciprocal altruism obviating most of the competitive aspects of maleness.

Many factors which favor simultaneous hermaphroditism could, under different circumstances, favor either gonochores with males fewer or smaller than females (as opposed to females which carry out male functions on the side), or monogamous gonochores whose males devote much of their effort to 'female functions' such as parental care. Synchrony of reproduction, which favors simultaneous hermaphroditism in plants, favors male participation in parental care among gonochores by depriving males of the prospect of mating with other females when they desert their offspring (Knowlton, 1979).

Particularly interesting in this respect is the role social convention sometimes plays in sex allocation. Just as social conventions promote male participation in parental care by closing off other avenues of increasing reproductive success, such conventions provide a striking advantage for simultaneous hermaphroditism among coral reef fish of the family Serranidae (Fischer, 1980, 1981, 1984; Fischer et al., 1987). Hamlets, *Hypoplectrus*, are simultaneous hermaphrodites which mate in pairs, switching sex roles after successive spawns: in effect, members of a pair trade eggs for each other to fertilize (Fischer, 1984). At first the prospective mates only offer each other a few eggs, then more as confidence builds up. This 'hesitancy' penalizes fish that move from mate to mate fertilizing eggs without offering any in return (Fischer, 1980), imposing a law of sharply diminishing returns on male function. Thus nearly fully functional females act as males 'on the side', with little sacrifice in egg production (Fischer, 1981).

The incidence of simultaneous hermaphroditism appears to agree at least roughly with theory (Fischer, 1980, 1981, 1984; Fischer et al., 1987; Ghiselin, 1969) although no survey of its incidence is complete enough to judge the accuracy of theory in this respect. Nor can we predict the circumstances under which it is feasible to evolve an otherwise desirable hermaphroditic habit. Finally, although it is difficult to measure precisely the effort hermaphrodites apportion to male versus female functions, a variety of opportunities exist for the testing of sex allocation theory (Fischer, 1980, 1981, 1984; Fischer et al., 1987).

*Sequential hermaphroditism*

The mathematics of simultaneous hermaphroditism also apply to animals which are born into one sex and change sex later on. Such animals divide their effort between male and female functions by changing sex at some size or age. Formally, we may imagine our hermaphrodite passing through successive ages, each more favorable, say, to producing male gametes rather than female, to be an animal in a heterogeneous environment moving through patches successively more favorable to producing males. In a constant environment, such an animal should produce only males in the patches most favorable for males, and vice versa, and a mixture of the two sexes in the one intermediate patch where increasing the proportion of females confers a proportionate gain in the total number of females produced in all patches, just balancing the proportionate loss in males (Bull, 1981). Likewise, sex change should be 'all or nothing', the hermaphrodite producing only female gametes until it is more profitable to produce males, then switching entirely to the latter.

Such sequential hermaphrodites seem to offer spectacular opportunities for testing sex allocation theory, and indeed, work on sequential hermaphrodites lends strong support to the proposition of the predominance of selection within populations (Leigh, 1977, 1983, 1987). The clear temporal distinction between male and female functions makes the apportionment between them much easier to discern than in simultaneous hermaphrodites. Moreover, since the age of sex change is usually an individual decision, it is more responsive to environmental circumstance than sex ratio among gonochores, or the apportionment of effort between male and female functions by simultaneous hermaphrodites. The most conspicuous difficulty is that sex ratio theory assumes that all offspring of a given sex from one mating are produced simultaneously and are equally valuable. However, the offspring of sequential hermaphrodites are generally produced over a long period. To apply the theory, we must assume that the numbers and age composition of a population of sequential hermaphrodites do not change with time. As we shall see, a more general difficulty is that the theory of sex change is a demographic theory, relating the age of sex change to growth, mortality and reproductive rates of fish of all ages. The theory then depends on an inconvenient multiplicity of variables.

When should sequential hermaphrodites replace gonochores? Ghiselin (1969) predicted that if one sex gained fertility more rapidly with age than the other, an animal should be born into the sex where youth hurts least, and change later to that sex where age benefits most. To explore the matter further, consider an animal which is born female and turns male at age $T$. Let us measure the proportion $1 - k(T)$ of its effort devoted to female function by the proportion of a newborn gonochore female's prospective egg output produced by age $T$. If $l(x)$ is the probability a female survives to age $x$, and if $b(x)$ is the egg output per unit time of a female aged $x$, then the average egg output $n_F(T)$ per newborn fish programmed to change sex at age $T$ is

$$n_F(\infty) [1 - k(T)] = \int_0^T b(x)l(x)dx,$$

where $n_F(\infty)$ is the egg output per female gonochore (female that never changes sex).

Let male fertility grow more rapidly with age than female. In particular, suppose that a male which changes sex at age $y$ ceases reproduction until age $y + t$, after which point he fertilizes $b(x)f(x)$ eggs per unit time when his age is $x > y + t$. Moreover, let a proportion $L(y + s)/L(y)$ of the animals turning male at age $y$ survive to age $y + x$, where $L(y)$ is the proportion of male gonochores that would survive to age $y$, and let $G(x) = L(x)f(x)/l(x)$

be an increasing function. Under these assumptions, the fraction $m(t)$ of a male gonochore's reproductive success realized by a fish that turns male at age $T$ is

$$\frac{\displaystyle\int_{T+t}^{\infty} dx\, b(x) f(x) L(x) l(T)/L(T)}{\displaystyle\int_{0}^{\infty} dx\, b(x) f(x) L(x)} = \frac{l(T) \displaystyle\int_{T+t}^{\infty} dx\, b(x) l(x) G(x)}{L(T) \displaystyle\int_{0}^{\infty} dx\, b(x) l(x) G(x)} < \frac{l(T) k(T+t)}{L(T)}$$

Let $T''$ be that age when a female exhausts half its prospective egg output, were it never to change sex. Then $1 - k(T'') = k(T'') = \frac{1}{2}$. If $dG/dx$ is large enough that

$$\frac{l(T) \displaystyle\int_{T+t}^{\infty} dx\, b(x) l(x) G(x)}{L(T) \displaystyle\int_{0}^{\infty} dx\, b(x) l(x) G(x)} \qquad k(t) = \frac{\displaystyle\int_{T}^{\infty} dx\, b(x) l(x)}{\displaystyle\int_{0}^{\infty} dx\, b(x) l(x)}$$

then $n(T'') > \frac{1}{2} = k(T'')$, and there is an age of sex change $T'' + s$, for which $m(T'' + s) > \frac{1}{2}$, $1 - k(T'' + s) > \frac{1}{2}$. Selection accordingly favors animals which change sex at age $T'' + s$ over gonochores. The more rapidly male fertility grows relative to female, the larger $dG/dx$, the greater $n(T'') - \frac{1}{2}$, the greater the advantage of changing sex, and the older animals should be when they do so (Leigh et al., 1976; Warner et al., 1975).

However, we risk forgetting the very provisional nature of our mathematics. To begin with, we have not established what conditions are necessary to render sex change advantageous. For example, a male's fertility need not increase indefinitely with his size to make sex change advantageous; all that is required is that, at some age, changing sex will increase the animal's prospective reproductive success over the remainder of its lifespan. More seriously, what we have represented as the age of sex change is often better represented as the threshold at which the animal 'decides' that the other males present are few enough, and that she herself is big enough, compared to them, to justify changing sex (Warner et al., 1975). In a population of stable age and size composition, fish with the same threshold of response will change sex at the same age, and the simplistic theory we presented will be more or less adequate. In a fluctuating population with seasonal recruitment, however, the age of sex change will depend on the number and the sizes of the males already present, which vastly complicates the theory. It is often best, however, to see how well a simple theory works before complicating the story; in that spirit, we shall proceed.

Pandalid shrimp have been used to test this theory. They show striking agreement with the theory's predictions (Charnov, 1979). These shrimp reproduce once a year. Males are assumed to fertilize the same number $bm$ of eggs each year, regardless of their age, whereas older females produce more eggs; the number $b(x)$ of eggs a female produces in her $x$th reproductive season is roughly $n[1 - e^{-k(x - x_o)}]^3$, where k, $x_o$ and n differ from population to population (Charnov, 1979). Thus theory predicts that these shrimp should be born male and become female later on. In most populations, all individuals are female by their second year, although some are born female. If an individual born male fertilizes $b(1)$ eggs, and if a proportion $L$ of these shrimp survive from one reproductive season to the next, then the proportion $W$ of shrimp born male should be such that

$$Wbm = Wb(1) = (1 - W)b(1) + b(2)L + b(3)L^2 + b(4)L^3 + \ldots$$

The relation between $W$, death rate $L$, and growth rate as reflected by $k$ and $x_0$ agrees roughly with this prediction (Charnov, 1979); too many inaccurately estimated variables are involved to hope for better. In some populations, all animals are born male. Since fishing increases mortality, theory predicts, and observation confirms, that in these populations, heavier fishing causes these shrimp to change sex earlier, and at smaller sizes (Charnov, 1982).

The best studied sequential hermaphrodite is the bluehead wrasse, *Thalassoma bifasciatum,* most of whose individuals are born female and transform later into 'gaudy' males with bright colors. A female bluehead spawns once a day, while a large male may spawn forty times a day; there is a distinct advantage to being male when large (Warner et al., 1975). Theory predicts that females should turn male at the age $T'$ when the proportionate gain in egg production from a slight delay in changing sex just balances the proportionate loss in male reproductive success, that is to say, when

$$\frac{b(T)\,l(T)}{\int_0^T dx\,b(x)\,l(x)} = \frac{b(T + t)f(T + t)L(T + t)\,l(T)/L(T)}{\int_{T+t}^{\infty} dx\,b(x)f(x)\,L(x)\,l(T)/L(T)}.$$

If we assume that all eggs are fertilized, that as many eggs are fertilized per newborn as are produced, and that all fish are born females, then the two denominators are equal and the fish should change sex at that age $T$ where

$$b(T) = b(T + t)f(T + t)L(T + t)/L(T);$$

that is to say, the age when, after sex change is complete, the fish is fertilizing $L(T)/L(T + t)$ as many eggs as it was producing just before changing sex.

Warner et al. (1975) measured the reproductive rates of small, recently transformed gaudy males, crediting a male with $\frac{1}{n}$ spawn when it and $n - 1$ others spawn jointly with one female. They found that such blueheads obtained an average of 1.5 pair spawns a day, in apparent agreement with theory. Now it appears that, after changing sex, a gaudy male is nonterritorial, and nearly nonreproductive, for roughly 90 days (Hoffman et al., 1985; Warner, 1984b), during which period the death rate is 0.0062 per day, so that a fish has probability 0.57 of surviving the 90 days. When they resume reproduction, they fertilize the equivalent of 3.7 spawns a day, while a female produces one a day. The disparity is over twice that which theory predicts. Here, the disadvantages of a demographic theory become apparent. Do small gaudy males mate with small females which have fewer eggs? Is the death rate right? Did we go astray because we assumed the wrasse population has a stable age composition when it almost certainly does not (Victor, 1986)? Is the time required to change sex, for which Hoffman et al. (1985) supply a minimum estimate, a gross underestimate? Or has the theory a more fundamental fault?

In many social systems where the largest males compete most successfully for females, small males may be able to escape notice and fertilize females 'on the sly' (Trivers, 1985; West-Eberhard, 1983). Some bluehead wrasses for example, are born male, but with female colors; these 'drab males' fertilize females by various forms of 'stealth' and groups of drab males mate with females quite openly when gaudy males are otherwise occupied (Warner et al., 1975). When a drab male is large enough, it assumes the bright colors of a gaudy male and joins the 'establishment' it previously 'subverted'. Direct measurements, and the similarity of the ratio of drab males to females (158 : 532) to the ratio of gaudy males born male to those born female (26 : 98) show that on large reefs the growth and death rates of females and similar sized drab males are the same (Warner, 1984b; Warner et al., 1975). Under these circumstances one would expect females and drab males to have similar reproductive success; indeed, one would expect the ratio of drab males to females would evolve toward the value where this was so (Warner et al., 1975). The daily reproductive success of drab males on large reefs was first reported to be roughly one spawn a day, indistinguishable from the female's production of one spawn per day (Warner et al., 1975). If the same conditions apply where gaudy males secure a fraction $Y$ of the matings, there should be $1 - Y$ drab males per female (Charnov, 1982). On one archipelago of Caribbean reefs, gaudy males were found to fertilize a mean proportion $Y = 0.55$ of all spawns, while there was a mean of 0.52 drab males per female (Warner and Hoffman, 1980). For an eastern Pacific congener of the bluehead, gaudy males fertilized far less than 1 % of the spawns, and, in fact, there were roughly as many drab males as females (Warner and Hoffman, 1980). In a variety of populations, the relation between the proportion of bluehead wrasses born male and the proportion of matings secured by gaudy males showed at least rough agreement with theory (Charnov, 1982).

Later work has revealed the imprecision of these 'tests', and the problems that prevent more precise testing. On small reefs, drab males do not mate, and grow twice as fast as females (Warner, 1984b). This does not reflect genetic differences between the drab males of small and large reefs; once they 'learn the ropes', drab males transplanted from large reefs to small behave like those which originally settled on small reefs. The calculations of Warner (1984 b) suggest that young drab males on small reefs would quintuple their lifetime reproductive output by migrating to large ones. On the other hand, revised calculations based on growth rates determined by Victor (1986), and the time elapsed between changing color and resuming reproduction as a fully territorial male from Table 3 of Hoffman et al. (1985), deny this. Perhaps more unsettling is Warner's (1984 b) conclusion that a drab male fertilizes 1.34 spawns per day. In assessing the evidence both for and against the theory's predictions being met, we should keep in mind that different populations censured differently will give different estimates for the relevant parameters. Further, it is likely that those parameters will change through time.

It appears that theory should account for two aspects of sex change. Judging by other fish, the physiological aspects of sex change need not delay reproduction by more than six days; bluehead wrasses apparently prolong the delay in order to grow larger (Hoffman et al., 1985). Thus we must jointly calculate the appropriate age or size at which to change sex, and the appropriate delay before resuming reproduction. These variates will reflect growth rates, which decline with increased population density (Victor, 1986). As recruitment varies seasonally (Victor, 1986), so must the population's density and size composition. It should therefore be possible to predict how age or size at sex change, and the delay before resuming reproduction, should change with the season. These predictions should be amenable to qualitative tests. Securing more precise estimates of size or growth rate, however, would disturb the populations under study. For the moment, at least, we must look elsewhere for precise, quantitative tests of sex allocation theory.

Warner and Robertson (1978), Robertson and Warner (1978), Warner (1984 a) and Ghiselin (1969) have reviewed the incidence of sequential hermaphroditism among fish and other animals. In general, we may conclude that sex allocation theory provides a necessary, but not sufficient, condition for sex change. Although as we have seen, animals which do change sex illustrate many facets of sex allocation theory, a number of animals whose males only reproduce when old and large are not sequential hermaphrodites. Birds such as manakins (Snow, 1985) and mammals such as gorillas (Jolly, 1985), where young males have almost no prospect of mating, do not change sex, even though the advantage that would accrue from doing so seems at first sight to be very great. The usual comment is that the structural specilization of the sexes in birds and mammals would make sex change rather difficult (Warner, 1978). One might further suggest that birds and mammals are almost the only animals where young clearly play (Jolly, 1985, p. 404), that in many of these animals, play is essential to learning one's role in life, and that one factor allowing structural specialization of the sexes in birds and mammals is that, for such complex animals, learning the signals associated with a particular sex role is a lifetime's work, so that there is no longer any point to changing sex. However, these suggestions are not meant to disguise our inability to predict with any accuracy those lineages for which sex change is an attainable alternative.

## Assumptions underlying sex allocation theory

In the following sections we will discuss the theoretical consequences of violating the assumptions of the basic model for predicting sex allocation outlined in the appendix and review some of the relevant evidence.

## Overlapping generations

We find (Appendix C) that whether different generations overlap or not should not alter the basic optimum sex ratio of 50:50. However, it is another matter if there is asymmetry between the sexes in their overlap across generations (Seger, 1983; Werren and Charnov, 1975; Werren and Taylor, 1984). Seger (1983) considered the case of bivoltine organisms in which individuals of one of the two sexes potentially reproduced in two successive generations. He showed theoretically that there exists selection to underproduce the sex which overlaps during the generation into which it overlaps. He further pointed out that the biasing associated with certain types of overlap provided conditions favorable to the evolution of eusociality. He found that data on the sex ratios of bivoltine solitary Hymenoptera and on the phylogenetic distribution of eusociality were consistent with both of these predictions.

## Survival and fecundity

What happens if the mother's survival prospects depend on the sex ratio of her offspring? We find that if the birth of one sex reduces mortality for the mother, selection favors mothers which produce an excess of that sex (Leigh et al., 1985). This creates a conflict of interest between the mother and the father because, unless the father's survival is also affected, the optimal sex ratio for increasing his representation is equal numbers of each sex.

While we do not know of any animals in which it is certainly true that offspring of one sex differentially affect the survival of the parent of one or the other sex, there are clear cases in a few vertebrate species in which offspring of one sex have a greater tendency to cooperate with their parents in the rearing of young (Clutton-Brock, 1986; Clutton-Brock and Iason, 1986; Gowaty and Lennartz, 1985; Maynard Smith, 1978). Therefore instead of leading to a decrease in expected future reproduction, the cooperative sex leads to an effective increase. Here, a disproportionate number of the cooperative sex is produced at birth. This is interpreted as the cooperative sex helping to 'pay' for itself, thereby reducing per capita parental investment. Theoretically, the bias in the sex ratio should accurately reflect the amount of contribution which the cooperative sex provides (Emlen et al., 1986). This is, potentially, a measurable quantity.

### Sex chromosomes and sex ratio distortion

Asymmetry of inheritance is particularly marked in the sex chromosomes. In diploids, the 'odd chromosome' of the heterogametic sex (for heterogametic males, the Y chromosome) shares no genes with its homogametic counterpart. Theory predicts that a Y chromosome gene can spread through its population by biassing segregation at meiosis in favor of its bearer chromosomes (Hamilton, 1967; MacArthur, 1961), even if the population dies out as a result. The homogametic sex chromosome can also spread through an appropriate meiotic bias against Y chromosomes, but a given bias spreads the X only a third as fast, as only a third of these chromosomes occur in the heterogametic sex (Hamilton, 1967).

As we might expect, selection on sex chromosome loci for the sex ratio bears no relation to the common interest of autosomal genes. Lyttle (1977; 1979; 1981) translocated a segregation-distorter locus onto the Y chromosome of *Drosophila melanogaster,* and observed instances where the distorter spread through entire populations, causing their extinction. Extrachromosomal factors also spread without regard to the common interest of the genome. In experimental populations of a parasitic wasp, a factor, inherited through the father, which causes his offspring to be nearly all male without reducing brood size, increases to high frequency when introduced (Werren et al., 1981). In this same wasp, a bacterium which is passed on from mother to daughter, kills males before they hatch, perhaps because this enhances female survival (Skinner, 1985). Finally, a maternally inherited extrachromosomal factor in this wasp causes her offspring to be entirely female (Skinner, 1982). The spectrum of asymmetrically inherited sex ratio factors in this wasp begins to remind one of the 'war of all against all'.

How is this anarchy resolved? In populations infected by sex ratio distorters, whether genetic or extrachromosomal, any autosomal modifier that suppresses the distortion effect spreads, because it assures *its* offspring an advantageous sex ratio (MacArthur, 1961), just as selection favors modifier alleles suppressing the effects on meiotic segregation-ratio of a distorter which is thereby spreading a phenotyic defect through the population (Prout et al., 1973). Moreover, selection must favor those *species* where autosomal loci have enforced the common interest of the genome with respect to sex ratio (Leigh, 1977). At least among those species living today, this common interest appears to have prevailed. Genes carried by sperm are usually inactive, possibly serving to prevent the distortion of meiosis (Crow, 1979; Hamilton, 1967). Y chromosomes, which would otherwise be most susceptible to selection for sex ratio distortion, are usually inert (Hamilton, 1967), even in recently evolved systems of sex determination (White, 1973). The absence, in many populations, of genetic variation in sex ratio (Williams, 1979) may itself reflect how effectively selection for the 'common interest of the genome' has eliminated ways of tampering with the sex ratio.

Indeed, there is experimental evidence for the efficacy of the common interest of the autosomes. When Lyttle (1977) translocated a distorter locus onto the Y chromosome of *Drosophila melanogaster* and inoculated these mutants into cages of normal flies, he found that populations which survived the introduction of this distorter Y responded in two ways. Most populations accumulated polygenic suppressors of the Y chromosome distorter, each suppressor slightly decreasing the strength of the sex ratio distortion (Lyttle, 1979). After 350 days sex ratios, though still very skew, were significantly closer to 1:1 than when the distorter had first spread through the population. In one population, aneuploid genotypes, XXY and XYY, appeared, and attained frequencies such that two females were now being born for every three males, even though the distorter Y had completely replaced the normal one (Lyttle, 1981).

## Haplodiploids

Haplodiploids, whose unfertilized offspring are haploid males and whose fertilized offspring are diploid females, provide numerous opportunities for testing sex allocation theory (Charnov, 1982). Their asymmetric inheritance causes a striking conflict of interest between mothers and daughters over sex ratio (Trivers and Hare, 1976). Only in social species do daughters provide essential help in raising offspring, and it is here where we might expect to find daughters influencing the outcome of this conflict of interest. Moreover, in many asocial Hymenoptera, there is clear evidence that mothers are able to adjust the sex ratio of their offspring by controlling fertilization and do so in response to a variety of factors (Charnov, 1982; Donaldson and Walter, 1984; Frank, 1985; Green et al., 1982; Hamilton, 1979; Herre, 1985; Orzack and Parker, 1986; Pickering, 1980; Seger, 1983; Waage, 1982; Waage and Lane, 1984; Werren, 1980). Such animals offer an unparalleled variety of tests of theory.

*The conflict of interest between mothers and daughters*

If, in outbreeding populations, mothers are inseminated by only one male, then daughters share the paternally inherited half of their genome with certainty since their father was haploid. They share the maternally inherited half of their genome with a certainty of $\frac{1}{2}$ since they have a probability of $\frac{1}{2}$ of inheriting the same allele at any given locus from their mother. Therefore, sisters are related to each other by a factor of $\frac{3}{4}$ and to their mothers by a factor of $\frac{1}{2}$ (Hamilton, 1964). Interestingly, sisters are also more closely related to each other than they are to their own offspring. This convoluted set of relationships has been cited as a factor contributing to the likelihood that in haplodiploid organisms offspring might forsake their own reproduction for the sake of rearing siblings, i. e. these relationships are thought to facilitate the evolution of eusociality (Hamilton, 1964). However, while mothers are equally related to both their sons and daughters, daughters are three times as related to their sisters as they are to their brothers. (The haploid brothers derive all their genome from their mother and so share none of their sisters' paternally inherited half of their genome with them. Of their maternally inherited genome they have a probability of $\frac{1}{2}$ of sharing any given allele with their sisters. Therefore, their relatedness to their sisters is $\frac{1}{4}$.) Therefore, although mothers should be under selection to invest equally in reproductive daughters and sons, daughters should be under selection to invest three times as much in their sisters as in their brothers (Trivers and Hare, 1976). The results of this intuitive argument have been reconfirmed by a variety of more explicit methods (Charnov, 1978; Leigh et al., 1985; Uyenoyama and Bengtsson, 1979, 1981).

Trivers and Hare (1976) first pointed out this conflict of interest and marshalled evidence to support the proposition that a female bias exists in the reproductive allocation in eusocial Hymenoptera in which the daughters raise their siblings. Here, they compared total weight of males with reproductive females rather than simple sex ratio. This view has proved controversial, and the alternative proposition – that the observed females are the result of local mate competition (which we will discuss later) – has been raised (Alexander and Sherman, 1977). It is difficult to distinguish between these two alternate interpretations because, by and large, they make nearly the same set of predictions. However, recent analyses suggest that the female biases are in fact due to the influence of daughters as Trivers and Hare proposed (Nonacs, 1986), although the question can by no means be considered settled.

## The effects of inbreeding

In inbreeding populations, mating among sibs leads to increased mother-offspring relatedness because the paternal genetic contribution to the offspring has a higher probability of being identical to the maternal genetic contribution by direct descent than in outbreeding populations. In diploid organisms, this increase in relatedness is symmetrical with respect to sons and daughters. However, in haplodiploid organisms, since the haploid males derive all their genome from the unfertilized eggs of diploid mothers, only the relatedness of daughters to mothers increases. This asymmetry in relatedness can select for female biased sex ratios independently of other effects (Frank, 1985; Herre, 1985; Nunney, 1985). Essentially, daughters become a more efficient vehicle for passing a mother's genes to future generations than sons. At present the only empiricial evidence suggesting this effect where inbreeding is not confounded with other influences is that presented by Herre (1985) for three species of fig pollinating wasps.

**Population structure**

In a central theoretical contribution, Hamilton pointed out that Fisher's argument for equal investment in the two sexes depends crucially on the structure of the mating population (Hamilton, 1967). He showed that selection for female-biased sex ratios exists when one or a few foundress mothers contribute offspring to isolated mating groups (broods) from which mated females disperse to found new broods. Hamilton originally termed the phenomenon 'Local Mate Competition' (LMC) and assembled a list of organisms whose life cycles fit the paradigm and showed the expected female-biased sex ratios (Hamilton, 1967). Slightly altered versions of the original model show that sex ratio bias can be selected for simply if there is differential dispersal between the sexes, with a sex ratio bias in favor of the sex with the greatest average dispersal (Bulmer and Taylor, 1980; Leigh et al., 1985; Stubblefield, 1980). Subsequent theoretical work has concerned itself with analyzing the roles that differential group productivity, individual selection, and inbreeding play in selecting for the female biases associated with LMC (Charnov, 1982; Colwell, 1981; Fischer and Harper, 1986; Hamilton, 1979; Herre, 1985; Nunney, 1985; Stubblefield, 1980; Werren, 1983). In an interesting contrast with selection for sex ratios in randomly mating populations, there is strong selection restricting variance in sex ratios. This opens the possibility of tests of variance in sex ratio. Green et al. (1982) find evidence supporting precise sex ratios in highly inbred wasps.

In the original model, Colwell (1981) has pointed to the fact that in order for female biased sex ratios to evolve from a sex ratio of $\frac{1}{2}$, there must be differential productivity of broods in terms of numbers of mated females that they produce. *Within* a brood, the best sex ratio is $\frac{1}{2}$, for this assures that no other female contributing to that brood has a higher relative fitness. However, *among* broods, the best sex ratio is female biased because broods producing more females are more productive, so much so that the among group advantage of producing female-biased offspring is able to outweigh the within group advantage of producing offspring in a 1 : 1 sex ratio. Principally on these grounds Colwell has maintained that the female-biased sex ratios which may be selected for in structured populations represent a case of a group selected trait. A number of workers find this perspective useful (Bulmer, 1987; Charnov, 1982; Frank, 1985; Herre, 1985; Taylor and Bulmer, 1980).

However, this interpretation has proven to be quite controversial (Leigh et al., 1985; Nunney, 1985; Werren, 1983). It has been attacked on two types of grounds. One is to demonstrate that female biases may be generated in different models which lack discrete groups, as is true in the case of the differential dispersal models. One problem with this approach to debunking the group view is that the demonstration that female biases may be generated in radically altered models without groups does not demonstrate that it is not differential group productivity which is the driving agent in the models in which the groups are intact. An orange does not prove that an apple is not an apple.

The second approach is to attempt to show that the female bias in the intact group model is caused by individual selection. For example, Nunney (1985) proposes that this selection comes via a process he terms 'Local Parental Control' (LPC). 'The hypothesis predicts that a sex ratio bias will occur if a parent can control the sex ratio of the mating group which contains its offspring and that increasing this local parental control (LPC) increases the potential for sex ratio bias', and 'The identification of LPC as the source of selection (for female biased sex ratios) allows an important point raised by Colwell (1981) and Wilson and Colwell (1981) to be addressed'. However, if group productivity of females is fixed (i. e. no differential group productivity) female bias will not evolve and local parental control exists without selection for female biased sex ratios (Colwell, 1981; Nunney, 1985; Wilson and Colwell, 1981). Clearly, LPC is not the 'source of selection'. In this case, Nunney points to differential changes in male vs female fitness as offspring sex ratio changes as being the key criterion. However, it is the between group component of selection which drives these fitness differentials (Charnov, 1982; Colwell, 1981). It thus appears that the group perspective gives the clearest, most parsimonious, and most useful view of the factors selecting for a female bias in the original LMC model.

The heuristic power of this perspective is illustrated in an important but seldom appreciated result from Taylor and Bulmer (1980). They find that the expected sex ratio depends not only on the number of foundresses per mating group (selection for female bias increases as the number of foundresses decreases) but also on the number of mating groups (for a fixed number of foundresses, the female bias decreases as the number of groups decreases). That is, the among-group advantage of producing female-biased sex ratios increases with both increasing number of mating groups and decreasing number of foundresses per mating group; the within-group advantage of producing a sex ratio of $\frac{1}{2}$ increases with both decreasing number of mating groups and increasing number of foundresses per mating group. In the limiting case of one group with two foundresses, a sex ratio of 1:1 is expected, despite high levels of LPC.

Perhaps the greatest value of sex ratio theory for structured populations is the fact that it provides quantitative predictions for the dynamics and equilibrium outcomes of the evolution of a character with a large and direct effect on fitness (Charnov, 1982; Herre, 1985; Leigh et al., 1985). Generally the empirical tests have given qualitative and often quantitative agreement with theoretical predictions. Several studies have shown female biases in populations in which LMC effects are expected (Charnov, 1982; Donaldson and Walter, 1984; Green et al., 1982; Hamilton, 1967; Waage, 1982), although there are cases of female biases where it is not obvious that they should be expected (Aviles, 1986). Moreover, an increasing number of instances of facultative adjustment of sex ratio have been reported (Frank, 1985; Herre, 1985; Werren, 1980). In an elegant set of experiments Werren (1980, 1983) showed that in a study population of the parasitic wasp *Nasonia vitripennis* females are able to shift sex ratio of their brood in response to changing intensities of LMC. The shifts conformed well to those predicted by theory. Data from fig pollinating wasps indicate that these wasps also adjust their sex ratios in response to different intensities of LMC again in qualitative and often quantitative accord with theoretical predictions (Frank, 1985; Herre, 1985). Judging by the available evidence, it is clear that the theory gives an accurate and powerful insight into the forces shaping sex ratio evolution.

Nonetheless, not all tests of theory give equally rosy results. The theory indicates that there is a powerful selective advantage for organisms that exercise precise control over their sex ratios and can match them appropriately to varying intensities of LMC. Yet several species only possess rudimentary ability to adjust their sex ratio, and others exhibit no facultative adjustment at all (Donaldson and Walter, 1984; Hamilton, 1979; Waage and Lane, 1984). Brood sex ratios in many test species show significant differences from those expected (Frank, 1985; Hamilton, 1979). Even within species, there are populational differences in sex ratio responses (Orzack and Parker, 1986; Orzack, 1986; Parker and Orzack, 1985). However, these discrepancies need not lead to dispair. Indeed they present an opportunity for an even more powerful generation of sex ratio studies.

At present the opportunities to test LMC sex ratio theory have only been exploited to the limited extent of testing predictions of equilibrium optimal sex ratios. The vast majority of sex ratio studies are correlational, relating observed individual brood or populational sex ratios with different population structures or life histories. These studies open the door to a new generation of experimental studies that will provide a complementary series of tests of theory. Orzack and Parker (1985, 1986; Orzack, 1986) have documented heritable variation in *Nasonia vitripennis* for brood sex ratios. Wasp population structures could be artificially manipulated and the dynamics of sex ratio evolution could be observed. The existing sex ratio theory should provide an accurate description of the selective forces acting on sex ratio evolution in subdivided populations. Do populations actually attain the expected equilibria? Do both number of foundresses per mating group and number of mating groups affect sex ratio evolution as Taylor and Bulmer (1980) predict? Do individuals that can adjust their sex ratio displace those that cannot? There are abundant possibilities for future experimental tests.

## Conclusions

Theory predicts and most evidence indicates that generally there should be equal investment in the sexes as a result of the within-population mating advantage of the rare sex. The types of organisms that are simultaneous hermaphrodites and their patterns of sex allocation are generally consistent with expectations. Allocation patterns in sequential hermaphrodites also generally give good agreement with predictions concerning age at which to change sex. The bulk of these predictions are rooted in the primacy of selection within populations (individual selection) and therefore give strong support for that proposition. In subdivided population structures the female biased sex ratios that are expected and observed may be usefully thought of as a special case in which selection for differential group productivity is able to override selection within subpopulations favoring a sex ratio of $\frac{1}{2}$. Overall, sex allocation theory is one of the most successful in all of evolutionary biology. Nonetheless, there are exceptions to the general patterns that challenge our understanding of the processes that lead to them and invite further study. Most present studies of sex ratio are correlational and interpretations are often obscured by potentially confounding influences. Organisms that lack constraining sex determining systems and possess heritable variation for sex ratios offer great promise for future experimental studies of the factors influencing sex allocation that the present correlation studies identify and so often powerfully support.

Acknowledgements. We thank Monica Geber, Elizabeth Stockwell, David Barker, Alexandra Basolo, Mike Ryan, and Jim Bull for discussion, and comments during the preparation of the manuscript. We also thank the Smithsonian Institute (all of us), the University of Iowa (EAH), the University of Washington (EAF), and the National Science Foundation (EAF) for support enabling us to accomplish this work.

## Appendices

### Basic theory of sex allocation

We derive models for selection on sex allocation in outbreeding organisms with separate sexes and discrete generations, assuming 1) sex ratio is determined by one locus, and 2) sex ratio is under polygenic control. We will then consider the case in which generations overlap and sex ratio is controlled by one locus. To address these questions, we consider sexual haploids, and assume that the probability an individual is male is governed by which sex ratio allele it carries. The diploid version complicates the mathematics without changing the essentials (Leigh, 1970; Uyenoyama and Bengtsson, 1979). The conditions under which selection on diploids favors a mutant so rare that it occurs only in heterozygous form are the same as for a rare mutant among sexual haploids. We also suppose that a sex ratio allele influences the probability that its bearers will abort before birth, as well as the probability that it will be male, in such a way that each mating produces $kS$ sons and $(1 - k)W$ daughters, where $k$ differs for different sex ratio alleles. This assumption restricts our attention to alleles whose reproductive effort per mating is the same. It is not our desire to confound the effects of changing sex ratio with those of changing reproductive effort. Our model of sex determination seems artificial, but again, it simplifies the mathematics. Moreover, were an allele prescribing a given sex ratio to act, not directly on the sex ratio of the bearers themselves, but on the sex ratio of the offspring of female bearers, genic selection would affect it the same way; female bearers of that allele would pass it on to their offspring in the sex ratio that allele would have prescribed had it affected the offspring directly. However, the selection would only be half as strong, as copies of the allele carried by males exert no effect. The same remark applies if sex ratio is controlled by fathers, or by both parents conjointly.

We will assume:

1) alleles affecting sex ratio affect nothing else, neither the survival nor the fecundity of their bearers or any of their kin.

2) each allele affecting sex ratio prescribes a fixed sex ratio, which cannot be adjusted to suit external circumstances.

3) mating is random: each female is equally likely to mate with any male in the population.

4) offspring are equally likely to inherit sex ratio alleles from either parent.

The reader interested in an explicit mathematical treatment of the consequences of violating these assumptions should consult Leigh et al. (1985).

*Appendix A: Separate sexes, discrete generations, one locus*

Consider a locus with two alleles, A and B, affecting sex ratio in a population of sexual haploids, and let $N_{BF}(t)$, $N_{BM}(t)$, $N_{AF}(t)$ and $N_{AM}(t)$ be the numbers of adult B females, B males, A females and A males at generation $t$. Suppose that each female mates only once, and let a mating of B with B produce $Sk$ B males and $W(1 - k)$ B females, a mating of A with A produce $S(k + d)$ A males and $W(1 - k - d)$ A females, and a mating of A with B produce $\frac{1}{2}Sk$ B males, $\frac{1}{2}W(1 - k)$ B females, $\frac{1}{2}S(k + d)$ A males, and

$\frac{1}{2}W(1 - k - d)$ $A$ females. If each generation mates at random, and if a fraction $F$ of the young females and a fraction $M$ of the young males mature, then $N_{BF}(t + 1)$ is

$$WF(1 - k)[N_{BF}(t)N_{BM}(t)/N_M(t) + \tfrac{1}{2}N_{BF}(t)N_{AM}(t)/N_M(t) + \tfrac{1}{2}N_{AF}(t)N_{BM}(t)/N_M(t)]$$

$$= WF(1 - k)\{\tfrac{1}{2}N_{BF}(t)[N_{BM}(t) + N_{AM}(t)]/N_M(t) + \tfrac{1}{2}N_{BM}(t)[N_{BF}(t) + N_{AF}(t)]/N_M(t)]\},$$

$$N_{BF}(t + 1) = WF(1 - k)N_F(t)[\tfrac{1}{2}N_{BF}(t)/N_F(t) + \tfrac{1}{2}N_{BM}(t)/N_M(t)]$$

$$= WF(1 - k)N_F(t)Q_B(t),$$

where $N_F(t) = N_{BF}(t) + N_{AF}(t)$ is the total number of adult females at generation $t$, $N_{BM}(t)$ is the total number of adult males, and $Q_B(t)$ is the average of B's frequency $N_{BF}(t)/N_F(t)$ among the females, and its frequency $N_{BM}(t)/N_M(t)$ among the males, of generation $t$. Similarly,

$$N_{BM}(t) = MSkN_F(t)Q_B(t);$$

$$N_{AF}(t) = FW(1 - k - d)N_F(t)Q_A(t);$$

$$N_{AM}(t) = MS(k + d)N_F(t)Q_B(t);$$

where $Q_A(t) = 1 - Q_B(t)$. Letting $k' = Q_Bk + Q_A(k + d) = k + Q_Ad$ be the average proportion of the population's reproductive effort spent on males, we find that

$$Q_B(t + 1) = \tfrac{1}{2}Q_B(t)\left(\frac{1 - k}{1 - k'} + \frac{k}{k'}\right) = \tfrac{1}{2}(n_{BF}/n_F + n_{BM}/n_M)Q_B(t) ,$$

$$Q_A(t + 1) = \tfrac{1}{2}Q_A(t)\left(\frac{1 - k - d}{1 - k'} + \frac{k + d}{k'}\right) = \tfrac{1}{2}(n_{AF}/n_F + n_{AM}/n_M)Q_A(t) ,$$

where $n_{BF}$ and $n_{BM}$ are the numbers of B daughters and B sons per B parent, $n_{AF}$ and $n_{AM}$ are the numbers of A daughters and A sons per A parent, and $n_F$ and $n_M$ are the numbers of daughters and sons per parent in the population at large. As Shaw and Mohler (1953) expected, genic selection increases A's frequency if $n_{AF}/n_F + n_{AM}/n_M$ exceeds 2. More specifically, $Q_A(t + 1) - Q_A(t)$ may be expressed as

$$\tfrac{1}{2}Q_A(t)\left(\frac{1 - k - d}{1 - k'} - \frac{1 - k'}{1 - k'} + \frac{k + d}{k'} - \frac{k'}{k'}\right)$$

$$= \tfrac{1}{2}Q_A(t)[1 - Q_A(t)][d/k' - d/(1 - k')] = Q_A(t)Q_B(t)V(k')d,$$

where $V(k') = (\frac{1}{2} - k')/k'(1 - k')$. If the proportionate gain in male off-spring from an increase $d$ in sex ratio outweighs the proportionate loss to the females, genic selection favors the change: the sex ratio is optimum when $k'$ is $\frac{1}{2}$ and the loss just balances the gain. The equations in Q apply whatever the values of $F$, $M$, $S$ or $W$, no matter how they change from generation to generation, so long as they are the same for A's and B's. Moreover, so long as the number of matings per female is not affected by the sex ratio locus, it does not matter how often they mate, or how their frequency of mating changes from generation to generation.

*Appendix B: Separate sexes, discrete generations, polygenic control of sex ratio*

Suppose now that other loci besides the one with alleles A and B affect sex ratio. How does selection at all these loci change the mean sex ratio $k'$? How precisely can we expect $k'$ to approach $\frac{1}{2}$? If the different loci assort independently, so that A and B have the same spectrum of genetic backgrounds, our equation for change in $Q_A$ applies. Since $k' = k + Q_A d$, the change in $k'$ from one generation's selection on A is

$$k'(t + 1) - k'(t) = [Q_A(t + 1) - Q_A(t)]d = Q_A(t)Q_B(t)V(k')d^2$$
$$= S_A^2 V(k'),$$

where $S_A^2 = Q_A(k - k')^2 + Q_B(k + d - k')^2$ is the variance in $k$ associated with the A locus (Fisher, 1930; 1941). Considering the joint effect of all sex ratio loci,

$$k'(t + 1) = k'(t) + S_T^2(t)V(k') = k'(t) + S_T^2(t)(\frac{1}{2} - k')/k'(1 - k'),$$

$$k'(t + 1) - \frac{1}{2} = [k'(t) - \frac{1}{2}][1 - S_T^2(t)/k'(1 - k')],$$

where $S_A^2(t)$ is the heritable part of the total variance in sex ratio at generation $t$. Polygenic control does not affect the direction of genic selection on sex ratio (Bulmer and Bull, 1982).

How does the variance $S_T^2(t)$ change over time? Since different loci assort independently, we expect that the proportion of individuals of generation $t$ which are programmed for sex ratios between $k$ and $k + dk$ will be (Falconer, 1981)

$$dk(1/2\pi S_T^2)^{1/2} e^{-(k - k')^2/2S_T^2}.$$

This expectation is reasonable only if $S_T^2$ is so small that individuals programmed for $k$ near 0 or 1 are rare. Individuals programmed for sex ratios between $k$ and $k + dk$ are $1 + (k - k')V(k')$, or roughly $e^{V(k')(k - k')}$, times as abundant in generation $t + 1$ as in generation $t$; thus, at generation $t + 1$, the proportion of such individuals in the population should be

$$Cdk(1/S_T^2 2\pi)^{1/2} e^{[(k - k')V(k') - (k - k')^2/2S_T^2]}$$

$$= dk(1/2\pi S_T^2)^{1/2} e^{[k - k' - S_T^2 V(k')]^2/S_T^2]},$$

where $C$ is a constant which makes the proportions sum to 1. Selection increases $k'$ by an amount $V(k')S_T^2$ per generation, as we expect, but it does not shrink the variance. Instead, variance is governed by the balance between mutation and the restricting effects of finite population size. In a finite population with $N$ equally reproductive haploid individuals, a fraction $1/N$ of the variance is lost each generation through the random sampling of $N$ mature individuals from the gametes formed by their parents, while each generation adds an amount, which we may call $U$, to the variance (Lande, 1976). Thus, at equilibrium, $S_T^2/N = U$, $S_T^2 = NU$, so long as $U$ is small enough that $S_T^2$ is far less than 1.

Finally, how precisely should $k'(t)$ approach $\frac{1}{2}$? If sex ratio alleles act directly on their bearers,

$$k'(t + 1) - \tfrac{1}{2} = (k'(t) - \tfrac{1}{2}) [1 - S_T^2/k'(1 - k')].$$

Selection thus decreases the mean square of the difference between $k'$ and $\frac{1}{2}$ by a factor $[1 - S_T^2/k'(1 - k')]^2$ each generation. In a finite population, however, sampling $N$ mature individuals from an array where $k$ has variance $S_T^2$ introduces a variance of $S_T^2/N$ in $k'$ (Lande, 1976). If $S_T^2$ is much smaller than 1, then, when the effects of genetic drift and selection on sex ratio are in equilibrium,

$$[k'(t) - \tfrac{1}{2}]^2[2S_T^2/k'(1 - k')] = S_T^2/N;$$

$$[k'(t) - \tfrac{1}{2}]^2 = k'(1 - k')/2N.$$

When sex ratio is maternally controlled, the mean square deviation $[k' - \frac{1}{2}]^2$ of $k'$ from $\frac{1}{2}$ would be four times greater. It would require a huge sample of the population's young to verify such a deviation. If each newborn had probability $k'$ of being male, the variance in the estimate of $k'$ from a sample of $n$ young would be $k'(1 - k')/n$: even for $n$ larger then the 'effective population size' $N$, this variance is of magnitude comparable to that in the deviation we would be trying to detect.

*Appendix C: Separate sexes, overlapping generations, sex ratio controlled by one locus*

Is the optimum sex ratio the same when generations overlap? Consider a locus with two alleles, A and B, which affects sex ratio, in a population of sexual haploids with overlapping generations, and assume that these alleles only influence sex ratio. How is A's frequency among the adults of year $t + 1$ related to that in year $t$? A's frequency in year $t + 1$ is a compound of A's frequency among the survivors from year $t$, and of A's frequency among the newly matured 'recruits' of generation $t + 1$. As survival prospects are unaffected by the sex ratio locus, A's frequency only changes through the newly matured recruits. But the difference in A's frequency among these recruits from that among their parents is just what it would be were the recruits and their parents of successive but distinct generations. Thus the optimum sex ratio is the same whether or not generations overlap.

Suppose, to be specific, that there are $N_{BF}(t)$ mature B females, $N_{BM}(t)$ mature B males, $N_{AF}(t)$ mature A females, and $N_{AM}(t)$ mature A males at year $t$. Let proportions $S_F$ and $S_M$ of the mature females and males of year $t$ survive to year $t + 1$; let mature females each mate once a year, and let matings of B with B each yields $kSB$ sons and $(1 - k)WB$ daughters, matings of A with A yield $(k + d)SA$ sons and $(1 - k - d)WA$ daughters, and matings of A with B yield $\frac{1}{2}kSB$ sons, $\frac{1}{2}(1 - k)WB$ daughters, $\frac{1}{2}(k + d)SA$ sons, and $\frac{1}{2}(1 - k - d)WA$ daughters. If a proportion $F$ of the daughters mature, taking $s + 1$ years to do so, while a proportion $M$ of the sons mature, taking $r + 1$ years to do so, then $N_{AF}(t + 1)$ is

$$S_F N_{AF}(t) + WF(1 - k - d)[N_{AF}(t-s)N_{AM}(t-s)/N_M(t-s) + \tfrac{1}{2}N_{AF}(t-s)N_{BM}(t-s)/N_M(t-s) + \tfrac{1}{2}N_{BF}(t-s)N_{AM}(t-s)/N_M(t-s)]$$

$$= S_F N_{AF}(t) + WF(1 - k - d)N_F(t - s)Q_A(t - s).$$

Similarly,

$$N_{AM}(t + 1) = S_M N_{AM}(t) + SM(k + d)N_F(t - r)Q_A(t - r);$$

$$N_F(t + 1) = S_F N_F(t) + WF(1 - k')N_F(t - s);$$

$$N_M(t + 1) = S_M N_M(t) + SMk'N_F(t - r).$$

$N_{AF}(t + 1) - S_F N_{AF}(t)$ is the number of adult A females recruited to the population between years $t$ and $t + 1$; the numbers of recruits of other types are defined similarly. The sum of A's frequencies among the female, and the male, recruits to the population in year $t + 1$ is

$$\frac{N_{AF}(t+1) - S_F N_{AF}(t)}{N_F(t+1) - S_F N_F(t)} + \frac{N_{AM}(t+1) - S_M N_{AM}(t)}{N_M(t+1) - S_M N_M(t)}$$

$$= Q_A(t-s)\left(\frac{1-k-d}{1-k'}\right) + Q_A(t-r)\left(\frac{k+d}{k'}\right).$$

Since recruits are the only source of change in A's frequency, it appears that A's frequency increases if $(1-k-d)/(1-k') + (k+d)/k'$ exceeds 2, that is to say, if $d(1-2k')$ exceeds 0. This is obviously true if $r = s$: if $r \neq s$, the argument is more delicate, but the conclusion still holds. The optimum sex ratio is not affected by the values of $W$, $S$, $F$, $N$, $S_F$ and $S_M$, or the yearly fluctuations therein, so long as the sex ratio locus does not affect these quantities. Indeed, regardless of the mode of sex determination, it makes no difference to the optimum sex ratio whether or not generations overlap, *if* sex ratio is influenced only by autosomal genes.

# A conflict between two sexes, females and hermaphrodites

P.-H. Gouyon and D. Couvet

The theory of the evolution of sex has suffered from the difficulties of untangling the effects of many unrelated phenomena (e. g. sexual reproduction, diploidy, outbreeding . . .). Much of the work on animals has been restricted to a subset of the problem, that is, dioecy. Plants, on the other hand, because they are subject to different constraints, allow us to broaden our perspectives. In particular, with regard to the differentiation of sexual functions within or between individuals, plants provide a remarkable model with gynodioecy, i. e. the coexistence of females and hermaphrodites. Although females exist extensively among hermaphroditic plant species (Cosmides and Tooby, 1981), they are present at a noticeable frequency in only a few species, which are termed gynodioecious. Females are actually hermaphrodites which have lost their male function, and they are referred to as male-steriles. This character is genetically determined, and females are similar to hermaphrodites in morphology with the exception of their flowers. Female flowers are generally smaller (Darwin, 1877 pp. 307–309), and this reduction in size has been related to the lack of anthers and, more precisely, to the male hormones present in them (Plack, 1958). Male-steriles do not produce pollen and, due to reallocation of resources, they are expected to produce more seeds (Darwin, 1877 pp. 280–281, 309; Charnov et al., 1976a), which could explain their maintenance in a population. The frequency of females can be very different among populations (e. g. Darwin, 1877, p. 301). The determination of the laws of variation of such a breeding system (the frequency of females can vary from 5 to 95 % in Thyme – Dommee et al., 1978) which oscillates from almost complete dioecy to total hermaphroditism gives insight into the variations of breeding systems in general.

## The explanations of the maintenance of gynodioecy

The maintenance of male-sterility has been a topic of interest in population genetics for a long time. Lewis (1941) showed that the conditions for maintenance of male-sterility depend on its mode of inheritance.

For nuclear single locus inheritance, whatever the dominance relationships between alleles, if $x$ is the frequency of females and $f$ their relative fecundity (production of seeds of females versus production of seeds of hermaphrodites), then at equilibrium:

$$x=(f-2)/(2f-2).\tag{1}$$

At equilibrium, females must have a fecundity more than twice the fecundity of hermaphrodites to be present in a population, and their frequency cannot exceed 0.5.

For cytoplasmic inheritance, if hermaphrodites do not receive more efficient pollen than females, then at equilibrium:

$$x=0 \text{ if } f<1, \text{ and } x=1 \text{ if } f>1 \text{ (undetermined if } f=1 \text{)}.\tag{2}$$

As a consequence, the selective pressure needed to maintain females among hermaphrodites at equilibrium differs greatly depending on the mode of inheritance of male-sterility. With regard to the origins of male-sterility, the mode of inheritance also has different consequences. Under nuclear inheritance, an original male-sterile mutant must be more than twice as fecund as an hermaphrodite to be retained (Charlesworth and Charlesworth, 1978). It seems implausible that a reallocation which doubles the production of seeds will appear, at a single stroke, in an original male-sterile mutant. In contrast, under cytoplasmic inheritance, a slight reallocation of resources toward female function is sufficient for male-steriles to spread (Lewis, 1941).

Few theoretical developments have altered the original results of Lewis (1941), although these results could not account for observations of gynodioecious populations. With cytoplasmic inheritance, no polymorphism is expected. With nuclear inheritance, females are expected to be less abundant or more fecund than has been observed in natural populations (see Fig. 1, and for a review Lloyd, 1976; Couvet et al., 1987).

Different sorts of models involving nuclear inheritance of male sterility have attempted to explain this discrepancy.

(i) Females can be assumed to have a higher survival than hermaphrodites (Lloyd, 1975; Van Damme and Van Damme, 1986). That would imply that the frequency of females increases with the age and/or the stability of a population, which is the inverse of what has been observed in *Origanum vulgare* (Elena-Rossello et al., 1976; Iestwaart et al., 1984), in *Silene maritima* (Baker and Dalby, 1981) and in *Thymus vulgaris* (Dommee et al., 1983). Moreover, the equilibrium reached when hermaphrodites are less viable than females is not shown to be an ESS (as defined by Charnov, 1982, p. 9): a mutant hermaphrodite investing more in survival and less in reproduction could possibly invade the population. This model implies that the optimal life-history (Charlesworth, 1980, pp. 231–251) differs between sexes.

(ii) Constraints on nuclear sex determinism have been introduced, for example the existence of a specific locus where the homozygotes are females and the heterozygotes are hermaphrodites (Gregorius et al., 1982), or the lethality of certain hermaphroditic genotypes (e. g. Jain, 1968). These constraints lead to higher frequencies of females at equilibrium, but seem to be a purely theoretical exercise since none of these constraints has been shown to exist in any gynodioecious species, and they do not explain the variation of frequency of females in wild populations. Moreover, the evolutionarily stable sex ratio of female and hermaphrodite phenotypes is as predicted by Lewis (Charnov et al., 1976a), and is not changed in these models. This means that sex determining alleles leading to this sex ratio will invade the population and will replace alleles subject to the constraints hypothesized (as stated by Bulmer (1986) in the case of Fisherian sex ratios).

Experiments have not helped to clarify these theoretical issues. Correns (1908) showed that male-sterility was maternally inherited in several plant species. Later, it was demonstrated that nuclear genes were involved in the determinism of male-sterility in *Origanum vulgare* (Lewis and Crowe, 1956) and in *Plantago lanceolata* (Ross, 1969). More recently, crosses among a wider range of parents showed that cytoplasmic genes were also involved in determining male-sterility in these two species (Kheyr-Pour, 1980 for *Origanum vulgare,* and Van Damme and Van Delden, 1984 for *Plantago lanceolata*). Experimental demonstrations of nuclear-cytoplasmic inheritance of male-sterility, as well as the failure of nuclear models to account for observations, have lead to the production of models of male-sterility involving both nuclear and cytoplasmic genes.

**The nuclear-cytoplasmic maintenance of gynodioecy**

The nuclear-cytoplasmic determination of male-sterility involves two (or more) types of cytoplasmic information. One of them, at least, can determine the female gender. However, the individuals bearing such cytoplasmic information can have their male fertility restored by particular nuclear genes specific to the cytoplasm. These nuclear genes are called restorers. The selective pressures on cytoplasmic and nuclear genetic units are different with regard to male-sterility. For genes which are not transmitted through male function (cytoplasmic), male-sterility has no deleterious effects, and can be advantageous if there is at least some reallocation of resources toward female function. For cytoplasmic genes, male-sterility does not lower fitness. In contrast, as shown by Lewis (1941), nuclear genes are usually selected to restore male fertility. The proportion of females at equilibrium for nuclear inheritance is given in formula (1), which means that when the proportion of females is higher than this value, selection on nuclear genes tends to decrease this proportion and restorer genes are favored. The situation is the same for cytoplasmic genes when the equilibrium is defined as in formula (2). These two different equilibria for cyto-

248

plasmic and nuclear genes define two curves, represented in Figure 1. These two curves divide the figure into three areas, each representing a different situation with regard to the outcome of male-sterility.

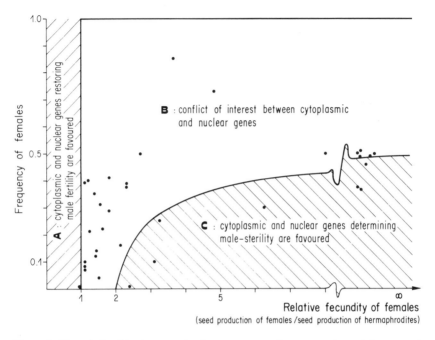

Figure 1. The relationships among selective pressures acting on sex determining genes in gynodioecious populations, the frequency of females, and their relative fecundity. The border between areas A and B is the equilibrium for cytoplasmic genes, the border between areas B and C the equilibrium for nuclear genes. Dots are data taken from the literature (see Table).

– In area A, females produce less seeds than hermaphrodites, both cytoplasmic and nuclear genes restoring male-fertility are favored and the species is expected to remain hermaphroditic.

– In area B, the frequency of females is such that nuclear genes determining male-sterility are now counter-selected while cytoplasmic genes determining male-sterility are still favored. There is, therefore, a conflict of interest concerning male-sterility between nuclear and cytoplasmic genes in this area. A review of the literature (Table) shows that natural populations are usually found in area B, the area of conflict. Nuclear-cytoplasmic models have been developed to determine what may happen, as a result of this conflict of interest.

– In area C, where females are more then twice as fecund as hermaphrodites, the frequency of females is below that which is expected at equilibrium, for both cytoplasmic and nuclear genes. Thus, cytoplasmic and nuclear genes determining male-sterility are both favored, and the frequency of females will increase.

Frequency and relative fecundity of females in 34 wild populations published in the literature (23 different species of 13 diffeŗent genera were observed).

| Species | Frequency of females | Relative fecundity of females | Reference |
|---|---|---|---|
| *Pimelea prostata* | .37 | 10.0 | Lloyd (1976) |
| *Pimelea traversii* | .39 | 2.3 | |
| *Pimelea sericeo-villosa* | .49 | 21.3 | |
| *Pimelea oreophila* | .50 | 18.5 | |
| *Cortaderia richardii* | .38 | 2.3 | |
| *Leucopogon melaleucoides* | .39 | 1.1 | |
| *Origanum vulgare* | .40 | 1.2 | |
| *Hebe subalpina* | .49 | 38.5 | |
| *Cirsium arvense* | .50 | 1916 | |
| *Cortaderia selloana* | .51 | 21.3 | |
| *Hirschfeldia incana* | .07 | 1.1 | Horovitz and |
| | .10 | 1.1 | Beiles (1980) |
| *Stellaria longipes* | .50 | 2.7 | Philipp (1980) |
| *Gingidia montana* | .30 | 6.2 | Webb (1981) |
| | .35 | 1.4 | |
| *Gingidia decipiens* | .12 | 1.4 | |
| | .25 | 3.2 | |
| *Gingidia trifoliata* | .14 | 1.5 | |
| *Gingidia baxteri* | .16 | 2.1 | |
| | .31 | 1.6 | |
| *Gingidia enysii* | .04 | 2.4 | |
| *Gingidia flabellata* | .003 | 0.9 | |
| | .21 | 1.3 | |
| *Scandia rosaefolia* | .28 | 1.8 | |
| | .46 | 33.3 | |
| *Lignocarpa carnosula* | .41 | 1.8 | |
| *Lignocarpa diversifolia* | .36 | 25.0 | |
| *Plantago lanceolata* | .04 | 1.5 | Van Damme |
| | .08 | 1.1 | (1984) |
| | .22 | 1.6 | |
| *Thymus vulgaris* | .10 | 3.1 | Couvet et al. |
| | .50 | 8.0 | (1987) |
| | .73 | 4.8 | |
| | .85 | 3.7 | |

As said above, in a nuclear-cytoplasmic model, there can be different cytoplasms determining male-sterility $S1, S2, \ldots Sn$. With each cytoplasm, $Si$, one or several nuclear loci may be associated, each with two alleles, a restored $Ri$ (it can be recessive or dominant) which restores fertility, and a maintainer $ri$ which maintains male-sterility. There may be several loci involved in determining sex for a single cytoplasm (as observed by Van Damme, 1983), with different types of epistatic effects. It is supposed that female phenotypes are more fecund (in terms of seeds) than those of hermaphrodites. To maintain a nuclear-cytoplasmic polymorphism at equilib-

rium, it is necessary to assume that restorer genes are favored in the cytoplasms where they restore male fertility, and are disadvantageous in the
cytoplasms where they are not active (Delannay et al., 1981; Charlesworth,
1981). This can be viewed as a cost of resistance, which is a well-known
phenomenon in bacteria, where resistance genes are counter-selected when
the disease is absent. Cytoplasmic genes can be perceived as a particular
type of pathogen which destroys male function. Nuclear restorer genes are
then akin to disease-resistance genes. The effects are as follows:

(i) The selective value of a given nuclear restorer gene $Ri$ depends on the
frequency of the corresponding cytoplasm $Si$ in the population. The more
frequent $Si$ is, the higher the fitness of $Ri$ becomes.

(ii) On the other hand, the selective value of a cytoplasm $Si$ depends on
what proportion of this cytoplasm is carried by females (because females
transmit more of this cytoplasm, as a result of their higher fecundity) and
thus depends on the frequency of the corresponding restorer gene $Ri$.

This interaction between the selective values of cytoplasmic and nuclear
genetic units can result in a frequency-dependent mechanism maintaining
the polymorphism at both the cytoplasmic and nuclear levels (Delannay et
al., 1981; Charlesworth, 1981; Gouyon et al., in prep.), allowing females to
be maintained in the population. When a stable equilibrium exists, this
equilibrium is reached slowly with oscillations of frequencies of the different genetic units and, consequently, the frequency of females (Charlesworth, 1981). With two cytoplasms and one restorer locus for each cytoplasm, stable limit cycles between the frequency of cytoplasmic and nuclear
genes are usually found (Gouyon et al., manuscript). Thus the frequency of
females oscillates endlessly and as a consequence the frequency of females
in a population has no significance.

The deleterious effects of restorer genes in cytoplasms where they are not
active – a hypothesis originally created for purely theoretical reasons and
which is critical for the maintenance of a nuclear-cytoplasmic polymorphism at equilibrium – has not yet been tested experimentally. However, this
hypothesis predicts that restorer genes will not be present when the corresponding cytoplasm is also absent. This effect has been shown to exist in
*Hordeum vulgare* (Ahokas, 1979), in *Brassica juncea* (Banga and Labana,
1985) and in *Plantago lanceolata* (Van Damme, 1985).

Crosses among individuals from different populations, resulting in frequent non-restoration of the male fertility of females in *Thymus vulgaris*
(Couvet et al., 1985) and in *Phacelia dubia* (Broaddus and Frosty, manuscript) suggest that there may be nuclear-cytoplasmic differentiation among
populations in these species (the hermaphrodites do not carry the restorer
genes of the cytoplasmic types of another population). Founder effects and
restricted gene flow may maintain population differentiation independently of, or in interaction with, any selective effect (Wright, 1946). In any
case, this differentiation between populations suggests that a nuclear-cytoplasmic polymorphism may be maintained at the scale of the metapopula-

tion, although it may be unstable within a population. That is, the conditions for maintenance of male-sterility at this level may be different from those applying to single populations. The maintenance of a nuclear-cytoplasmic polymorphism within a population depends on the respective selective values of cytoplasmic and nuclear genes (see above). The maintenance of a nuclear-cytoplasmic polymorphism in a metapopulation depends on other factors such as the respective migration rates of nuclear and cytoplasmic genes, the life span of populations and the diversity of the genetic units (cytoplasmic and nuclear) among the founders of a population (which itself depends on the number of founders).

To illustrate this last point, high frequencies of females are found, in the main, in young and/or disturbed populations for several gynodioecious species (Couvet et al., 1986). In *Thymus vulgaris,* the frequency of females is significantly related to the age of a population (Dommee et al., 1983; Belhassen et al., manuscript). This is interpreted as a consequence of nuclear-cytoplasmic interactions. When founders originate from different populations, females will often not be restored by hermaphrodites (see above). A non-restored cytoplasm spreads very rapidly in a new population due to its higher selective value, and as a consequence the frequency of females rises. When the specific restorer gene for this cytoplasm reaches it, the restorer will invade and the proportion of females will decrease until equilibrium is reached. The probability of a cytoplasm being present in a new population without its corresponding restorer gene(s) is thus critical.

## Some evolutionary consequences of gynodioecy

From its discovery, gynodioecy has been discussed as (i) a mechanism promoting outbreeding and (ii) an intermediate stage on the way from hermaphroditism to dioecy (Darwin, 1877 pp. 278–279). These two points are not exclusive, particularly because it has been argued that dioecy itself is an outbreeding mechanism. Note however that, as early as 1877, Darwin himself seems to favor the hypothesis of resource allocation rather than an outbreeding mechanism. Darwin (1876) claimed that differentiation of sexual functions was important for cross-fertilization, but that it could be attained by many mechanisms other than dioecy that are actually present in hermaphrodite individuals (like dichogamy or monoecy). Moreover, he expressed strong doubts about the possibility of an evolutionary process leading from a selfing species to an outbreeding one through the evolution of dioecy (or gynodioecy): "As we must assume that cross-fertilization was assured before an hermaphrodite could be changed into a dioecious plant, we may conclude that the conversion has not been effected for the sake of gaining the great benefits which follow from cross-fertilization". We now develop theoretical and experimental arguments on these two aspects of gynodioecy.

*Outbreeding*

The existence of females which are obligate outbreeders among a population of self-compatible hermaphrodites necessarily increases the level of outcrossing of the population. The consequences are twofold, occurring at the level of the individual and at the level of the population.

*Heterozygosity: the individual level.* The increase in the outbreeding rate necessarily increases the proportion of heterozygotes in the population. The way this is achieved is relatively complicated since, in gynodioecy, there is a sort of 'heritability' of the sexual form. Contrary to what is found in other species, in gynodioecious species, the knowledge of the gender of the parents gives information about the gender of the progeny. Roughly, an individual is likely to have the same sex as its mother. One consequence is that the proportion of heterozygotes may be higher within females than among the hermaphrodites (Gouyon and Vernet, 1982; and see Gouyon and Vernet, 1980 for observations). The exact relationship between the different parameters acting on the proportion of heterozygotes can be modelled in the following way.

The parameters of the model:

$FH =$ Proportion of females in the progeny of hermaphrodites
$FF =$ Proportion of females in the progeny of females
$f =$ Female advantage: ratio of the female fitness of females over the female fitness of hermaphrodites
$s =$ Selfing rate in hermaphrodites
$h =$ Heterosis: ratio of the survival or fecundity of outbred over inbred individuals
$X_n =$ Proportion of females at generation $n$

If we define:

$$a = fh / (h + s - hs)$$

which is a sort of global female advantage including heterosis, the proportion of females at generation n+1 is then given by (see Appendix 1).

$$X_{n+1} = (X_n (a\, FF - FH) + FH) / (X_n (a-1) + 1)$$

This series converges quickly towards a value which corresponds to the solution of the second degree polynomial given by equalizing $X_{n+1}$ and $X_n$. In fact, this formula is complicated and it is simpler to reiterate the above recurrence relation to get the equilibrium value *(X)* of the proportion of females. The value of $X$ is necessarily intermediate between *FH* (the value obtained if only the hermaphrodites produced seed) and *FF* (the value obtained if only the females produced seeds).

Given this result, it is possible to calculate the deficit of heterozygotes (or fixation index) for a locus with two neutral alleles, in females *(F_f)*, hermaphrodites *(F_h)*, and in the whole population *(F_t)* (see Appendix 2). Gouyon

(1978) has made numerous computer simulations of these formulae. One example of the results is given in Figure 2. The different sets of values which have been studied generally give the same shape of response. Note that these formulae can be used only 'locally', because *FF* and *FH* may change between generations, as can *X* (Couvet et al., 1987). In conclusion, the proportion of heterozygotes in the total population seems to increase roughly linearly with the proportion of females.

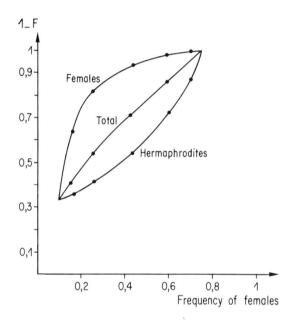

Figure 2. Heterozygote deficiency for females, hermaphrodites and the whole population, depending on the frequency of females, for s=0.8, h=1, FH=0.1, FF=0.75 (see text for definition).

*Adaptability and ecological success: the group level.* The presence of outbreeding females has other consequences on a more ecological level.

1) The mixing of different genetic information is increased as a direct consequence of the increase of heterozygosity. The ability of the progeny of each individual to adapt, as well as that of the population, increases.

2) The neighborhood area (Gliddon et al., 1987) is increased by the decrease of selfing and adaptations of females to disperse their seeds more efficiently (Belhassen et al., manuscript). The basic reproductive units are increased in size by the existence of females. Each reproductive unit is thus more likely to sample a variety of environments; the effective grain of the environment is then decreased and the level of local polymorphism should increase. One might also suggest that, by increasing the neighborhood area, the presence of females increases the effective number of indi-

viduals in the neighborhood. However, when the proportion of females is high, the increase in area could be compensated, from the point of view of effective number of individuals, by the decrease in effective size due to unequal gender proportions (Gliddon et al., 1987). As far as we know, no precise investigation of the combined roles of changes in area, in homozygosity, and in variance of progeny number on the effective size of a population has been carried out.

3) The reproductive output of the population is increased by the higher seed production of females or their longer life span. This is also true only as long as the proportion of females is not such that pollen becomes limiting (obviously a population composed of 100% females would have a null seed output).

These three different aspects may have one general effect: they are very likely to make the presence of females beneficial for the group. We have seen in the preceding sections that, at a very proximal level, this group advantage could not be considered to be the cause of the existence of females in variable proportions in gynodioecious species. The proximate cause is the existence of a nuclear-cytoplasmic conflict. However, as pointed out by Gouyon and Couvet (1985), these group effects can be very important for the ecology of the species. They influence the whole constitution of its genetic system and the kind of habitats in which some populations are able to establish. As a consequence, they influence the kind of environment which will be encountered by the species and thus (i) the kind of selection which the individuals will experience and (ii) the dynamics of the populations and, in particular, the dynamics of the nuclear-cytoplasmic conflict. In that sense, the group effects can be considered as belonging to the general causes of the existence of females. One can even imagine that these effects are absolutely necessary to permit the continued existence of the species in its habitat. One could, for instance, think that Thyme would not be able to invade disturbed sites if it was only composed of hermaphrodites (hermaphrodites are not very good at producing seeds, because, when isolated, they exhibit a large amount of selfing . . .). In this case, the adaptation to the environment through high female investment and/or outcrossing would become an ultimate cause of the existence of high proportions of females in certain types of sites. To continue the example, the colonization of disturbed sites is necessary for the nuclear-cytoplasmic dynamics to produce females in high numbers. If, as supposed above, high proportions of females are necessary for the species to live there, the necessity of outbreeding in disturbed areas can be considered an ultimate cause of the existence of females there, while the nuclear-cytoplasmic conflict is the proximate cause (Gouyon and Couvet, 1985).

The proportion of females may thus have an important ecological significance, but this will not always or even often be true. As said above, in a model with two cytoplasms and a restorer locus for each, limit cycles seem to be the rule rather than the exception (Gouyon et al., manuscript). If such systems happen in nature (as could be true in *Plantago*), one should expect to find huge variations in the frequency of females that would be due to position in the cycle at the time of observation and would not be correlated at all with any ecological variation.

## The path to dioecy

In his seminal work on the different forms of flowers on plants of the same species, Darwin (1877, p. 278–279) proposed that a hermaphroditic species could become dioecious through a process starting with the loss of stamens in some plants (which would then be females) followed by the loss of female organs in the remaining hermaphrodites (transforming them into males). The intermediate stage in this process is obviously the co-occurrence of females and hermaphrodites, that is, gynodioecy. Why did Darwin not propose the reverse scheme? The reason could be that, as stated later (Darwin, 1877, p. 299), androdioecious species are "extemely rare or hardly exist". Given this, it could be considered surprising that Darwin refers to gynodioecy only for taxa which "rarely show any tendency to be dioecious, as far as can be judged from their present condition and from the absence of species having separate sexes within the same groups". This last statement applies clearly to the *Labiatae* and to the genus *Plantago* (which are the two groups extensively cited in Darwin, 1877). However, Darwin's conclusions would have been modified if he had looked closely at groups like the genera *Cirsium* (Delannay, 1979), *Silene,* or *Fuschia* (Raven, 1979).

The question of the possibility of gynodioecy leading to dioecy is thus left open and has received very little attention. Basically, the available data indicate that:

1) Subdioecious species are often gynodioecious species where hermaphrodites usually produce very few seeds (e. g. figs; Valdeyron and Lloyd, 1980), and the gender determination is nuclear and simple (one locus, two alleles).

2) Stable cases of gynodioecy which have been extensively studied are cases where the determination is nuclear-cytoplasmic (see above).

Theoretical studies on the question are even scarcer:

(i) Nuclear gynodioecy. Charlesworth and Charlesworth (1978) have demonstrated, in a formal model, that nuclear gynodioecy could lead to dioecy. There are two necessary conditions.

First, a locus bearing an allele which makes hermaphrodites lose their female function must be linked to the locus of male sterility. Secondly, the compensation curve (or fitness set) between the male and female fitnesses of an individual must be convex (in the mathematical sense of the second derivative being positive, which is the reverse of the terminology used by Charnov et al., 1976a). This can be due to the resource allocation system of the species or to inbreeding depression in populations where selfing regu-

256

larly occurs in hermaphrodites. Where the fitness set is convex, Charnov et al. (1976a) showed that the Evolutionarily Stable Strategy was dioecy (Fig. 3b); they also showed that, in order to be stable, nuclear gynodioecy needed a concavo-convex fitness set (Fig. 3c).

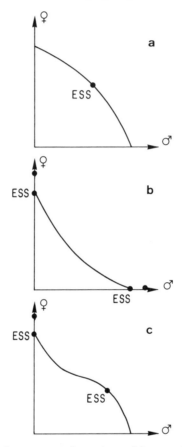

Figure 3. The shape of the fitness-set as a determinant of the reproductive system (the fitness-set represents the trade-off between male and female reproduction in a simultaneous hermaphrodite), a: hermaphroditism, b: dioecy, c: gynodioecy (from Charnov et al., 1976).

To argue that gynodioecy is a path to dioecy, one needs to propose a mechanism for the fitness-set to change its shape from concavo-convex to convex. Such a situation seems to be met in figs (Kjellberg et al., 1987). Within the genus *Ficus* the allocation of a female flower to female (production of a seed) or male (production of pollen) function is regulated by a simple change in the style morphology. Resource allocation is thus very easily changed. It seems that in species living in areas where hermaphroditism is selected against (such as seasonal climates where the synchronization of fig production occurs, leading to a high level of selfing), a process comparable to that described by Charlesworth and Charlesworth (1978) has occurred. Indeed, in *Ficus carica,* the locus of male sterility is linked with a locus changing style morphology so that hermaphrodites devote all their female flowers to male function and become functionally male (Valdeyron and Lloyd, 1980).

(ii) Cytoplasmic gynodioecy. Ross (1978) stated that "stable gynodioecy often shows cytoplasmic inheritance, and it probably does not evolve towards dioecy because linkage cannot develop between nuclear female sterility and cytoplasmic male sterility". As for genetic determination, it seems that later authors have tangled together every non-nuclear type of determination. The argument from Ross (1978) does not hold for nuclear-cytoplasmic male-sterility because, in this case, there are nuclear determinants of hermaphroditism (called restorers) to which a locus of female-sterility can be linked.

(iii) Nuclear-cytoplasmic gynodioecy. In a verbal model, Gouyon and Couvet (1985) have proposed a mechanism by which nuclear-cytoplasmic gynodioecy could lead to dioecy. The basic idea of this model is that the fitness set curve is itself subjected to selection (Fig. 4). The presence of females due to cytoplasmic information then creates selection in favor of particular adaptations to females which make the fitness set concavo-convex. If the proportion of females is high enough, the ESS for nuclear genes can be the transformation of the remaining hermaphrodites (the restored genotypes) into males through fixation of an allele for female sterility linked to the restorer gene. A formalization of this model is difficult and is critically dependent on the kind of nuclear-cytoplasmic determination which is present.

As shown at the beginning of this paper, nuclear inheritance of male-sterility is rather elusive. However, nuclear inheritance is always the rule in dioecious species. In contrast, nuclear-cytoplasmic inheritance is the rule in gynodioecious species. It would be dangerous to conclude from this that nuclear gynodioecy is unstable and leads to dioecy, while nuclear-cytoplasmic gynodioecy is stable. Actually dioecy must have a nuclear inheritance (because all individuals carry the cytoplasm of the females) and if nuclear inheritance is hard to find in gynodioecious species, it may be because it does not happen, rather than because it leads to dioecy. Precise studies of groups where dioecy and gynodioecy coexist, together with theoretical models, are now required to allow a general scheme for the evolutionary importance of gynodioecy to be constructed.

Gynodioecy is an example of the result of a conflict between two compartments of the genome. We have seen that this conflict may have very different results, the consequences of which can be important enough to determine the habitat or the reproductive system of the species. This interaction between forces acting at the molecular and the ecological level exemplifies the fact that selective pressures acting at different levels of the system are involved in the evolution of sex.

258

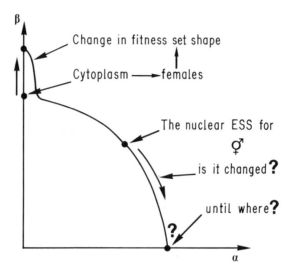

Figure 4: The evolution of the fitness-set and the ESS for nuclear genes under nuclear-cytoplasmic inheritance of male-sterility.

## Summary

We have reviewed models dealing with the maintenance of male-sterility under different modes of inheritance. For females to remain among hermaphrodites requires cytoplasmic information determining male-sterility and nuclear information restoring male-fertility. Such a nuclear-cytoplasmic polymorphism can be maintained at equilibrium given certain assumptions. However this equilibrium result may be irrelevant because gynodioecious populations are usually not at equilibrium, and because founder effects may give rise to high frequencies of females, due possibly to the fact that populations are often very different from one another genetically.

The genetic structure of such a species depends on the frequency of females. Females are obligate outbreeders and usually produce more seeds than hermaphrodites. These two traits are beneficial to a group where females are present, in terms of effective size and seed output, although they are not, themselves, causes of the occurrence of females. Nuclear-cytoplasmic gynodioecy may lead to dioecy.

## Appendix 1

$FH =$ Proportion of females in the progeny of hermaphrodites
$FF =$ Proportion of females in the progeny of females

$f =$ Female advantage: ratio of the female fitness of females over the female fitness of hermaphrodites

$s =$ Selfing rate in hermaphrodites

$h =$ Heterosis: ratio of the survival or fecundity of outbred over inbred individuals

$X_n =$ Proportion of females at generation $n$

At generation $n+1$, the total number of individuals is

$$W_{n+1} = X_n f h + (1-X_n)[(1-s)h + s]$$

The proportion of females at generation $n+1$ is

$$X_{n+1} = [X_n f h\, FF + (1-X_n)[(1-s)h + s]\, FH]/W_{n+1}$$

Defining

$$a = \frac{f h}{s + h - hs}$$

gives

$$X_{n+1} = \frac{X_n\,(a\,FF - FH) + FH}{X_n\,(a-1) + 1}$$

This harmonic series converges if and only if an equilibrium value exists, i. e., there exists an $x$ such that

$$x = \frac{x\,(a\,FF - FH) + FH}{x\,(a-1) + 1}$$

or the polynomial

$$P(x) = x^2\,(a-1) + x\,(1 - a\,FF + FH).$$

has real roots.

Given that $P(0) = -FH \leq 0$ and $P(1) = a\,FH \geq 0$, there is necessarily a root and the series converges.


## Appendix 2

Given that the population is at equilibrium for the proportion of females $(x)$, we will calculate the proportion of heterozygotes for a neutral marker with two alleles in proportions $p$ and $q$ in hermaphrodites, then in females and finally in the whole population. Hermaphrodites are supposed to reproduce in panmixis when outcrossed.

### 1) Heterozygosity in hermaphrodites

Let $z_n$ be the proportion of heterozygotes in hermaphrodites at generation $n$ and $W$ be the total number of individuals in the next generation. Among hermaphrodites, the proportion of individuals issuing from selfing is: $s(1-FH)/W$, their (hermaphrodite) parent being heterozygote with a probability $z_n$, the proportion of heterozygotes among them is $z_n/2$.

The number of hermaphrodites resulting from outcrossing is the total number of hermaphrodites at generation $n+1$ minus the preceeding number. They are heterozygous with probability $2pq$ so that

$$z_{n+1} = \frac{z_n}{2} \, \frac{s\,(1-FH)}{W} + 2pq \, \frac{W - s\,(1-FH)}{W}$$

with

$$W = xfh + (1-x)\,[(1-s)\,h + s]$$

This series converges because $s(1-FH)/2W$ is obviously less than $1/2$. The limit $z$ is given by equalizing $z_n$ and $z_{n+1}$.

$$z = 2\,p\,q\,[1 - \frac{s\,(1-FH)}{2\,xfh + 2\,(1-x)\,[(1-s)\,h + s] - s\,FF}]$$

so that the deficit of heterozygotes in hermaphrodites is

$$F_n = \frac{s\,(1-FH)}{2\,xfh + 2\,(1-x)\,[1-s)\,h + s] - s\,FF}$$

### 2) Heterozygosity in females

Let $y_n$ be the proportion of heterozygotes in hermaphrodites at generation $n$. Among females, the proportion of individuals issuing from selfing is: $sFH\,(1-x)/W\,x$, their (hermaphrodite) parent being heterozygous with a probability $z_n$; the proportion of heterozygotes among them is $z_n/2$.

The number of females from outcrossing is the total number of females at generation $n+1$ minus the preceeding number. They are heterozygous with probability $2pq$ so that

$$y_{n+1} = \frac{z_n}{2} \frac{s\,FH}{W} + 2pq\,\frac{W - s\,FH}{W}$$

At equilibrium, the proportion of heterozygotes among females is thus

$$y = 2pq\,[1 - \frac{1-x}{x}\,\frac{s\,FH}{2xfh + 2\,(1-x)\,[(1-s)\,h + s] - s\,(1-FH)}]$$

and

$$F_f = \frac{1-x}{x}\,\frac{s\,FH}{2xfh + 2\,(1-x)\,[(1-s)\,h + s] - s\,(1-FH)}$$

In the whole population, the proportion of heterozygotes is

$$F_t = \frac{s\,(1-x)}{2xfh + 2\,(1-x)\,[(1-s)\,h + s] - s\,(1-FH)}$$

# Parallels between sexual strategies and other allocation strategies

D. G. Lloyd

## Introduction

Biological strategies concerned with deploying the various adaptive mechanisms operating in a population are usually considered according to the nature of the strategy. Hence we examine strategies dealing with life histories, sex ratios, foraging behavior, flowering periods, seed dormancy, sperm size, etc. But if we wish to explore the shared features of diverse strategies and the common principles underlying them, we need to consider these strategies according to the kinds of decisions being made. Several major classes can then be distinguished, including phenology strategies, size-number compromises and packaging strategies (e. g. putting seeds into a particular number of fruit). The most diverse and widespread class of strategies is the class of allocation strategies, in which an individual apportions an limited resource, such as energy or time, among two or more competing activities. Much of the pioneering work on allocation strategies has dealt with sex allocations. These take two forms (Charnov, 1979b). In dioecious (gonochoristic) populations, the allocations involve the proportions of male and female offspring. Explanations of sex ratios began with the well-known (but, it will be argued below, still under-appreciated) argument of Fisher (1930), who concluded that parents optimally invest equal resources in sons and daughters. The second major class of sex allocation concerns investment in paternal and maternal reproduction by cosexes (equivalent to hermaphrodites in the animal kingdom, but encompassing a variety of genetically monomorphic sex conditions in plants, only one of which is hermaphroditism). Various allocation ratios may be selected (Charlesworth and Charlesworth, 1981; Charnov, 1979b, 1982; Lloyd, 1984a).

In addition to sex allocation strategies affecting male versus female offspring or gametes, a wide range of allocation strategies involve morphological, genetical or behavioral alternatives. These include:

1) In repeated behavioral sequences, animals may perform different behaviors randomly with fixed probabilities (a mixed *s. s.* or stochastic strategy) or confine themselves to one behavior (a fixed strategy) or choose either behavior according to the circumstances (a conditional strategy – Maynard Smith, 1979, 1982).

2) The open developmental system of plants allows them to produce their organs repeatedly. Plants frequently produce at one time two kinds of structures that perform the same function – a multiple strategy. Examples include simultaneous heterophylly, heteromorphic seeds or fruits, sexual and asexual embryos, etc. (Lloyd, 1984c).

3) When alleles of autosomal loci undergo unequal segregation, modifying genes that alter the segregation ratio may be selected (Crow, 1979; Lloyd, 1984b; Prout, 1953). That is, resources put into gametes and offspring are allocated among alleles.

4) Foraging strategies are concerned primarily with the proportions of different food items or time spent in different patches in natural environments (cf. 9) (Kamil, 1983; MacArthur and Pianka, 1966; Pyke et al., 1977).

5) An animal testing different types of food may have a learning strategy by which it samples alternatives (allocates food-seeking efforts) according to past rewards (Harley, 1981; Maynard Smith, 1984).

6) An altruist may dispense its favors in varying proportions to different classes of relatives with which it shares its genes with different probabilities (Altmann, 1979; Schulman and Rubenstein, 1983).

7) Colonies of social insects produce varying frequencies of distinct sterile castes (Oster and Wilson, 1978; Wilson, 1971).

8) In a spatially heterogenous environment, individuals can occupy different patch types with varying frequencies, e. g. ideal-free distributions (Fretwell, 1972).

9) In psychology, animals in operant conditioning experiments choose between alternative reinforcements available in a limited period of time (cf. 4) (Herrnstein and Vaughan, 1980; Staddon, 1980, 1983).

10) Individual humans buying goods from a market are engaged in choosing the proportions of various products within a limited budget (Kagel et al., 1980; Rachlin, 1980).

This list of alternative allocations is by no means exhaustive. Moreover, other biological allocation strategies of a somewhat different nature are concerned with the amounts of investment in one kind of reward at different times (e. g. strategies for reproductive effort or germination time), or with allocations to two (or more) activities which are both required before one function succeeds (e. g. diaspores require allocations for dispersal *and* for subsequent growth). We are concerned here only with allocations between alternatives which provide separate rewards.

With few exceptions, studies of the various allocation topics have developed independently of each other. In recent years, some psychologically and zoologically oriented behaviorists and other biologists have actively forged connections between their previously separated approaches (Kamil, 1983; Pulliam, 1981; Staddon, 1980, 1983). Both groups have also developed connections with economics, particularly in methodology (compare for instance, the formulation of fitness sets, feedback functions and budget lines – Kagel et al., 1980; Levins, 1968; Rachlin, 1980). Parallels between the empirical results of economics and biology are limited, however, because economists are largely concerned with the effects of varying the prices of competing goods, whereas biologists rarely concentrate on variation in unit costs.

The theme of the present paper is that the similarities between allocation strategies dealing with different topics of plant and animal behavior are so fundamental and extensive that these diverse allocation topics should be brought into a single framework. Shared phenomena call for a common explanation. Towards this end, the primary elements of sex allocation theory are first presented, and then an analogous model which utilizes parallel features of other allocation topics is developed. The selection models are deterministic phenotypic models that assume selection operates on individuals in populations with discrete generations. Some of the principles and patterns that are shared by diverse allocation topics are subsequently discussed.

## A model of sexual strategies

Various methods of analyzing the theory of sex ratios and cosexual allocations have been developed (Charnov, 1982), including fitness set analysis, the Shaw-Mohler equation, and maximization of the product of male and female fitnesses. Here I examine the basic features of sex ratio strategies using the simplest and most general method available. This entails summing contributions from male and female offspring weighted by their reproductive values, and comparing the absolute fitnesses of phenotypes that differ in the sex ratio of their progeny.

Since the sexual functions of males and females are complementary, the contributions that sons and daughters make to the fitness of their parents are non-substitutable alternatives that can be added together. The fitnesses of the sons and daughters of a parent may be unequal, however, so the numbers of the two sexes cannot be simply added to provide a measure of parental fitness. To make male and female progeny commensurate, the numbers of each sex are weighted by their reproductive values, which measure the expected contributions to the 'ancestry of future generations' (Fisher, 1930). If parent $i$ produces $s_i$ sons and $d_i$ daughters (counted at fertilization or later) with reproductive values $v_{si}$ and $v_{di}$, the parent's fitness,

$$w_i = d_i v_{di} + s_i v_{si}. \tag{1}$$

The exploration of sex ratio strategies then centers on specifying the effects of various factors on the reproductive values of sons and daughters, which can be split into as many components as necessary. One convenient factor is the individual 'fitnesses' of sons and daughters, $w_{si}$ and $w_{di}$, including appropriate viability and fertility factors, but not their mating success.

Using Bateman's Principle (Bateman, 1948) that the fitness of females is usually limited by their nutritional resources, whereas that of males is characteristically limited by their ability to fertilize eggs, we define the reproductive value of *mature* daughters as one and express the reproductive value of mature sons in female units as the number of females they fertilize, weighted by the proportion of each female's eggs that they fertilize. If a son of parent $i$ has an expected 'competitive share', $c_{ij}$, of mate $j$'s eggs, then

$$w_i = d_i w_{di} + s_i w_{si} \sum_j c_{ij}. \tag{2}$$

When all males have the same fitness and all females are equally fit, the reproductive value of a mature male is simply the ratio of females to males in the population (i. e. $\sum_j c_{ij} = F/M$).

To determine the outcome of selection on the sex ratio, an ESS is found by examining the fitness advantage that a rare mutant in a population has over the prevalent phenotype that has a slightly different sex ratio. The ESS occurs where neither phenotype has a fitness advantage. Consider the situation in which sons and daughters are equally expensive to produce and survive at equal rates during the period of parental expenditure, but have different post-independence 'success' rates, $p_d$ and $p_s$, which may reflect differential survival or fertility. If sons and daughters compete with the progeny of numerous parents, an individual's own sex ratio has a negligible effect on the sex ratio experienced by its progeny. When individuals of the prevalent phenotype, 1, in a population produce $n$ offspring, in any proportions $r_1$ and $1 - r_1$ of sons and daughters at independence, and a rare mutant of phenotype 2 produces the sexes in proportions $r_2$ and $1 - r_2$, the fitnesses of the competing phenotypes, using (2), are

$$w_1 = n\left[(1 - r_1)p_d + r_1 p_s \cdot \frac{(1 - r_1)p_d}{r_1 p_s}\right],$$

and

$$w_2 = n\left[(1 - r_2)p_d + \frac{r_2(1 - r_1)p_d}{r_1}\right].$$

The fitness advantage,

$$w_2 - w_1 = np_d\left[(1 - r_2) + \frac{r_2(1 - r_1)}{r_1} - 2(1 - r_1)\right],$$

and

$$\frac{\delta(w_2 - w_1)}{\delta r_2} = np_d\left[-1 + \frac{1 - r_1}{r_1}\right].$$

Neither phenotype has a fitness advantage when $\delta(w_2 - w_1)/\delta r_2 = 0$, i. e. when

$$\frac{r_1}{1 - r_1} = 1. \tag{3}$$

The second derivative is zero, indicating that there is an equilibrium at which $r_2$ may have any value when the population as a whole produces equal numbers of males and females (Fisher, 1930; Shaw and Mohler, 1953). The model shows that when the number of offspring produced by parents varies independently of the sex ratio, the sex ratio that is selected is not affected by the total number of offspring or their individual size or cost or proficiency. In these circumstances, sex ratio strategies can be decoupled from reproductive effort and other life history strategies.

The costs of raising sons and daughters to maturity may differ. If parents spend $R$ resources on their families and each son and daughter costs $E_s$ and $E_d$ respectively, the number of offspring that can be produced, $n = R \div [r_i E_s + (1 - r_i)E_d]$. Substituting for $n$ in the above derivation gives at equilibrium,

$$\frac{r_1}{1 - r_1} = \frac{E_d}{E_s} \text{ , or } r_1 E_s = (1 - r_1)E_d. \tag{4a}$$

The rates of survival of sons and daughters during the period of parental investment may also differ. Suppose that fractions $s_s$ and $s_d$ of the sons and daughters respectively survive, and that average fractions $f_s$ and $f_d$ of the full investments in offspring are spent before the proportions $(1 - s_s)$ and $(1 - s_d)$ of sons and daughters die. Then
$n = R \div \{r_i[s_s + (1 - s_s)f_s] + (1 - r_i)[s_d + (1 - s_d)f_d]\}$.
Substituting for $n$ in the above derivation gives at equilibrium, when

$$\frac{r_1}{1 - r} = \frac{s_d + (1 - s_d)f_d}{s_s + (1 - s_s)f_s}$$

or

$$r_1[s_s + (1 - s_s)f_s] = (1 - r_1)[s_d + (1 - s_d)f_d]. \tag{4b}$$

Examination of (3) and (4a, b) shows that in all cases parental investments in the two sexes are equal, whether their costs are equal or unequal. The numerical ratio of sons and daughters invested in is affected by sex-differential expenditure or survival during the period of parental investment but not by differential survival after parental investment has ended, as Fisher argued verbally in 1930.

These derivations of Fisher's results differ from most sex ratio models in two respects.

First, the use of absolute fitnesses weighted by reproductive values permits a very simple and versatile formulation of selection, obviating the need for more complex procedures such as the Shaw-Mohler equation or fitness set analysis. Parental fitness can be expressed in terms of the expected contributions of the immediate progeny, rather than in terms of the number of grandchildren, as is usually done. (The number of post-investment grandchildren is not a valid measure of fitness when the costs of male and female progeny are unequal.) Taylor (1985a) has also pointed out that the coefficients $1/f$ and $1/m$ in the Shaw-Mohler equation should be regarded as measuring the relative value of placing a gene in a female or male offspring.

Although Fisher used the concept of reproductive value in several context, including sex ratios and comparisons of the products of self- and cross-fertilization, his example has rarely been followed (e. g. Uyenoyama, 1984). The concept of reproductive value has been used primarily for comparisons of different age classes, and is often defined as the age-specific expectation of future offspring. The concept of reproductive value raises difficulties when used as an index of the intensity of selection in age-structured population (Charlesworth, 1980), but these difficulties do not apply to the situations considered here.

Second, the procedure of examining the fitness advantage of a rare mutant is a variation of the popular method of finding an ESS by examining the fitness of a mutant. The use of a fitness advantage, which seems to have been introduced by Hamilton (1967), is superior in several respects. It portrays selection realistically as a process of competition (Clark, 1981) rather than optimization and thus emphasizes that the primary outcome of selection is the persistence of certain types based on their sustained competitiveness, not the attainment of optimality on any absolute scale. Hence the principal merit of the ESS concept, the notion that selection produces strategies that persist, is incorporated explicitly into the derivation of an ESS. In addition, fitness advantage formulations are able to predict the full course of selection, not just the ability of a rare mutant to spread, and are therefore able to predict equilibrium frequencies in a polymorphism (Lloyd, 1983; Taylor, 1985a).

The gender allocation strategies of cosexes can be studied by an analogous method. If a cosexual individual, $i$, allocates resources to paternal and maternal reproduction so that it produces $m_i$ female units and $l_i$ male units with reproductive values $v_{mi}$ and $v_{li}$ respectively, then

$$w_i = m_i v_{mi} + l_i v_{li}. \tag{5}$$

If the female units (eggs, zygotes, or seeds, etc.) are given reproductive value of one, the reproductive value of male units (sperm, pollen grains, etc.) can be expressed as the number of eggs they are able to fertilize, weighted if necessary by their subsequent behaviors. If the maternal fitness of mate $j$ of individual $i$ is $m_j$, and $i$ fertilizes a competitive share, $c_{ij}$, of that fraction of $j$'s eggs that it is eligible to fertilize, $e_{ij}$, then

$$w_i = m_i + \sum_j m_j e_{ij} c_{ij}. \tag{6}$$

Equation (6) can be used to derive gender allocations (Lloyd, 1983, 1984a), just as (2) was used above for sex ratios.

### Factors affecting sex allocations

Theoretical studies have shown that a considerable number of factors can affect the selection of sex ratios or gender allocations. These are listed in the Table. All deterministic factors (all those except sex ratio homeostasis) can be analyzed using the method outlined above. The factors may be grouped into five classes according to how they affect the male or female fitness curves which describe how the fitness obtained from male or female investment increases as the allocation increases. *'Intrinsic'* factors that affect the number, cost, or efficiency of all male or female units do not affect the sex ratio selected (above) or the selected allocations in cosexes (Lloyd, 1984a).

*Size* factors determine the maximum fitness that can be obtained from male or female investment if all reproductive resources were committed to one function (Figure). The unique and omnipresent feature of sex in anisogamous species is the obligatory interdependence of male and female functions that arises from fusion of their gametes or nuclei. In the absence of complicating factors, this complementarity causes equal investments in male and female functions and serves as a baseline to compare other factors against. Although male and female gametes contribute autosomal genes equally in the population as a whole, varying circumstances may cause particular parents to have unequal opportunities for fitness through their male and female gametes or offspring (Bull, 1981; Lloyd and Bawa, 1984; Trivers and Willard, 1973). In cosexual plants for instance, when pollen and seeds are dispersed equal distances the selected ratio of allocations is equal to the ratio of the viabilities of an individual's progeny from pollen and seeds (Lloyd and Bawa, 1964) i. e.

$$\frac{a}{1-a} = \frac{v_p}{v_s}. \tag{7}$$

Factors affecting sex allocations

| Factor | Direction of deviation, effect on stability | | References |
|---|---|---|---|
| | Dioecism sex ratio | Cosexuality gender allocation | |
| **A) Size factors** | | | |
| 1) Sex complementarity | ♂ = ♀ | ♂ = ♀ | (Fisher, 1930; Maynard Smith, 1971 b; Shaw and Mohler, 1953) |
| 2) Variable parental status or progeny success | > ♂or♀ | > ♂or♀ | (Bull, 1981; Lloyd and Bawa, 1984; Taylor and Sauer, 1980) |
| 3) Self-fertilization | –[a] | > '♀' | (Charlesworth and Charlesworth, 1981) |
| **B) Shape factors** | | | |
| 1) Finite population | No deviation (S)[a] | No deviation (S) | (Lloyd, 1983; Taylor and Bulmer, 1980) |
| 2) Local mate competition | > ♀(S) | > ♀(S) | (Hamilton, 1967; Taylor and Bulmer, 1980) |
| 3) Local resource competition | > ♂(S) | > ♂(S) | (Clark, 1981) |
| 4) Unilateral fixed cost | No deviation (S) | > ♂or♀[c](D) | (Charnov, 1979b; Lloyd, 1984 a) |
| 5) Unilateral fitness limit | No deviation (D) | > ♂or♀ (S) | (Charnov, 1982; Lloyd, 1984 a, Strathmann et al., 1984) |
| 6) Interference between sex functions | –(S) | > ♂or♀ (D) | (Lloyd, 1984 a) |
| 7) Facilitation between sex functions | –(D) | > ♂or♀ (S) | (Lloyd, 1984 a) |
| **C) Uncertainty of fitness returns** | | | |
| 1) Sex ratio homeostasis | No deviation (S) | No deviation (S) | (Taylor and Sauer, 1980; Verner, 1965) |
| **D) Constraints on allocation** | | | |
| 1) Segregation constraint | Limits deviation | – | (Maynard Smith, 1980) |
| 2) Assessability constraint | Limits variation | Limits variation | (Bull, 1981, Lloyd and Bawa, 1984) |
| **E) Allocation-independent intrinsic factors[b]** | | | |
| 1) Sex-differential costs | No effect | No effect | (Charlesworth, 1977; Fisher, 1930) |
| 2) Sex-differential proficiency or survival of gametes or progeny | No effect | No effect | (Fisher, 1930; see text) |
| 3) Reproductive effort | No effect | No effect | See text |

[a] – = not relevant. (S) = stabilizes sex ratio or gender allocation. (D) = destabilizes sex ratio or gender allocation. [b] Sex differential costs or success affect the numbers of the two sexes or gametes, but not the allocations. [c] Unilateral fixed costs affect gender allocations if they are included in the allocations (Lloyd, unpubl.).

*Shape* factors modify the shape of a fitness curve (Figure). If the paternal and maternal fitnesses of a cosex are proportional to their respective allocations raised to the powers $y$ and $z$ respectively, then (Charlesworth and Charlesworth, 1981; Charnov, 1979b, 1982; Lloyd, 1984a)

$$\frac{a}{1-a} = \frac{y}{z}. \tag{8}$$

If the allocation curves decelerate ($y < 1$, $z < 1$), the equilibrium at (8) is fully stable. The best-known shape factors are local mate competition (Charnov, 1982; Hamilton, 1967) and local resource competition (Clark, 1981), in which the male or female fitness returns respectively are limited by competition among the products of a restricted number of parents. Unilateral local competitions occur sporadically among animals, but are virtually ubiquitous among seed plants because of the limited distances that seeds and pollen travel (Bulmer and Taylor, 1980; Lloyd, 1984a). Most of the other shape factors affect the selection of gender allocations but not that of sex ratios. Nevertheless they can favor, and help to stabilize, either combined or separate sex conditions (Table).

The only *stochastic* factor that has been examined to date is the chance fluctuations in the sex ratio that occur when mating groups are small. The offspring sex that happens to be in the minority in a group will have a higher reproductive value than the other sex. Verner (1965) and Taylor and Sauer (1980) have shown that selection for sex ratio homeostasis leads to a fully stable sex ratio at equality.

A number of *constraints* may restrict the ability of size and shape factors to modify sex ratios or gender allocations. The chromosomal machinery imposes a severe 'segregation constraint' on deviations in the ratios of genetically determined sexes whenever deviations are caused by eliminating a fraction of one sex (Maynard Smith, 1980). The eliminations impose a cost on deviation that is permissible only if the forfeited investment is a small fraction of the total investment required per offspring. Another constraint, an 'assessability constraint', arises from the inability of parents to judge whether male or female offspring, or those from maternal or paternal investment, are likely to be fitter. This is particularly important in plants, which have only limited means of sensing their environment, and restricts the number of plants that choose their sex according to the conditions (Lloyd and Bawa, 1984).

More factors affect the gender allocations of cosexes than affect the sex ratios of dioecious organisms (Table). Together with the segregation constraint that severely restricts deviations in sex ratios, this means that the gender allocations of cosexes offer much more scope for analyzing a variety of sex allocation strategies. Nevertheless, most sex allocation studies to date have examined the less pliable sex ratios of dioecious populations.

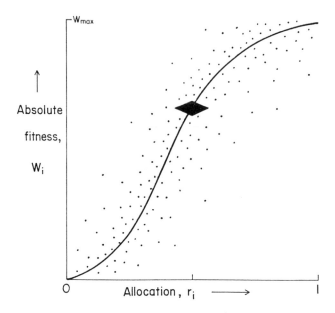

Four kinds of factors affecting a reward curve that describes how the rewards from investment in an activity increase as the proportional allocation increases. The shape of the curve illustrating a deterministic factor (solid line) is one that accelerates at first, then decelerates as it approaches the maximum reward when the allocation is 100 % (the potential 'size' of the rewards). The scattered points represent stochastic variation in the reward. The diamond figure in the center represents the limited deviation in sex ratios and other segregation ratios that is imposed by a 'segregation constraint' (see text for further explanation).

## A general model of allocations

A number of features that occur repeatedly in various allocation decisions can be explained by a general theory. Every allocation topic has unique aspects, and some subjects already have more complex theories than that provided below. The present aim is not to cover every eventuality in specific topics, but to demonstrate the generality of certain features that are shared by seemingly unrelated subjects.

Suppose there are two activities, A and B, that make separate contributions to the total reward which an individual receives. The alternative activities may be morphological structures, the segregation of genes or gene combinations, or biochemical, physiological or behavioral acts. The individual allocates proportions $a$ and $1 - a$ of a single limited resource (energy, a nutrient, time, etc.) to A and B respectively. The contributions from A and B, $c_a$ and $c_b$, are functions of the allocations to the activities.

The contributions from A and B are measured on a single scale of rewards (reproductive fitness, rate of food intake, reinforcement rate, utility etc.)

by assigning reward values, $v_a$ or $v_b$, to units of the two activities. Then the total reward for individual $i$,

$$R_i = c_a + c_b = f(a)v_a + f(1-a)v_b.$$

To find the stable strategy, the relative allocations that do not have a competitive disadvantage over any other, we examine the situation where a single individual of type 2, with allocations to A and B of $a_2$ and $1 - a_2$, competes with all other individuals of type 1 with allocations $a_1$ and $1 - a_1$. Then the reward for the prevailing type,

$$R_1 = f(a_1)v_a + f(1-a_1)v_b.$$

The reward for the single individual of type 2,

$$R_2 = f(a_2)v_a + f(1-a_2)v_b.$$

The advantage of type 2 changes with its allocation according to

$$\frac{\delta(R_2 - R_1)}{\delta a_2} = f'(a_2)v_a + f'(1-a_2)v_b.$$

Neither type has an advantage when $\delta(R_2 - R_1)/\delta a_2 = 0$, i. e.

$$f'(a_2)v_a + f'(1-a_2)v_b = 0. \tag{9}$$

The stationary point represents a stable equilibrium if the second derivative is negative,
i. e.

$$f''(a_2)v_a + f''(1-a_2)v_b < 0,$$

which includes the situation when both reward curves are decelerating.

Equation (9) is a marginal value theorem that applies to allocations between any alternatives that provide additive rewards. At the ESS allocation, the absolute values of the marginal gains for the two activities, weighted by their reward values, are equal.

The Shaw-Mohler equation (Charnov, 1982) for sex allocations is a special case of (9). In considering allocations to male and female gametes, or to sons and daughters, pollen grains and seeds, etc., the total fitnesses of males and females in a population are equal, i. e. $v_{\male}\bar{m} = v_{\female}\bar{f}$. Substituting into (9) and rearranging gives the Shaw-Mohler equation,

$$\frac{dm}{\bar{m}} + \frac{df}{\bar{f}} = 0.$$

Equation (9) shows that the ESS for alternative allocations depends on the shapes of the reward curves and on size factors represented by the reward values. The effects of size and shape factors can be seen more clearly when they are represented by specific functions. Suppose that when all resources go to A (or to B), the individual produces $n_a$ (or $n_b$) units. If both activities are conducted, a linear combination of numbers is produced. The units of each activity experience allocation-independent success rates $s_a$ or $s_b$, associated with their post-investment survival rate, efficiency at locating appropriate environments, etc. Hence the effective numbers of A and B activities are $n_a a s_a$ and $n_b (1 - a)s_b$. The contributions that the two activities make to the rewards are proportional to the numbers raised to a power, $y$ for A or $z$ for B, which express the shapes of the reward curves.

Then

$$f(a) = (n_a a s_a)^y \text{ and } f(1 - a) = [n_b(1 - a)s_b]^z,$$

so

$$f'(a) = y(n_a s_a)^y a^{y-1}$$

and

$$f'(1 - a) = - z(n_b s_b)^z(1 - a)^{z-1}.$$

The reward values can be specified in a competition model that assumes that there are restricted numbers of opportunities, $O_a$ or $O_b$ per parent, for the activities to contribute to the rewards, because of limits in the available space to grow or numbers of food items, time requirements, etc. The reward values of A and B units are then the number of available opportunities divided by the number of competing units. The competition pools for $O_a$ and $O_b$ are considered here to be completely separate; that is

$$v_a = \frac{O_a}{[n_a a s_a]^y} \text{ and } v_b = \frac{O_b}{[n_b(1 - a)s_b]^z}.$$

Substituting for $v_a$, $v_b$, $f'(a)$ and $f'(1 - a)$ in (9) gives

$$\frac{a}{1 - a} = \left(\frac{y}{z}\right)\left(\frac{O_a}{O_b}\right). \tag{10}$$

Also

$$\frac{\delta^2(R_2 - R_1)}{\delta a_2^2}\bigg|_{a_1 = a_2 = a}$$

$$= y(y-1)a^{-2}O_a + z(z-1)(1-a)^{-2}O_b.$$

A number of conclusions from the model are parallel to those presented above for sex allocations. First, both mixed strategies (do two things in combination) and pure strategies (do only one thing in any given circumstance) are possible. The strategy that is most competitive depends on the shape of the reward curves. When the curves decelerate ($y < 1$, $z < 1$), the second derivative is negative and there is a mixed strategy at a stable equilibrium specified by (10). When the curves are linear ($y = 1$, $z = 1$), an equilibrium is obtained in which any combination of activities is equivalent when the average ratio of investment in the population is equal to the ratio of opportunities, $O_a : O_b$ (cf. (3, 4)). If the curves accelerate, a mixed strategy is unstable and restriction to one or other activity provides the most competitive reward.

Second, the shapes also affect the exact allocation in mixed strategies. The activity for which the rewards decelerate most quickly is allocated less (cf. (8)).

Third, the differences in the costs of single A or B activities (incorporated into $n_a$, $n_b$) and the success rates of A and B ($s_a$ $s_b$) do not affect the most competitive allocation ratio (cf. (3, 4)).

Fourth, the allocation ratio of a mixed strategy depends on the size factors, $O_a$ and $O_b$. The property of the model that A and B activities offer rewards that are separately limited introduces a frequency-dependent element into their rewards (compare the effects of $O_a$ and $O_b$ in (10) with the ratio $(1 - r_1)/r_1$ in the derivation of (3)).

Fifth, if the shapes of the allocation curves are identical ($y = z$), the ratio of allocations is equal to the ratio of opportunities (cf. (3, 7)). This is not so when the shapes are not identical.

Sixth, the separate size and shape factors have multiplicative effects on the equilibrium allocation ratio. It has similarly been shown that a number of combinations of factors have multiplicative effects on gender allocation ratios (Lloyd and Bawa, 1984) or on other mixed strategies for plants (Lloyd, 1984c), provided the factors vary independently of each other.

This general model is analogous to that presented above for sex allocations, but it differs in allowing the total opportunities for the two activities to vary independently of each other. The general model could be extended by the inclusion of numerous specific factors, as has happened in models of sex allocation. If would be relatively easy to add some factors to the model, e. g. local competitions, fixed costs, upper limits to one or other reward, or certain types of constraint. Other potential additions to the model, e. g. three or more alternative activities, the simultaneous operation of different limiting resources such as energy and time, partially substitutable activities, or randomly varying rewards for a given allocation, may require more complex mathematical procedures such as linear or dynamic programming. The inclusion of any of these additional factors would increase the ability of the general theory of allocations to provide a unified explanation of the diverse types of allocation decisions.

## Observed patterns of allocation strategies

The following brief survey notes some of the features common to sexual strategies and a variety of other allocation strategies for which the competing alternatives provide separate rewards.

### Classes of strategies

There are several principal classes of allocation strategies (Lloyd, 1984c; Maynard Smith, 1979, 1982).

1) *Fixed* strategies: individuals choose the same option under all permissive conditions. The complementarity of male and female gametes precludes parents from following a fixed sex allocation strategy; they produce either two types of gametes or two sexes of progeny. For many biological functions, fixed strategies are the norm.

2) *Conditional* strategies: individuals choose different options according to the circumstances. Conditional strategies include sex-choosing, which is more common among animals (Bull, 1981; Charnov, 1982) than among plants (Lloyd and Bawa, 1984), and occur sporadically for other subjects, such as choices between open and cleistogamous flowers or sexual and asexual embryos in plants (Lloyd, 1984c), and some alternative mating strategies and other environmentally determined 'polymorphisms' such as winged versus non-winged or horned versus non-horned individuals (Austad, 1984; Eberhard, 1982; Harrison, 1980).

3) *Mixed* strategies: individuals carry out two or more activities with non-zero probabilities. Mixed strategies occur in some instances in each of the ten topics listed in the introduction. The subclass of multiple strategies occurs when individuals carry out both activities at the same time, as in simultaneous hermaphroditism, seed heteromorphism, etc. (Lloyd, 1984c; Schoen and Lloyd, 1984; Venable and Lawlor, 1980). The subclass of stochastic strategies occurs when alternative behaviors are performed in succession with constant probabilities, as in some foraging and learning strategies (Maynard Smith, 1982; 1984).

*Mixed strategies are associated with separately limited rewards*

In the case of sex allocation strategies, male and female gametes have complementary roles that cannot be substituted for each other as long as sexual reproduction is retained. The ways in which the alternative activities of other mixed strategies provide separately limited opportunities are being investigated in many fields e. g. flowers with adaptations for self- or cross-fertilization, seed heteromorphisms, seed and vegetative reproduction (Lloyd, 1984c; Uyenoyama, 1984). The separate advantages of distinct sterile castes in social insects are being explored (Oster and Wilson, 1978; Wilson, 1971). In foraging strategies, the concept of resource depression describes situations where predators acquire separately diminishing gains from patch types because prey are depleted or take evasive action (Charnov et al., 1976b; Krebs and McCleery, 1984). Operant behaviorists have recognized that their experiments frequently contain separate time limitations on alternative reinforcements (Staddon, 1980). Conversely, some fixed strategies are associated with a lack of separate limitations on the rewards from different activities e. g. learning experiments with constant reward rates (Maynard Smith, 1984; Staddon, 1980), single syndromes for externally mediated pollination or dispersal (Lloyd, 1984c).

*Matching rules*

In operant conditioning experiments, particularly those with variable-interval schedules, a 'simple matching law' (or rule) has been repeatedly observed (Herrnstein and Vaughan, 1980; Staddon, 1983). The rule is that the ratio (or proportions) of responses (investments) to two signals is equal to the ratio (or proportions) of positive reinforcements (rewards) obtained from the two signals. The matching rule of psychologists may be generalized to cover situations in any topic of allocation strategies in which the relative rates of investment in alternative activities, $a_1$ and $a_2$, match the relative rewards from the activities, $R_1$ and $R_2$, i. e.

$$\frac{a_1}{a_2} = \frac{R_1}{R_2} \quad , \quad \text{or} \quad \frac{a_1}{a_1 + a_2} = \frac{R_1}{R_1 + R_2}. \tag{11a}$$

The matching rule may also be expressed as a statement that the total benefit to cost ratios of two activities are equal. By transposition from (11a),

$$\frac{R_1}{a_1} = \frac{R_2}{a_2}. \tag{11b}$$

Matching rules occur in diverse allocation strategies. The classical conclusion of Fisher in 1930 that parents in a dioecious population are selected to invest equally in sons and daughters follows the generalized matching rule, since males and females contribute (autosomal) genes equally in the population as a whole, as Fisher noted. A comparable matching rule applies to the maternal and paternal allocations of cosexes (Maynard Smith, 1971b). Moreover, when there is variation within a cosexual population in the ability of parents to produce offspring via male and female gametes, or in the relative opportunities for the two classes of offspring, the selected ratio of investments is equal to the ratio of parental or offspring opportunities (cf. (7) above), provided the allocation curves have the same shape (Lloyd and Bawa, 1984). The males of gynodioecious plant populations also produce some seed, and the relative investments in pollen and seed match their fitness returns (Lloyd, unpubl.). It can also be shown that the increase in seed investment by a cosexual plant as the frequency of self-fertilization increases (Charlesworth and Charlesworth, 1981) exactly matches the proportionate increase in the fitness the parent obtains from expenditure on seeds rather than pollen (Lloyd, unpubl.). Matching rules are also observed in 'ideal free distributions' of animals settling in a patchy environment (Fretwell, 1972), and are predicted by theory as a learning rule for animals sampling rewards from different behaviors (Harley, 1981) and for genes modifying the segregation ratio of certain autosomal loci (Lloyd, 1984b). Similarly, in one of two digger wasp colonies observed, (Brockman et al., 1979), females dug new burrows or used old ones in frequencies such that the average fitnesses derived from the two behaviors were equal.

Some observations do not conform to the matching rule. In a second digger wasp colony, the two nesting strategies did not provide equal fitnesses (Brockman et al., 1979). In operant conditioning experiments, variable-ratio schedules do not generally follow the rule (Herrnstein and Vaughan, 1980; Staddon, 1980). Some outcrossing plants do not spend equal resources on pollen and seeds (Lloyd, 1984a).

The models presented above for sexual and other types of allocations predict that the matching rule will be followed in some situations but not in others. With only two alternatives, the marginal rewards from the two activities are necessarily equal at equilibrium. This follows because the total reward is the sum of rewards from the two activities, so when there is no net change a small change in one must be opposed by an equal and opposite change in the other. This is 'molecular' or 'local' matching in the terminology of psychologists. But there is no necessity for the total or overall benefits from two activities to match – for 'molar' or 'global' matching. The models show that this occurs only when the shapes of the two allocation curves are identical. If they are not, as in the presence of unilateral fixed costs or local competitions or any other unilateral shape factors, the ratios of rewards to costs for different activities are expected to be unequal at equilibrium. Peter Taylor has discovered a general form of the Shaw-Mohler equation that is valid for local mate competition and can be regarded as a generalized matching rule (pers. comm., 1985).

*Constraints on allocations*

Although each type of allocation is likely to have its own combination of constraints limiting the allocation strategies that can be achieved, a number of constraints have parallel effects on different subjects. Three examples will suffice. First, a segregation constraint severely restricts the ability of 'outlaw' genes to segregate preferentially (Lloyd, 1984b), just as a segregation constraint on sex ratios is a major barrier to the selection of deviant ratios (Maynard Smith, 1980). Secondly, the generally poor ability of plants to assess the prospects of alternative activities is a major constraint on strategies involving a variety of vegetative and reproductive functions (Lloyd, 1984c; Lloyd and Bawa, 1984). Similarly, Great Tits do not restrict their foraging to the most rewarding prey but show a 'partial preference' for lower-ranking prey (i. e. a mixed strategy) in part because they cannot discriminate accurately among the various types of prey (Rechten et al., 1983). Thirdly, in some plant species, limited resources associated with smaller size or poorer conditions affect the ability of a plant to function as a maternal parent (Lloyd and Bawa, 1984) or to produce chasmogamous (open) flowers (Schoen and Lloyd, 1984), and thus impose a cost constraint.

*Risk sensitivity*

The benefits obtained from any given allocation are likely to fluctuate. If variation in potential rewards cannot be predicted before an investment is made, a response to the particular conditions experienced by an individual is precluded, and the variation is perceived as effectively random. Stochastic models suggest that organisms that take account of random variability by risk-sensitivity strategies improve their fitness (Caraco, 1981; Gillespie, 1977; Real, 1980a, 1980b; Rubenstein, 1982). Jensen's Inequality, a theorem of statistical decision theory, compares the expected utility from a constant benefit with randomly varying benefits that have a mean value equal to the certain returns. It shows that when reward curves decelerate, individuals do better by engaging in diversified activities – a risk-averting mixed strategy. When reward curves accelerate, individuals do better by concentrating on the potentially more rewarding activity – a risk-prone fixed strategy (Caraco, 1981; Keeney and Raiffa, 1976; Rubenstein, 1982). Hence the stochastic effect of the shapes of reward curves operates in the same general direction as the deterministic effect of shape factors discussed above.

When the local breeding populations are small and subject to random variation in the ratio of males to females, selection for risk-aversion leads to sex ratio homeostasis (Taylor and Sauer, 1980). Comparable risk-sensitive strategies have been discussed for a considerable variety of allocation subjects, including reproductive effort (Cohen, 1966; Goodman, 1984; Lacey et al., 1983; Schaffer, 1974), dormancy (Cohen, 1968; Venable and Lawlor, 1980), foraging (Caraco, 1981) and others.

**Concluding remark**

Although each type of allocation has its own characteristics, both theoretical models and empirical results demonstrate that diverse allocation topics share an impressive number of features. In particular, several features of sex allocations that have been discovered over the past 50 years have direct parallels in other fields. These shared features warrant common explanations. There are considerable heuristic advantages to be gained from uniting these hitherto largely unconnected topics. Efforts to develop widely applicable principles should reduce the duplication of labor involved in rediscovering the same concept independently in different fields, as has happened repeatedly in the past. Broadening the applications of general principles will increase the explanatory power of the theories that have been applied to particular types of allocation. The total range of phenomena will be explained more economically by general theories than by a series of parallel but unconnected theories for each specific subject.

**Summary**

Allocation strategies in which a limited resource is apportioned among alternative activities are applicable to diverse structural, genetical and behavioral topics, including male versus female investments. In a model of sex allocation strategies, the absolute fitnesses of individuals are calculated by summing the production of male and female gametes or offspring, each weighted by its reproductive value. The ESS is obtained by examining the fitness advantage of one phenotype over another. An analogous method is used to obtain a general model of allocation strategies that incorporates some widespread features. Allocation strategies are affected by the sizes and shapes of the reward curves, stochastic factors, and constraints on the allocations permitted. A number of parallels among diverse types of allocation strategies, including the occurrence of fixed, conditional and mixed strategies, and matching rules, are discussed.

Acknowledgments. I am grateful to C. Aker, C. Lively, S. Stearns, P. Taylor and C. Webb for their helpful comments on a draft of the manuscript.

# Quantitative genetic models of sexual selection: A review

S. J. Arnold

## Introduction

The aim of this paper is to survey recent quantitative genetic models for evolution by sexual selection. These models treat the evolution of continuously distributed traits such as tail length in the males of a lekbreeding bird and female mating preference based on such an attribute. The quantitative genetic models have some advantages over two- and three-locus models of sexual selection, recently discussed by Andersson (1986a, 1986b), O'Donald (1980), Kirkpatrick (1982), Maynard Smith (1985) and Seger (1985). Most traits of interest to students of sexual selection are continuously distributed with polygenic inheritance. Traits composed of two or three discrete classes with single factor inheritance are relatively uncommon. Secondly, the parameters of inheritance and selection used in the quantitative genetic models can be measured in natural populations, whereas the crucial gene frequencies and selection coefficients in the oligolocus models are virtually always inaccessible to measurement. The polygenic models may also permit sustained evolution, but the oligogenic models may quickly reach limits imposed by gene fixation, which may obscure some evolutionary phenomena. In defense of the two- and three-locus models, we may note that their analysis is sometimes more tractable. In any case, the principal conclusions are the same using both types of models.

A review of the quantitative genetic models is timely for three reasons. Models of this type have recently proliferated and it is easy to become overwhelmed by new results. Secondly, much of the controversy in sexual selection theory hinges on which aspects of fitness are thought to be correlated with the male traits that are the objects of mate choice (Heisler, 1985). One camp (the 'good genes' school) has argued that the traits used in mate choice are viability or fecundity indicators, while another camp (the 'arbitrary' trait school) has argued that the traits need only be mating – success indicators. Quantitative genetic models are just beginning to connect these two viewpoints (Kirkpatrick, 1985, 1986; Heisler, 1985). Perhaps they will help to move the controversy into constructive avenues of research. Thirdly, the model family is now large enough to provide some generalizations about equilibria and stability conditions that instruct our intuition about evolutionary phenomena in general, as well as our understanding of sexual selection. For these reasons I will focus on the common denominators in the models and on a comparative analysis of their conclusions.

Several terms are used here in a formal sense that sometimes differs from popular usage. *Sexual selection* refers to selection acting via effects on the mating success of surviving males (or females) whereas *natural selection* refers to effects on other components of fitness (e. g., viability). This usage is common to all the theoretical models discussed here. The use of mutually exclusive terms rankles some, despite the fact that Darwin (1859, 1874) used the 'natural' and 'sexual selection' in just this way (Arnold, 1983; Wade and Arnold, 1980). Some prefer to see sexual selection as a subset of natural selection. The important point is that viability, or other forms of, selection may act in a different direction than sexual selection, as when the peacock's tail makes him vulnerable to predators but attractive to peahens. Separate terms are useful to describe such opposition of selective forces, but it may be a sterile semantic issue to quibble about the precise labels.

In the models discussed here *fitness* is measured as a zygotic progeny count (Crow and Kimura, 1970). The number of zygotes produced by an individual is taken as that individual's fitness, *not* the quality of offspring or the number surviving to sexual maturity. The point in so defining fitness is that inheritance and selection can be explicitly separated in the models. A count of surviving offspring is a problematic fitness measure because it confounds the reproductive success of parents (zygotic progeny count) with the viability of offspring. A better approach is to treat parental care as a parental attribute, viability as an offspring attribute, and fitness as a zygotic progeny count in both generations. The effects of parental care on offspring viability can then be treated as separate inherited and direct parental effects (Cheverud, 1984).

## Common denominators

Certain features are common to all the models. These common denominators include: assumptions about the frequency distributions and mode of inheritance of the characters, the form of viability selection and the modes of female choice that produce sexual selection. The principal differences among the models lie in the number of evolving characters and in the modes of selection on the characters (Table 1).

All of the models deal with the evolution of a male secondary sexual trait, sometimes referred to as a handicap and sometimes simply as the male trait. The extraordinary development of such characters in some species (e. g., the peacock's tail, the newt's crest, the cricket's mating call) led Darwin (1859, 1874) to propose the concept of sexual selection. The models deal with only one type of Darwinian sexual selection, mate choice, and do not treat the evolution of characters used in male-male contest or combat.

The characters are assumed to show normal (Gaussian) frequency distributions. This is an assumption that greatly facilitates getting mathematical results, because, for example, a Gaussian character distribution acted on by a Gaussian selection function yields a Gaussian character distribution after selection. It is also an assumption with solid empirical foundation. Many characters are normally distributed or can be transformed to normality, often with simple logarithmic transformations (Wright, 1968). Evolution is then modeled on the transformed scale.

Character inheritance is assumed to be polygenic with numerous loci contributing to genetic variation in each character. Continuously distributed characters of the kind used in the models often show polygenic or multifactorial inheritance (Falconer, 1981; Wright, 1968).

In all the models the male trait is acted on by both natural (viability) and sexual selection. In other words, the trait influences both the male's survivorship and his mating success. Viability selection is assumed to be of a Gaussian form, with selection acting against extremes and favoring intermediate development of the male trait (Fig. 1). A Gaussian form is a noncontroversial assumption because any strength or peakedness of the function is allowed (the width of the selection function is a parameter in the model). Furthermore when the population mean lies far from the intermediate optimum, the population mainly experiences directional selection favoring, say, smaller tails or softer calls. Very weak or no viability selection can be accommodated by making the function nearly or actually flat.

All of the models specify one of three types of female mate preference and many of them treat all three types (Fig. 2). Hardly any empirical work has focused on individual differences in mate preference within populations (a critical issue in evolutionary models) and so the modes are meant to represent three diverse possibilities. Surprisingly, the different modes of preferences have relatively little effect on the qualitative results of the models (but see Seger, 1985). *Absolute preferences* describe a situation in which each female is most prone to mate with a certain male phenotype (Fig. 2a). Females vary in most preferred mate, with a Gaussian frequency distribution of the characteristic used in mate choice. *Relative preferences* differ from absolute preferences in that each female's most preferred mate is some characteristic distance from the average male phenotype instead of being an absolute (Fig. 2b). As in absolute preferences, the tendency of the female to mate falls off as a Gaussian function in both directions away from her most preferred mate. The most preferred mates of all females in the population form a Gaussian distribution of the characteristic used in mate choice. *Psychophysical* or *open-ended* preferences describe the common psychological phenomenon that perception increases exponentially with stimulus intensity. The tendency of a particular female to mate is an exponentially increasing function of male phenotype (Fig. 2c). The exponent of the mating function varies among females with a Gaussian distribution.

Mate choice, as modeled for example by the three functional forms just discussed, is unfortunately often visualized as a cognitive process in which

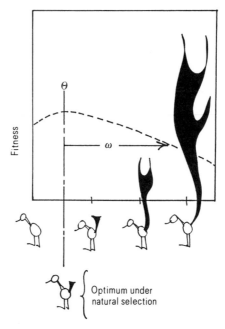

Figure 1. Viability of the male is affected by a hypothetical character, tail size. The curve specifying the relationship is a Gaussian function, shaped like a normal curve, with an optimum at the tail size specified by θ and a characteristic width of ω. Reproduced with permission from Arnold (1983).

the male phenotypes are under conscious scrutiny and evaluation. Such a cognitive process is only one, narrow interpretation of the actual possibilities and is probably an uncommon vehicle for sexual selection via mate preferences. Broadly construed, 'mate preferences' capable of generating sexual selection include any morphological, behavioral or physiological differences that produce differences in mating success in the opposite sex: differential reaction to male aphrodisiacs, variation in genitalia, differences in auditory reception or processing, sensitivity to tactile stimuli, etc. The models discussed here could represent the joint evolution of any such female attribute with corresponding male stimuli or structures.

## Anatomy of the models

The models share a basic structure, because they all employ the same evolutionary equation. The basic equation was derived by Lande (1979) and is the multivariate version of the quantitative genetic equation that has been used for several decades by plant and animal breeders (Lush, 1945). The equation gives the predicted change per generation in averages of an

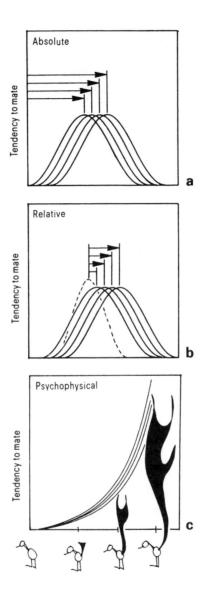

Figure 2. Three modes of female mate preference each based on the hypothetical male attribute of tail size. Individual variation is illustrated for each mode. The frequency of male tail sizes in the population is shown by the dashed curve in the middle figure. *a* Females with absolute preferences are most inclined to mate with males with a tail of the size indicated by the vertical line at the ends of the arrows. For each female, mating tendency falls off as a Gaussian curve in both directions from that most preferred mate. *b* Females with relative preferences most prefer to mate with a male whose tail is some characteristic size larger than the male average. *c* The mating tendency of females with psychophysical preferences increases exponentially with male tail size (shown on a logarithmic scale). Females differ in rate of increase of mating tendency as a function of male tail size.

arbitrarily long list of phenotypic characters. The prediction is based on only three ingredients: (1) inheritance, (2) trait variation and covariation and (3) selection. The patterns of selection and trait variation transform the set of potential parents into the set of actual parents. The pattern of inheritance transforms the set of actual parents into a set of descendants in the next generation. The transformations are then iterated to extrapolate the population into the future, generation by generation, and to predict its phenotypic composition at equilibrium.

The hereditary transformation consists of a matrix of genetic parameters. These parameters describe resemblance among relatives in the phenotypic traits. The parameters are called additive genetic variances and covariances and measure, respectively, the resemblance between offspring and parents in a particular character or the association between one character in offspring and some other character in parents (Falconer, 1981). Arnold (1983) gives a graphical account of the main genetic parameters used in the models.

The matrix of additive genetic variances and covariances is treated as a constant in the models. Empirical work and some theoretical studies provide support for the supposition of constancy. Studies of polygenic characters have shown substantial influx of variation each generation from mutation. Theoretical models indicate that a stable equilibration of genetic variance and covariance can be achieved at which the production of variation and covariation by mutation and recombination is balanced by losses due to selection (Bulmer, 1980; Lande, 1975, 1980b; Turelli, 1984). The present models assume that this equilibration of genetic variation has been achieved, so that the same hereditary matrix can be used generation after generation to predict evolutionary response to selection.

O'Donald (1983) objected to one of Lande's models on the grounds that it merely posits the existence of a critical genetic parameter, a genetic covariance that describes genetic coupling between a male trait and female mating preferences based on that trait. O'Donald (pers. comm.) overlooked a critical section of Lande's (1981) paper that modeled the maintenance of genetic variances and covariances. Constancy of genetic parameters is not a bald assumption, as O'Donald claimed, but the result of a model nested within the main model.

In contrast to the constancy of the equilibrated parameters of genetic variation, selection is perpetually changing in the evolving population. Selection is described in the models by the selection differential which gives the shift in the phenotypic mean of each character that is induced by selection. Usually two forms of selection act on the male secondary sexual trait: viability selection (a form of natural selection) and sexual selection induced by female mate choice. The total selection differential on the male trait is specified by writing the expression for the shift in mean due to viability selection and adding to it the shift in mean due to sexual selection. The relative strengths of those two selective forces change according to the value of the male trait average before selection each generation, which affects the strength of viability selection, and the average value of mate preference in the female population, which affects the strength of sexual selection.

In many of the models the mate choice of the females is assumed not to affect her fitness (measured as a zygotic progeny count). In such models there is no direct selection on mating preferences, so they evolve as a correlated response to direct selection on the male trait. In the models the selection gradient on mate preference, which measures the direct force of selection, is set to zero. Lande (1979), Lande and Arnold (1983) and Arnold and Wade (1984) discuss the concepts of selection differentials and gradients.

A major goal in using the models is to solve for the equilibrium composition of the population. What is the average value of the male and female traits when the population has stopped evolving? Equilibrium is not specified by exhaustion of genetic variation since that variation equilibrates at nonzero values, as described above. Instead the equilibrium is specified by the vanishing point for directional selection. Thus for the male trait described above, total directional selection will disappear when natural (viability) and sexual selection exactly balance. One solves for this equilibrium by setting the expressions for the total selection differential equal to zero. The resulting expressions give the composition of the equilibrium population in terms of the phenotypic averages of the male and female traits.

A second major goal is to analyze the stability properties of the equilibrium. Is the equilibrium stable and what is the rate of approach to or departure from the equilibrium? These questions can be answered by setting up a new system of dynamic equations which give the change in average character values measured as distance from the equilibrium. In order to determine whether the system is stable, one asks for the conditions under which the distance from the equilibrium will shrink each generation. The analysis consists of solving for the eigenvalues of the matrix of coefficients corresponding to the new set of linearized dynamic equations. The signs of the real parts of the eigenvalues indicate stability, their magnitudes indicate rates of approach to or departure from equilibrium. The corresponding eigenvectors give the direction of evolution in the vicinity of the equilibrium. Roughgarten (1979) gives a readable, elementary account of stability analysis.

## The evolution of sexual dimorphism

Lande's (1980a) model is in some ways the most general of the quantitative genetic models of sexual selection. The model is multivariate and features an arbitrary number of male and female characters. The focus of the model is on the joint evolution of homologous male and female characters. In contrast to later models, however, the sexual selection process is specified only in general terms. Mate choice and its impact on sexual success are not specified. Consequently the model does not illustrate evolutionary feedback. It does provide a general perspective on the evolution of sexual dimorphisms.

The main actors in the model are an arbitrary number of characters that are sexually homologous in the sense that they may be expressed to varying degrees in both sexes. An arbitrary pattern of genetic correlation is allowed both within the sexes and between the sexes. Natural selection is allowed to act on any or all of the characters. In addition the model specifies sexual selection on a subset of the male characters. A form of sexual selection is assumed such that the advantage of a phenotype in mate choice (or in combat) is a function of its deviation from the phenotypic mean. Such relative sexual selection yields a constant force of sexual selection from one generation to the next.

Several maladaptive features of sexual selection are revealed by the model. At evolutionary equilibrium the opposing forces of natural and sexual selection on male secondary sexual characters perfectly balance. Lande shows that this balance point corresponds to a site downhill from an adaptive peak (see also Lande, 1976, 1979). Thus mean fitness in the population is lowered by sexual selection (Fig. 3). The magnitude of maladaptation at equilibrium can be measured as the vertical distance to the peak of the adaptive landscape from the equilibrium point. The distance corresponds to a loss in population mean fitness, a *sexual selection load*. In principle, strong sexual selection could impose such a large load that local populations could be vulnerable to extinction. Whether sexual selection actually promotes extinction in natural populations is an outstanding empirical issue.

Sexual selection can also cause maladaptation in ordinary male characters that are not sexually selected. Sexually selected characters will generally deviate from their adaptive optimum at equilibrium. That maladaptation in the sexually selected characters can be translated into a maladaptation of the ordinary characters if the two sets of characters interact in their effects on fitness. As a hypothetical example, let the peacock's tail represent the sexually selected characters and let wing length represent the ordinary character. (Actually the peacock's wing may be sexually selected, since it is vibrated to produce sound during the sexual tail display, but let us suppose that the wing is only exposed to natural selection). We imagine that the tail is maintained at its extraordinary size by a balance between sexual selection favoring even longer tails and natural selection favoring smaller tails. Because the wing and tail work together during such aerodynamic activities as takeoff, flight and landing, the wing and tail sizes probably interact in determining non-sexual aspects of fitness. Differently put, there may be an aerodynamic premium on a particular ratio of wing and tail sizes, a premium that could be measured by computing mesures of correlational selection for the wing and tail (correlational selection acts directly on trait combinations and consequently changes the covariance between traits). Lande (1980a, eq. 14) shows that maladaptation in the peacock's tail will

Female character (Ȳ)     Female character (Ȳ)

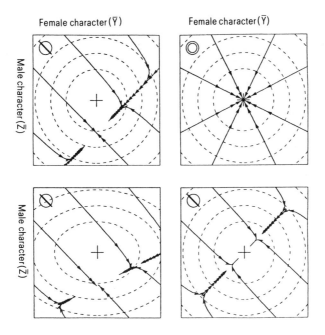

Figure 3. Lande's (1980a) model for the joint evolution of sexually homologous characters. Solid lines with arrowheads show the evolution of the population averages of the male and female characters. Dashed lines indicate fitness contours of an adaptive topography determined by viability. The adaptive peak (cross) is situated at the center of each figure. The solid circle in the lower left-hand corner of each figure has a radius of one phenotypic standard deviation. The inner ellipse indicates the pattern of additive genetic variation for the characters: a circle indicates no genetic correlation between the sexes, a narrow ellipse indicates a high genetic correlation between the sexes. The evolutionary trajectories were computed assuming a heritability of 0.5 for each of the two characters. Arrowheads are shown at intervals of 200 generations, except near the equilibria, and the total elapsed time for each trajectory is 2000 generations. *a* Upper left. With no sexual selection and no genetic correlation between the sexes, populations evolve directly towards the adaptive peak. *b* Upper right. With no sexual selection and high genetic correlation between the sexes, populations make a very slow final approach to the adaptive peak. Sexual dimorphism evolves during this slow phase. *c* Lower left. With sexual selection and high genetic correlation between the sexes, the population equilibrates off the adaptive peak, in the direction favored by sexual selection. A constant force of sexual selection was assumed in computing the trajectories (a sexual selection gradient of one phenotypic standard deviation per generation). *d* Lower right. Same conventions as in the previous figure but with natural (viability) selection on females half as strong as on males. Reproduced with permission from Lande (1980a).

induce maladaptation in any interacting body part, such as the wing. The induced maladaptation in the wing may be permanent. Temporary maladaptation in other characters may arise due to genetic correlation with sexually selected characters, even in the absence of correlational selection.

Temporary but long lasting effects may be exerted on genetically correlated male characters.

Consider next the nonobvious consequences on female traits. Sexual selection on the male will also cause temporary maladaptation in the female. The effect arises because pleiotropic gene effects tend to be similar in the two sexes, causing high genetic correlations between homologous traits in males and females (Lande, 1980a; Lande and Arnold, 1985). Thus, because of a large genetic correlation between male and female body size in *Drosophila*, we will increase female body size even if we only select larger males each generation, while choosing female parents at random.

Genetic correlations induce correlated responses to selection in unselected characters (Falconer, 1981, chap. 19). In Lande's (1980a) model, with natural selection on both male and female characters, but sexual selection on only the male character, the genetic correlation between the sexes causes a correlated evolutionary response in the female to sexual selection on the male. The net consequence of this correlated response to sexual selection, together with the direct response to natural selection on the female, is a curved evolutionary trajectory (Fig. 3). The evolving female population can actually evolve away from its optimal phenotype as it follows a curved trajectory (e. g., Fig. 3c). Departure from the optimum or adaptive peak is only temporary, but the final approach to the optimum may require hundreds of thousands of generations. The reason for this slow approach is the necessary antagonism between genetic correlation between the sexes and genetic variance for sex dimorphism. When the genetic corelation between the sexes is high (as is usually the case) and genetic variances are comparable in the two sexes, then genetic variance for sex dimorphism is vanishingly small. Because of this antagonism, the evolution of the average of male and female traits (toward upper right and lower left in Fig. 3c) is much faster than the evolution of sex dimorphisms (toward upper left and lower right in Fig. 3c). Thus Lande's results confirm Fisher's (1930, p. 157) argument that the evolution of sexual differences will be prolonged compared with the evolution of the sexual average and will depend on the slow build-up of genes with sex limitation to their expression.

Slatkin (1978) presents quantitative genetic models for sexual dimorphism evolving in response to ecological pressures rather than in response to sexual selection. Slatkin makes the important point that a genetic correlation between the sexes in a homologous trait can affect the evolutionary outcome, as well as the trajectory, if there are two or more stable equilibria.

### Fisher's runaway process

Lande's (1981) paper is a formal model of Fisher's (1915, 1930) account of a positive feedback process of sexual selection, although Lande discovered features apparently unappreciated by Fisher. The model describes the joint evolution of two traits: a male secondary sexual trait and female mating preferences based on that male trait. The male trait is subject to both natural and sexual selection and is not expressed in females. Thus the male trait affects both male survivorship (viability) and mating success. In contrast, the fitness of the female is not affected by her mate choice: even females with the most extreme mating preferences are eventually mated and there is no penalty for any incurred delays. In this sense mating preferences are selectively neutral. Maynard Smith (1982) and Arnold (1983) give explications of the model.

The main results of the model are that a variety of evolutionary outcomes are possible and the equilibrium may be stable or unstable. Remarkably, the qualitative results are unaffected by mode of female mating preference. The unstable case corresponds to the runaway process anticipated by Fisher (1930). In this case, the population evolves away from a line of unstable equilibria at ever increasing speed. The population may evolve towards elaborated or diminished male character depending on its initial condition (Fig. 4a). Instability or the triggering of such a runaway process is promoted by a large genetic covariance between the sexes, weak natural selection on the male character and strong mating preference (narrow curves in Figs. 2a and 2b). Relative preferences are more prone to trigger the runaway than absolute mating preferences.

The stable case (Fig. 4b) was not described by Fisher and is promoted by the converse of the conditions just listed: small genetic correlation between the sexes, strong natural selection, weak mating preferences and absolute rather than relative mate choice. In the stable case the population evolves towards any of a large number of possible stable combinations of average male character and female mate preference. The set of stable combinations forms a line, a so-called line of neutrally stable equilibria. The existence of a line of equilibria can be appreciated from the fact that sexual selection must balance natural selection on the male trait at equilibrium and for every strength of natural selection there is conceivably a strong force of mate preference that will exactly counterbalance it. Thus the force of natural selection increases as the population moves away from the optimum (dashed vertical line), say, to the right in Fig. 4b. A stronger force of sexual selection is required to balance an increasingly strong force of natural selection and so the line of equilibria slopes upward, towards more extreme female preference.

Populations that should drift off of equilibrium will be driven back towards it in the stable case. In contrast, populations are free to drift along the line, unopposed by selection (hence the line is neutrally stable). Lande (1981, eq. 20) has worked out the conditions that will promote diversification by sampling drift of populations in the vicinity of the line of equilibria. Diversification in male traits by drift per se will be promoted by small effective population size, strong female mating preferences, high heritability of the preferences and weak natural selection on the male trait. Genetic vari-

294

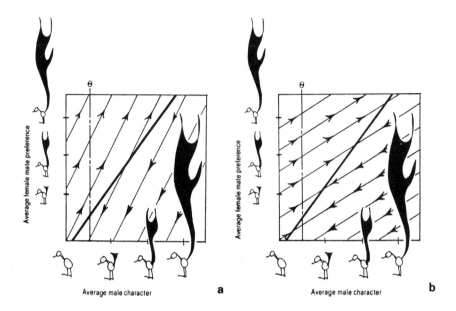

Figure 4. Lande's (1981) model for the joint evolution of female mating preference and a sexually selected male character that is not expressed in females. The tail size of males most preferred as mates is shown on the vertical axis. Thus each male caricature on that axis corresponds to the average most preferred mate of a female population with absolute mating preferences. Evolutionary trajectories are shown with solid lines with arrowheads at arbitrary intervals. The heavy solid lines indicate lines of equilibria. The average tail size in a male population that is at the viability optimum is specified by the vertical dashed line. *a* Runaway sexual selection occurs when the slope of the evolutionary trajectories exceeds the slope of the line of equilibria. In this case populations evolve away from the line of equilibria at ever increasing speed. *b* Stable outcome to sexual selection occurs when the slope of the evolutionary trajectories is less than the slope of the equilibrium line. Populations decelerate as they approach the line. Reproduced with permission from Harvey and Arnold (1982).

ance for mating preferences and small effective population size will promote differentiation in female mating preferences. In addition, drift can promote population diversity by interacting with selection. Thus sister populations that drift to opposite sides of the stable line of equilibrium (Fig. 4b) will be moved back towards the equilibrium line in opposite directions.

## Joint evolution of mate choice and sexual dimorphism

Lande and Arnold (1985) treat the simultaneous evolution of three characters: a male secondary sexual trait, the homologous female trait and female
mate choice based on the male character. The model thus combines
features of Lande (1980a) which treated the evolution of homologous traits
in males and females and Lande (1981), which modeled the joint evolution
of a sex-limited male trait and female choice based on that trait. Primary
motivations for the three-charcter model were to see whether expression in
females of the male sexually selected trait might collapse the line of equilibria (Lande, 1981) to a point and to see whether bisexual expression might
impede or eliminate the runaway process. Surprisingly, female expression
of the male trait had relatively little effect on evolutionary dynamics and the
feature of indeterminant equilibrium was retained.

Both the male and female (homologous) traits are under stabilizing
natural selection, but a sexual difference in optima is allowed. Thus natural
selection could be more stringent on females than on males, for example.

At equilibrium the female trait is found to have evolved to its optimum
specified by natural selection. The possible combinations of average male
trait and female preference form a line at equilibrium that is identical to the
equilibrium line in Lande's (1981) model for sexual selection on a sex-
limited male trait (Fig. 5). Thus expression of the sexually selected trait in
the female sex does not affect the outcome of evolution.

Both stable and unstable equilibria are possible, just as in the simpler
two-character case. Furthermore, the conditions for stability are identical
in the two- and three-character models (assuming in the three-character
model that there is no pleiotropy between female preferences and the
homologous traits in either sex). Thus natural selection on the homologous
female character does not make the conditions for unstability more stringent or affect those conditions in any way. In the unstable case, the
homologous female trait, as well as the sexually selected male trait, evolves
at ever increasing speed.

The feature of a rapid phase of average character evolution followed by a
gradual evolution of sexual dimorphism is present in the three-character
case just as it was in the model of sexually homologous traits (Lande,
1980a). Thus Figure 3c and Figure 4 can be taken as two-dimensional pro-
jections of evolutionary trajectories occurring in three dimensions.

## Onset and cessation of the runaway

The onset of the runaway process and its cessation are not explicitly modeled in Lande (1981, 1982), Lande and Arnold (1985), Kiester et al. (1984),
or Kirkpatrick (1985, 1986). The initial stage in the evolution of female sexual preference is problematic because when novel preferences are rare they

296

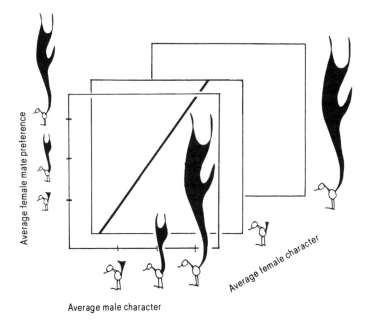

Average female mate preference

Average male character

Average female character

Figure 5. The equilibrium set for female mating preferences, a sexually selected male character and the homologous female character. Absolute female mating preferences and other conventions are the same as in Fig. 4. The line of equilibria (heavy diagonal line) lies on a plane at which female tail size is at a viability optimum.

will produce a trivial force of sexual selection. Fisher (1915, 1930) proposed that a novel preference for some male feature (e. g., tail size) might yield more viable progeny for the female and so produce an indirect force favoring evolution of the preference. Heisler (1984, 1985) has modeled this critical early stage in the evolution of sexual preference.

Heisler (1985) shows that an indirect force arises if the preference induces a genetic covariance between preference for a trait and viability. Thus the expected gain in progeny viability from choosing a particular male phenotype is proportional to a genetic covariance between preference and viability. This genetic covariance in turn depends on the genetic covariance between the male trait and viability. Consequently if female preference is based on multiple traits, the optimal weight given to each trait is proportional to its genetic covariance with viability. Thus the indirect force favoring the early stage of preference evolution depends on whether the preferred traits are genetical indicators of viability [so-called 'good genes' – Ed.]

The early evolution of sexual preference will also depend on an initial departure from equilibrium in the population, so that one or more male trait is favored by directional viability selection. Such a circumstance would arise if the environment suddenly changed so that the average values of one or more male trait are some distance from the viability optimum. In this circumstance novel preferences could evolve as a correlated response to viability selection on the male traits. The particular form of multivariate preference that is favored will depend on the genetic covariances just described as well as on genetic correlations among male traits.

Consequently the rank order of weightings used by females in choosing males may differ considerably from the rank order of viability selection on the traits. When the novel sexual preferences have increased in frequency, they may then produce a force of sexual selection on the male traits and trigger the runaway process or evolution to a stable equilibrium.

Another possibility for a trigger is that alleles causing selectively neutral mating preferences (those that do not affect the female's progeny count) might drift upward from mutation frequency and then start the runaway process (Kirkpatrick, 1982; Lande, 1981). Pleiotropy provides yet another possible scenario for origination of preferences. Genes affecting mating preference for one character are likely to have pleiotropic effects so that mating is biased with respect to other male characters as well. Once genes affecting preference for one trait have increased in frequency, pleiotropic effects may become the vehicle for a new epoch of sexual selection on another character. Thus pleiotropy could lead to a continuing series of sexual selection episodes.

Fisher (1930) argued that the runaway process might stop either because the mating preferences of females would become so extreme that they would fail to find mates or because of the intervention of strong counterselection against extreme males. In the latter regard Lande (1981) has pointed out that in order to stop the runaway, viability must fall off faster than a Gaussian curve as males depart from the viability optimum. In addition, genetic parameters might change during the runaway so that the population reverts to a region of genetic stability (Kirkpatrick, 1985, 1986).

## A fecundity-indicator model

Kirkpatrick's (1985) model is designed to test the logic of Weatherhead and Robertson's (1981) 'sexy son' hypothesis. The hypothesis applies to species in which mate choice is based upon the care which is given to mates or offspring. Weatherhead and Robertson supposed that females might evolve preferences for attractive males even though such males provide inferior care of offspring. The losses in the current generation would be offset in the next generation because the sexy sons of the attractive males would have many offspring. Kirkpatrick showed that such 'deficit financing' of fitness does not work.

Kirkpatrick adapted Lande's (1981) model to the problem by letting the male character, which is the focus of female choice, affect the fecundity of the male's mates as well as his viability and mating success. Thus the male trait plays the additional role of being a fecundity-indicator. While this additional component of fitness may seem a minor addition to the assumptions of the original model, it considerably complicated the mathematics. In Kirkpatrick's model selection acts directly on female mating preferences because female fecundity is affected by choice of mate.

Kirkpatrick also explored two possible relationships between the mating success and fecundity components of male fitness. In the first possibility, termed *unlimited male reproductive potential,* the average fecundity of mates is unaffected by the male's mating success. Thus in scorpionflies or other species in which males offer their mates nuptial feeding gifts (Thornhill, 1981; Thornhill and Alcock, 1983), a male's past sexual successes are unlikely to affect the size of future nuptial gifts and hence the fecundity of his mates. (In the case of unlimited male reproductive potential, Kirkpatrick assumed an intermediate fecundity optimum for male care or protection, with Gaussian stabilizing selection). In contrast, in polygynous species in which mates reside on the males territory, it is conceivable that average fecundity is smaller in larger harems. Kirkpatrick termed this possibility *limited male reproductive potential* and modeled it by letting average male fecundity fall off as a negative exponential function of male mating success.

The mode of interaction between mating success and male fecundity had a major impact on the type of evolutionary equilibrium. In contrast, different modes of female choice produced qualitatively similar evolutionary results. Kirkpatrick used the three types of mate choice devised by Lande (1981): relative, absolute, and psychophysical.

Selection on mating preferences had a dramatic effect on the evolutionary equilibrium. Instead of a line of equilibria, as in Lande's (1981) model, the equilibrium is a point, a unique combination of average male trait and average female preference (Fig. 6). As in Lande, however, the equilibrium may be stable or unstable, with strong genetic coupling between the male trait and female preference promoting instability. Thus a runaway process is possible even when preference is based on an attribute that directly affects female fecundity.

The location of the equilibrium is affected by the existence of a tradeoff between male mating success and male fecundity. When there is no tradeoff (unlimited male reproductive potential), the male trait equilibrates at a level that maximizes mate fecundity. In the case of an evolving nuptial gift, the gift in the equilibrium population would maximize female fecundity, even if such a gift were deleterious to male survivorship. Females in the equilibrium population would prefer males with extreme gifts, but because this force of sexual selection exactly balances the opposing force of viability selection, which favors less extreme gifts (Fig. 6) the equilibrium gift-giving level is maladaptive for males, in the sense that it decreases their survival.

In contrast, when there is a tradeoff between male mating success and male fecundity, the population equilibrates at the male trait level that maximizes male survivorship. Female preference for this male trait value also

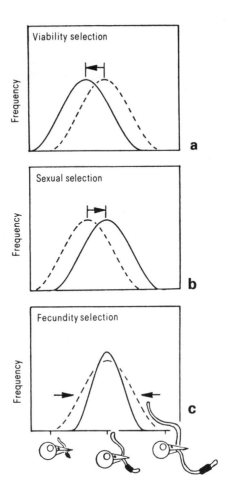

Figure 6. The equilibrium in Kirkpatrick's (1985) fecundity-indicator model when male reproductive potential is unlimited. Three episodes of selection act on paternal performance. The figures show the selection patterns when the population has evolved to an equilibrium. *a* High paternal performance lowers the male's survivorship (e. g., the larger the worms the male tends to pursue, the more vulnerable he is to predation during prey capture). The effect of viability selection is to shift the distribution of paternal performance towards lower values. *b* Females preferentially mate with males inclined to show high paternal performance. At equilibrium this directional force exactly balances the opposing force of viability selection. *c* Males with intermediate paternal performance raise larger families with each mate (e. g. small worms yield inferior growth but so do unwieldy large worms). In equilibrium populations average paternal performance lies at an intermediate fecundity optimum.

maximizes female fecundity. In the equilibrium population the average male will offer the least risky territory size or level of care and the average female will most prefer to mate with such an average male, and so will maximize her fecundity. Thus a tradeoff between mating success and male fecundity dramatically changes the nature of equilibrium populations. Remarkably, it is the fact of tradeoff and not the magnitude of tradeoff that matters. No matter how small or large the tradeoff, the population will equilibrate at the same trait values.

Populations do not equilibrate at a level of male care or protection that is deleterious to females. Thus the particular maladaptive outcome predicted by Weatherhead and Robertson (1981) is contradicted by Kirkpatrick's (1985) results. Sexy sons in the next generation will not compensate for low fecundity arising from poor mate choice in the present generation. Thus the 'sexy son' fails because it violates an important principle in evolutionary biology: at equilibrium the forces acting on genes and phenotypes must balance within each generation (Kirkpatrick, 1985). [Could this follow simply from the single-generation definition of fitness as progeny zygote count? – Ed.]

Weatherhead and Robertson (1981) and Heisler (1981) attempted to bolster the 'sexy son' argument by a 'good genes' argument in which fitness is tabulated by counting grandchildren. Neither paper, however, modeled the inheritance of mating preference and the sexually selected trait and their joint evolution. Thus the contradiction of the 'sexy son' proposal by Kirkpatrick's explicit genetic model is a severe blow to the unbridled use of 'good genes' logic.

## Viability-indicator models

Zahavi (1975) proposed a novel explanation of sexual selection processes. Zahavi embraced Darwin's (1874) and Fisher's (1930) view that male secondary sexual characters must often be deleterious for male survival and termed these characters *handicaps*. The central problem addressed by Zahavi is the adaptive significance of female mate preference. Rather than accept the view that mating preferences simply evolve as a corollary of the sexual evolution they cause, Zahavi sought an adaptive explanation for mate preference. He proposed that mating preferences based on elaborate characters, handicaps, evolve because such characters are indicators that males have passed a viability test. "Females which select males with the most developed characters can be sure that they have selected from among the best genotypes of the male population" (Zahavi, 1975).

Maynard Smith (1976c) argued that Zahavi's mechanism was unlikely to work because the sons of discriminating females inherit the handicap as well as the genes that promote viability. Maynard Smith built a three locus model to test Zahavi's proposition and concluded that the handicap as an indicator of male quality cannot cause mating preferences to evolve, even when the handicap is expressed only in males. Nevertheless, adaptive explanations for the evolution of mating preferences have much intuitive appeal and a series of authors have continued to explore modifications of Zahavi's proposal (Andersson, 1982; Bell, 1978b; Dominey, 1983; Eshel, 1978; Kodric-Brown and Brown, 1984; Nur and Hasson, 1984; Thornhill and Alcock, 1983; Zahavi, 1977). General conclusions from the three locus models are that the mechanism cannot increase the frequency of preferences when they are rare and that the mechanism works only in conjunction with a Fisherian (pure sexual selection) process (Maynard Smith, 1985). Recent work with three locus models indicates that the mechanism may, under special circumstances, enhance evolutionary rates (Andersson, 1986a, 1986b).

What exactly is meant by the 'handicap mechanism'? The unique feature is *not* the notion of a handicap, since hardly anyone doubts that some sexually selected traits, such as the peacock's tail, actually expose the male to predatory risk. The key feature in Zahavi's proposal is that the handicap is an indicator of male genetic quality. Thus 'indicator mechanism' is a more revealing label. The indicator aspect of the handicap is supposed to play an important role in the evolution of mating preferences. Following Andersson (1986b) we can ask, does the presence of a quality trait, for which the handicap is a marker or indicator, affect the equilibrium, its stability properties, or the approach to equilibrium? These questions have been pursued by two similar quantitative genetic models of the indicator mechanism (Heisler, 1985; Kirkpatrick 1986).

Kirkpatrick (1986) addressed these issues by modeling the joint evolution of three quantitative traits: (1) a male secondary sexual trait (the viability-indicator), (2) a viability trait (overall quality) expressed in both sexes and (3) a female mating preference based on the indicator. To allow the male trait to be an indicator of genetic quality, the model assumes a correlation between the male trait and the viability trait. Normally in thinking about the indicator mechanism, the correlation between the male trait and the viability trait would be positive and genetic, but the model is more general, allowing a correlation of arbitrary sign or magnitude that might be partly or wholly nongenetic.

The model assumes stabilizing viability selection on the male trait towards an intermediate optimum. Sexual selection then acts via female mate choice among the survivors, using the indicator as a criterion. The model is developed using absolute mate preferences, as described by Lande (1981), but relative and psychophysical choice functions gave qualitatively the same conclusions. As in previous models, there is no selection on female mating preferences.

The model allows two forms of selection on the viability trait. In the first case, there is stabilizing selection towards an intermediate optimum. The viability trait in this instance might be overall running or flying ability. In this case, the model allows the possibility of correlational selection on the handicap and viability trait. Thus there may be a premium on particular trait combinations, such that the subset of surviving males shows an increased correlation between handicap and viability trait. In the second case, there is no intermediate optimum for the viability trait and fitness increases monotonically with trait value. This case corresponds to equating the viability trait with the viability component of total fitness or to some feature, such as parasite resistance, for which selection is always directional. The mode of selection on the viability trait has a major impact on the nature of the equilibrium. We will first consider the equilibrium when an intermediate optimum exists for the viability trait.

A major conclusion from Kirkpatrick's (1986) model is that when mate choice is based on an indicator of male quality (viability), the mating preferences do not evolve to a unique equilibrium. Thus preference based on an indicator trait does not constrain the evolution to a single adaptive outcome of the sort found when mate choice directly affects female fecundity (Kirkpatrick, 1985). Instead Kirkpatrick (1986) finds a line of equilibrium: many different combinations of the indicator, preference and viability traits can occur at equilibrium. In this sense the equilibrium is the same as in other models with no selection on female preferences (Lande, 1981; Lande and Arnold, 1985). Furthermore, as in those models, the equilibrium may be stable or unstable, with strong genetic coupling between the handicap and mating preferences promoting a runaway process.

Contrary to the expectations of some proponents of the 'good genes' school of sexual selection, mate choice based on a viability-indicator does not lead to the evolution of maximally adapted males. Instead Kirkpatrick (1986) finds that males in equilibrium populations are pulled off of their adaptive, viability peak for the indicator by the force of sexual selection produced by female choice.

In the preceding discussion, the viability trait was assumed to be under stabilizing selection and the model specifies an equilibrium. Alternatively, when fitness is a continually increasing function of the viability trait, no equilibrium exists. Instead, the population continues to evolve so long as there is genetic variance for the viability trait, which is now equivalent to a fitness component. Kodrick-Brown and Brown (1984) implicitly invoked such a non-equilibrium model when they proposed that mate choice might be based on the ideal indicator of male quality. As in Kirkpatrick's model, their Figure 1, which graphs genetic quality of the male (or, more precisely, his breeding value for the viability component of fitness) as a linear function of the indicator trait, specifies a perpetually evolving, nonequilibrium system.

Kirkpatrick (1986) explores the nature of equilibrium when such a system exhausts genetic variance for the viability trait. As before, the equilibrium is a line with innumerable possible combinations of indicator, viability and preference trait.

Kirkpatrick (1986) also used his model to explore some variations of the original indicator mechanism and to evaluate its effects on the qualitative aspects of the equilibrium. Thus one can ask whether the indicator mechanism would work in the absence of sexual selection (Bell, 1978b). The equilibrium is unaffected. With mate choice but no sexual selection the equilibrium is the same as with natural selection alone. Dominey (1983) proposed that the indicator mechanism would work if variation in the indicator were purely environmental, but again Kirkpatrick finds that the viability trait equilibrates as it would under natural selection alone. Finally, Zahavi (1977) proposed that his process might work if the handicap (indicator) was expressed in both sexes. Expression in both sexes seems the worst case for the process, because both sexes incur a liability (Maynard Smith, 1976c). Indeed, Kirkpatrick found that an indicator expressed in both sexes yields a line of equilibria as before.

Although the indicator mechanism (and its various modifications) give a line of equilibria just as in Lande's (1981) model of the Fisherian process, we can ask whether the indicator causes a qualitative change in the equilibria and the conditions for stability. Indeed, the line of equilibria for preference and the male trait is quantitatively affected by the indicator mechanism. The effect is due to correlational selection on the indicator and viability traits and to phenotypic correlation between preference and viability traits in females. Thus, as Kirkpatrick (1986) notes, when there is no correlational selection between male indicator and viability traits, the equilibria is the same as in Lande's (1981) model. Further comparison of the equilibria indicates that the indicator mechanism will yield a steeper equilibrium slope than the pure Fisherian process whenever correlational selection on the male indicator and viability traits is positive and greater than the phenotypic correlation between preference and viability traits in females. Under these conditions it appears that the indicator mechanism has the effect of making the equilibrium more stable and so lessens the possibility of a runaway.

Additional insight on the effects of an indicator trait on the Fisherian process can be gained from Heisler's (1985) results. Heisler modeled a special case of the indicator situations considered by Kirkpatrick (1986). In Heisler's (1985) model there is only sexual selection on the indicator, the viability trait is under Gaussian selection, and there is no correlational selection on the male indicator and viability traits. As in Kirkpatrick's model, Heisler finds a line of equilibrium. She goes on to analyze stability conditions and finds that with psychophysical and relative preferences a runaway process seems inevitable. Comparing Heisler's conditions for stability with Lande's (1981) conditions, it appears that within these two modes of preference, an indicator trait promotes instability of the Fisherian process and hence the runaway. Likewise with absolute preferences, the indicator trait promotes instability of the Fisherian process whenever the indicator trait has greater genetic variance than the viability trait. Viability selection on both sexes (rather than just on males) promotes stability.

The discovery of lines of equilibria when mate choice is based on a quality indicator led Kirkpatrick (1986) to re-examine the haploid three-locus models that had been purported to support the indicator mechanism by Bell (1978b) and Andersson (1982). He found that those genetic models gave the same qualitative results as the quantitative genetic models: evolution of the viability trait to an ecological optimum, the evolution of preference and handicap dependent on initial conditions and towards a line of equilibrium.

In summary, quantitative genetic models for the indicator mechanism, working in conjunction with the Fisherian process, indicate that it results in a line of equilibria rather than a unique evolutionary outcome. The indeterminacy of outcome is a consequence of no direct selection on female preferences. A second major conclusion is that an indicator trait may enhance or decrease the stability expected under a pure Fisherian process. Additional analytical and numerical work is needed to contrast evolutionary rates and stability conditions when sexually selected traits are and are not indicators of overall viability.

In light of the need for additional work on trajectories and stability, it is difficult to contrast the results of the quantitative genetic and three-locus models of the indicator mechanism. It does appear that the indicator mechanism can speed up the sexual selection process under special conditions in both modeling realms (Andersson, 1986a, 1986b – but see Maynard Smith, 1985). Whether those conditions prevail in nature is an outstanding empirical issue.

**Sexual selection in a cline**

Lande (1982) made the important contribution of building a sexual selection model for the process of clinal speciation. Fisher (1930, pp. 139–143) proposed that a gradient in environmental conditions would cause tension in a species continuously distributed along the gradient, triggering the evolution of divergent mating preferences and eventually speciation. Mayr (1963, p. 525) objected to Fisher's model of 'semigeographic speciation' on the grounds that migration and genetic cohesion would disrupt the process and argued that there was no possibility that the mechanism would work. Lande's (1982) formulation of Fisher's suggestion refutes Mayr's contention and shows that clinal or semigeographic speciation is a real possibility. Furthermore, the model builds a bridge between discussions of sexual selection and speciation, closely related topics that have often been artificially separated (see also West-Eberhard, 1983).

The issue of whether clinal speciation actually happens in nature cannot, of course, be resolved by a theoretical model. Lande's model does define the conditions for clinal speciation and these could be useful in designing field observations that might detect the process in nature.

Lande's (1982) model describes the joint evolution of geographic variation in a male secondary sexual trait and female mating preferences based on that trait. The model is a more general vision of Lande (1981), incorporating the same assumptions but adding a spatial dimension. In the earlier model, female preferences evolved as a correlated genetic response to sexual selection on the male trait; here geographic variation in preferences evolves as a genetic consequence of evolving geographic variation in the male trait.

Lande's model describes the outcome of the following scenario. Suppose that a continuously distributed species is suddenly exposed to a new spatial pattern in selection imposed by changing environmental conditions. Suppose too that the changed conditions affect natural (viability) selection on a male secondary sexual trait. If heritable female mating preferences for that trait exist, how will the male trait evolve and what new kind of equilibrium clinal variation in the male trait will be established? In particular could very sharp clines develop, corresponding to the initial stage of speciation?

Alternatively, the model gives the equilibrium clines that would evolve after a species suddenly invades a region with some new spatial pattern in viability selection. In the model the male trait is under spatially varying stabilizing selection: the optimum varies in space, the width of the selection function is a constant. Whether the environment changes suddenly or the species invades a new region rapidly, the initial condition for the species is no geographic variation in the male trait or in mating preference, with uniform population density and with migration. These conditions are the least favorable for the evolution of sharp clines and speciation because there are no initial differences between populations (e. g., no premating or postmating isolation mechanisms) and no barriers to gene flow. Clinal speciation demonstrated under these conditions might prevail under much more general circumstances in nature.

Thus the idea of the model is to impose a particular new spatial pattern of selection on the male trait and investigate the subsequent evolution of geographic variation in the trait. The model permits exploration of a large number of spatial patterns in selection, but all of them are one-dimensional. Ecologists or naturalists familiar with the ubiquity of multi-dimensional gradients in environmental variables should not be dismayed by the one-dimensional feature of the model. One-dimensional gradients are mathematically the most tractable. The results can be mapped directly onto species occupying linear habitats (e. g., stream courses), but they also instruct our expectations for more complex environmental gradients.

A major result of the model is that sexual selection exaggerates geographic differences in the male trait. Thus sexual selection amplifies the steepness of equilibrium clines (Fig. 7). Both genetic variance in mating preferences and the strength (stereotypy) of those preferences tend to amplify clines. Migration and strong stabilizing natural selection at each location along the cline oppose amplification and act to diminish the steepness of the equilibrium cline.

The amplification result extends the general conclusion that sexual selection has a tremendous potential to create population diversity. Earlier mod-

306

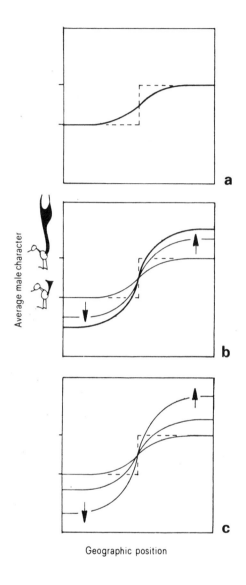

Figure 7. Lande's (1982) model for stable and unstable clines in a male character. A sudden change in the environment favors a small tail size in one geographic region and a larger tail size in an adjoining region. The geographic change in optimum tails size favored by viability selection is shown by the dashed line. *a* In the absence of sexual selection, a stable cline forms (heavy line). *b* Sexual selection amplifies the character cline and may yield a new stable cline (heavy line). *c* Alternatively, a runaway sexual selection process may occur simultaneously across the two geographic regions, with an ever-increasing exaggeration of the original cline.

els stressed diversification arising from indeterminate equilibria and initial conditions, sampling drift interacting with selection in populations at or near equilibrium, slow approach to equilibrium caused by genetic coupling and, alternatively, runaway processes. To this catalog of diversifying mechanisms we can now add exaggeration of clinal variation, which can occur even without any initial, phenotypic differences between populations.

In contrast to their tendency to exaggerate the height of clines, strong and genetically variable preferences tend to lengthen the geographic width of the cline. In Slatkin's (1984) terminology, they increase the characteristic length of the cline. At distances much shorter than the characteristic length (a distance considerably greater than individual dispersal distances), the population fails to track spatial differences in selection and no clines evolve. Ironically, strong, variable preferences promote tracking failure over short geographic distances. A second nonobvious finding is that the genetic variance and strength of mating preferences act to retard the approach to equilibrium. Again the effect as exerted through the relationship with characteristic length.

Another major result is that runaway sexual selection can occur simultaneously over a broad geographic area. As in previous models, equilibria may be stable or unstable. In the unstable case the cline evolves exponentially so that the height of the cline becomes increasingly exaggerated (Fig.7c). The condition for stability is identical to the case without geographic variation (Lande, 1981): genetic variance in preferences (acting through the genetic coupling between preference and sexually selected trait) must exceed a critical value that increases with the strength of natural selection and decreases with stereotypy of preference.

The amplification result suggests that premating isolation could evolve rapidly along a cline that straddles an environmental discontinuity, just as Fisher (1930) argued. Although Lande's model treats only the stages of speciation before coexistence, it does not confirm the popular notion of character displacement. Lande found that equilibrium clines were always monotonic and did not show greater character divergence in parapatry than in allopatry.

## Coevolution of plants and their pollinators

Kiester et al. (1984) present a sexual selection model for the coevolution of plants and pollinators. In many species of euglossine bees, for example, males collect fragrances from orchid flowers and apparently use the fragrances to attract mates (Dodson, 1975; Dressler, 1968, 1981, 1982; Schemske and Lande, 1985). The intent of the model, as applied to such a system, is to trace the evolution of three traits: (1) amount (or mixture) of some chemical fragrance produced by the orchid, (2) preference of male bees for the orchid fragrance and (3) mating preferences of female bees based on the fragrances collected and carried by male bees. The model is interesting from a number of viewpoints, but for the present discussion perhaps the most relevant focus is on how mate preferences based on attributes derived from a second coevolving species affect evolution by sexual selection.

Aside from the plant character, the only major departure from Lande's (1981) model is that the viability of the male is determined by the relative number of plants that he visits rather than by a Gaussian natural selection function. Two kinds of pollinator preference are modeled. Pollinators with *absolute* preferences most prefer a certain value of the plant character and preference falls off as a Gaussian curve in either direction from that most preferred value. 'Preference' is meant to cover a whole spectrum of plant-pollinator interactions. Thus, tongue length of the pollinator can formally be modeled as absolute pollinator preference because the pollinator will have most success with flowers whose corolla length matches his tongue length. In contrast, pollinators with *relative* preferences most prefer a value of the plant character that lies some characteristic distance from the plant population mean. This preference mode implies that the pollinator surveys the plant population and perhaps forms a search image for plants with, say, greater than average concentrations of fragrance.

The second force acting on male pollinators is sexual selection. Females are assumed to base their mate choice on the average of phenotype of plants visited by the male. In the orchid bee example, mate choice would be based on the average fragrance of the orchids successfully visited by an individual male bee. Kiester et al. (1984) employ an open-ended or psychophysical model of mating preference in which the tendency of a particular female to mate with a male is an exponentially increasing function of her characteristic ability to discriminate among males and the average of plant phenotypes (e. g., fragrances) visited by that male.

Female mating preference is assumed to be selectively neutral, as in some previous models (Lande, 1981). The assumption seems plausible for euglossine bees in which males neither protect nor provision their mates or offspring.

The model predicts both stable and unstable evolutionary outcomes. As in previous models, stability depends on the amount of heritable variation in female mating preference. That variation exerts its destabilizing influence through the genetic coupling between mate preference and sexually selected character. Strong effects of plants on the viability of male pollinators increase stability; large phenotypic variance in the plant character (e. g., fragrance) promotes instability. In the unstable case, the plant character, pollinator preferences and mating preferences evolve at ever increasing speed, at least initially. In the stable case, the female pollinator population tends to lose net mating preference so that at equilibrium there is no average mate preference exerted in the population. This unique female outcome contrasts with the variety of female outcomes that occur in the absence of a plant-derived male trait (Lande, 1981).

The equilibrium combination of plant and male pollinator attributes, however, is not unique. Instead an infinite number of outcomes is possible, the set forming a line of equilibria. With either absolute or relative pollinator preferences, at equilibrium the pollinators most prefer the average plant phenotype.

## Discussion

A major conclusion from the present survey is that quantitative genetic models often reveal critical features hidden or incorrectly evaluated in verbal accounts of sexual selection processes. Weatherhead and Robertson (1981) argued, for example, that females might evolve mating preferences for mates that provide inferior parental care, this liability being offset by the production of 'sexy sons'. The argument sounds plausible, at least to some ears. Nevertheless, a formal genetic model of the process contradicts their conclusions and shows that females will evolve preferences for mates with the best parenting abilities (Kirkpatrick, 1985). 'Good genes' arguments can lead to other erroneous conclusions. Following Zahavi's (1975) logic, one might expect the evolution of unique, optimal female mating preferences when mate choice is based on a trait that is an indicator of male genetic quality. Instead Heisler (1985) and Kirkpatrick (1986) find that mating preferences do not evolve to a unique endpoint: there is no optimal mate choice. Why do these two apparently plausible accounts of sexual selection lead to indefensible conclusions?

The key is found by focusing on whether mate choice yields immediate fitness effects. When the female's mate choice does not affect her progeny count, as in Zahavi's (1975) original proposal, there is no tendency for the evolution of unique, adaptive mate choice. When mate choice does affect the female's progeny count, as in the 'sexy son' model, females evolve unique, adaptive mate preferences.

In contrast, a sure route to erroneous conclusions is to treat the attributes of offspring as fitness currency and follow the logic of optimization used in 'good genes' arguments. If we argue that evolution will maximize the sexual success of sons or the quality of offspring, we are bound to go astray for two reasons. Offspring attributes are not a simple fitness currency. Unlike an individual's zygotic progeny count, the viability or mating success of progeny depends on inheritance from the individual and its mates. Verbal arguments that gloss over this inheritance issue or fail to account for genetic effects on evolution can easily lead to false conclusions. Secondly, sexual selection is by nature a frequency-dependent process and in general such processes do not result in fitness maximization (Lande, 1979; Wright, 1969). Maynard Smith (1982) has discussed both points in stressing the need to use evolutionarily stable strategies, or other approaches that do not depend on fitness maximization, to model evolution under frequency-

A summary of quantitative genetic models of evolution by sexual selection and their equilibrium properties

| Name of model | Reference | Characters in addition to the male trait ('handicap') | Selection in addition to selection on male trait | Dimensionality of equilibrium |
|---|---|---|---|---|
| Sexual dimorphism | Lande (1980a) | Homologous female trait | On female trait | Point |
| Mate choice | Lande (1981) | Female mate choice | None | Line |
| Sexual dimorphism and mate choice | Lande and Arnold (1983) | Homologous female trait, female mate choice | On female trait | Line |
| Sexy son | Kirkpatrick (1985) | Female mate choice | On female mate choice | Point |
| Handicap | Kirkpatrick (1986) | Male viability trait, female mate choice | On male viability trait | Line |
| Sexual selection in a cline | Lande (1982) | Female mate choice | None | Line |
| Coevolution of plants and pollinators | Kiester, Lande and Schemske (1984) | Plant trait, female mate choice | On plant | Line |

dependent selection. He also points out that explicit genetic models are needed to treat sexual selection processes because the evolutionary dynamics depend on genetic phenomena, such as the genetic coupling between male trait and mate preference induced by assortative mating. Thus, while verbal accounts of sexual selection processes may often be provocative, they are unlikely to be the last word. Formal genetic modeling will usually be necessary to test expectations and mitigate disputes.

Indeterminacy of equilibrium is a feature common to many of the models (Table). Thus whenever selection does not act directly on female mate choice, there is a line of equilibria, with many possible combinations of mating preference and sexually selected male traits as evolutionary outcomes. The same feature holds in two- and three-locus models (Kirkpatrick, 1982, 1986; Seger, 1985). The expectation that evolutionary outcome will depend on initial conditions is one of the major contributions of Lande's (1981) model and its descendants. Indeterminancy of outcome was alluded to by Fisher (1930) but largely ignored in later discussions of sexual selection. When multiple characters are the focus of mate choice, the variety of possibilities at equilibrium is enormous with the equilibrium set forming a hyperplane (Lande, 1981).

The family of sexual selection models illustrates a general rule for predicting the number of dimensions that constitute the equilibrium set. The dimensionality of the equilibrium equals the number of genetic degrees of freedom (the number of phenotypic characters with less than perfect genetic correlation) minus the number of independent selective constraints (Kirkpatrick, 1986; Lande and Arnold, 1985). Thus in the sexual dimorphism (Lande, 1980a) and 'sexy son' (Kirkpatrick, 1985) models with selection acting on both male and female characters, the equilibrium set is a point (dimensionality is zero). In the other models (Table), selection does not act on one of the characters (female mating preference) and so the equilibrium is a line (dimensionality is one).

The existence of stable and unstable evolutionary outcomes is another major result from the models. The unstable case is of particular interest, because it corresponds to the runaway process first described by Fisher (1930), with accelerating evolution of male traits and female preference. The conditions favoring instability or runaway processes are identical or extremely similar in all the quantitative genetic models. Thus genetic variance in female mating preferences promotes instability as does stereotypy of mate choice and weak natural selection on the male trait. In many of the models, genetic variance in mating preferences, expressed as a genetic regression of mating preference on male trait, must exceed a critical ratio of mate choice stereotypy to strength of viability selection on the male trait in order to trigger the runaway. Homologous female traits may also evolve at accelerating rates during the runaway due to genetic correlation with sexually selected male traits (Lande and Arnold, 1985). Furthermore the runaway may occur over a broad geographic area (Lande, 1982).

The need for empirical research in a number of areas is highlighted by the models. In particular we need new information on the nature of mate choice, on the inheritance of mate choice and sexually selected traits, and on the intensities of selective forces acting in nature. Estimates of critical variables in natural populations could tell us whether runaway sexual selection is plausible, whether mate choice acts on arbitrary male attributes and could enable us to explore many other issues.

Typology has prevailed in past studies of mate choice, with a focus on average behavior rather than on variation within populations. Consequently we have a poor understanding of how individual females react to the spectrum of male phenotypes in natural populations and how females vary in those reactions. Such phenotypic studies are especially important because they lay the groundwork for genetical studies of mate choice, a virtually unexplored field. Promising starts at characterizing phenotypic variation in mate choice have been made by Boake (1985, 1986) and Houck et al. (1985). The basic issue of the form of mate preference (e. g., absolute, relative or open-ended) is also critically important because the mode of choice also affects the possibility of a runaway process.

The importance of a series of genetical parameters is revealed by the models. Perhaps the most critical of these genetical issues is the existence and magnitude of genetic covariance between mating preference and sexually selected traits. This genetic covariance plays a critical role in predictions of both evolutionary trajectory and instability, yet no estimates of it have been made in either experimental or natural populations. The covariance is expected to originate and be perpetuated as a consequence of linkage disequilibrium arising from assortative mating (Lande, 1981). Breeding designs of the kind usually employed to estimate such a genetic parameter (Falconer, 1981) will impose a relaxation of normal assortative mating and so the linkage disequilibrium and genetic covariance will gradually decay from generation to generation. Thus special care must be taken to estimate the parameter soon after genotypes are sampled from nature and to correct for decay in covariance due to relaxation of assortative mating.

Decay in genetic correlation between the sexes during the evolution of sex-limitation is an unresolved empirical issue. The supposition of high genetic correlation between the sexes in homologous traits during the early stages of sexual differentiation is supported by genetic studies of species with slight or modest sexual dimorphism (reviewed in Lande, 1980a). Pleiotropic gene action is undoubtedly responsible for the observed correlations between expression in male and female relatives. The correlation approaches zero as one sex evolves to the point of not expressing the trait and presumably the correlation passes through intermediate values during the elaboration of extreme sexual dimorphism that precedes sex-limitation (Lande, 1980a). Remarkably, no genetical studies have measured the genetic correlation in traits with extreme sexual dimorphism. Thus we cannot make quantitative assessments of correlated female responses to sexual selection on males, as modeled by Lande and Arnold (1985), during the interesting phase when sexual dimorphism is pronounced.

Boake (1985, 1986) has stressed the need for estimates of genetic correlation between male mating success and other components of fitness. Quantitative genetic theory provides a means of measuring male 'genetic quality', for it can be viewed as the male's breeding value for fitness or its major components (e. g., viability), measured by scoring fitness in large samples of progeny as described by Falconer (1981). Thus a key test of the 'good genes' outlook is to seek correlations between breeding values for sexually selected traits and non-sexual components of fitness. Genetic correlations are correlations in breeding values. Boake (1985, 1986) outlines a strategy for estimating the critical parameters and illustrates it with an example.

Phenotypic correlation between male fitness components plays a key role in Kirkpatrick's (1985) model for sexual selection when males provide parental care. Thus a negative correlation between male mating success (e. g., harem size) and the average fecundity of males constrains the equilibrium state of paternal care to a value that maximizes male survivorship. Kirkpatrick (1985) refers to this tradeoff situation as limited male reproductive potential. In contrast, when male reproductive potential is unlimited, mating success is not correlated with mate fecundity and paternal care equilibrates at a value that maximizes mate fecundity but that value may be maladaptive for male survival. Downhower and Armitage (1971) pursued the critical tradeoff issue in a field study, but many subsequent studies of polygynous populations have neglected the issue. Arnold and Wade (1984) show how the tradeoff can be measured as a weighted covariance that also represents part of the total opportunity for selection in the male population.

A balance between sexual and natural selection at equilibrium is a prediction of all the quantitative genetic models and one anticipated by Darwin (1874). I know of no quantitative test of this expectation in natural populations. A variety of studies have detected or measured sexual selection on male attributes, but studies documenting viability selection on sexually-selected traits are rare (nonexistent?). There is a great need for studies that measure sexually selected traits, mating success, and survivorship in the same population so that the strengths of sexual and natural selection can be compared.

Lande's (1982) model of sexual selection in a cline suggests that ethological isolation can evolve rapidly over a wide geographic area in the absence of post-zygotic reproductive isolation. In many taxa post-zygotic barriers do not accompany pre-zygotic isolation (e. g. Twitty, 1961), contrary to the view that ethological isolation is mainly elaborated by direct selection in secondary contact zones. Thus Lande's results suggest that conspecifics exerting sexual selection, rather than sister taxa, may be the main selective agents in the evolution of reproductive isolation. The Fisher-Lande model for stasipatric speciation points to the need for clinal studies of sexual selection, an almost entirely neglected field of enquiry.

A number of sexual selection processes have never been analyzed with formal genetic models and might be the focus of future theoretical work: (1) Sexual selection arising from male contest or combat is an unexplored area. Models of sexual selection via mate choice suggest that male-male interaction could cause the elaboration of weapons deleterious to the survival of the weapon-bearing male. (2) Male-male interactions commonly consist of offensive and defensive tactics (Arnold, 1976). The joint evolution of such tactics might involve interesting dynamics. (3) Fisher (1930) suggested that the male displays might intimidate rivals as well as impress females. He predicted that a process of 'war propaganda' among males might lead to the evolution of displays but that these would not be as elaborate as the products of selection via mate choice. Borgia (1979) has discussed this possibility but the process has not been formally modeled. (4) Many authors have proposed that females should evolve preferences for attributes used by males in sexual combat. A formal model would involve episodes of viability selection, combat and then sexual selection via mate choice. (5) Many species possess epigamic characters that appear to be sexually selected but with no sexual dimorphism (Huxley, 1914, 1938). Darwin (1874) and Huxley (1914) proposed a process of *mutual sexual selection* for such species, with each sex exerting mate choice on the other. Could such a process lead to runaway sexual selection? (6) Fisher (1930) proposed that evolutionary elaboration of sexually selected traits would occur suddenly with periods of stasis between epochs of rapid evolution. Rapid development of one trait might be followed by sudden elaboration of another trait in the next epoch, perhaps with deterioration of the traits that were sexually selected in the early epochs. A multivariate model of mate choice might test Fisher's idea of an uneven tempo for evolution by sexual selection.

## Summary

Quantitative genetic models of sexual selection have disproven some of the central tenets of both the handicap mechanism and the 'sexy son' hypothesis. These results suggest that the 'good genes' approach to sexual selection may often lead to erroneous results.

Runaway sexual selection seems possible under a wide variety of circumstances. Quantitative genetic models have revealed runaway processes for sexually selected attributes expressed in both sexes and for attributes of parental care. Furthermore, the runaway could occur simultaneously in a series of populations that straddle an environmental gradient. While the models support the feasibility of runaway processes, empirical studies are needed to evaluate whether runaways actually happen. Estimates of critical genetic parameters are particularly needed, as well as measures of natural and sexual selection acting on the same population.

The models also show that sexual selection has tremendous potential to produce population differentiation, particularly in epigamic traits. Differentiation is promoted by indeterminancy of evolutionary outcome, transient differences among populations during the final slow approach to equilibrium, sampling drift among equilibrium populations, and the tendency of sexual selection to amplify geographic variation arising from spatial differences in natural selection.

Recent work with two- and three-locus models of sexual selection has produced results that parallel the results of the polygenic models (Kirkpatrick, 1982, 1985, 1986; Seger, 1985). Thus the feature of indeterminate equilibria (outcome dependent on initial conditions) is common to both types of model.

Acknowledgments. I am grateful to J. Felsenstein, R. Huey and M. Slatkin for hospitality during the writing of the manuscript and to M. Kirkpatrick and R. Lande for helpful discussions and comments on the manuscript. The preparation of this manuscript was supported by U. S. Public Health Service grant 1-K04-HD-00312-01 and by N. S. F. grants BRS 81-11489 and BSR 85-06766.

# The evolution of plant reproductive characters; sexual versus natural selection

D. Charlesworth, D. W. Schemske and V. L. Sork

## Introduction

"unarmed, unornamented, or unattractive males would succeed equally well in the battle for life and in leaving a numerous progeny, if better-endowed males were not present." (Darwin, 1871, p. 258)

"The exercising of choice in mate selection is exhibited throughout the plant and animal kingdoms." (Jones, 1928, p. 1)

Over the last decade, the rôle of sexual selection in the evolution of plant characters has received considerable attention (Bawa and Webb, 1984; Queller, 1983; Stephenson and Bertin, 1983; Thomson and Barrett, 1981; Willson, 1979; Willson and Burley, 1983). Much of the history of empirical results in this area has been reviewed by Stephenson and Bertin (1983). In the present paper, our aims are 1) to define sexual selection in plants, taking account of their frequently bisexual nature, and suggest criteria by which to distinguish between natural and sexual selection in plants, and 2) to discuss examples of plant characters that may have evolved by sexual selection. We will focus on the selective forces that act on plant breeding behaviour and on the question of how total fitness is influenced by these characters, rather than on semantic issues about which forces should be called sexual selections.

The concept of sexual selection was fully laid out in Darwin's book "The descent of man and selection in relation to sex" (Darwin, 1871). Darwin's development of the theory of sexual selection in animals was motivated by the observation that characters in one sex, usually males, were often so exaggerated as surely to cause a reduction in the individuals' ability to survive. The selective value of such characters was due to "the advantage which certain individuals have over others of the same sex and species in exclusive relation to reproduction". Darwin stressed the difference between natural and sexual selection, but made it clear that it is often difficult to distinguish between the two in practice. For example, he states (p. 257): "if the chief service rendered to the male by his prehensile organs is to prevent the escape of the female before the arrival of other males, or when assaulted by them, these organs will have been perfected by sexual selection, that is by the advantage acquired by certain individuals over their rivals. But in most cases of this kind it is impossible to distinguish between the effects of natural and sexual selection." He also (p. 256) stressed that the primary sexual organs are under the influence chiefly of natural selection. This makes it difficult to apply Darwin's definition of sexual selection to plants, since flowers contain the primary sex organs and also the organs of attraction for animal pollinators. It may be significant that despite his many contributions to the evolutionary biology of plant reproductive characters, Darwin's only discussion of sexual selection in plants is in a footnote (Darwin, 1871, p. 260).

The definition of sexual selection commonly used for animals and referred to above is inapplicable to plants because they are frequently bisexual (88 % of all genera and 83 % of species of flowering plants are estimated to be either hermaphroditic or monoecious – Yampolsky and Yampolsky, 1922). We will define sexual selection in bisexual species as competition between individuals for mates caused by traits which yield unequal fitness gains through male and female function, where the selective value of a trait is not due to its effect on survivorship of the sporophyte itself. Clearly, for there to be selection on such a trait, it must increase total reproductive success, i. e. sum of male and female components of fitness as they enter into the correct formulation for total fitness (Lloyd, 1974; also see below). This definition differs from that of Stephenson and Bertin (1983), in that it stresses that individuals may have both male and female sexual function, and does not include all traits expressed after sexual maturity which increase reproductive success. As with animals, the relevant categories of effects can be grouped under two headings:

1) Male-male competition: direct interference with male function of other individuals,

and 2) Female choice between different types of male parents.

Because plants are sedentary and depend on wind and animals for pollination, there is probably little opportunity for them to choose males on the basis of characters that affect only mating success, e. g. nectar production or floral display. These traits are therefore usually discussed in plants in terms of male-male competition. There is also an opportunity for male-male competition between pollen tubes growing in the same pistil. Female choice is usually thought of as a post-pollination phenomenon. It might occur in the process of fertilization if gametes from some donors are 'preferred' to those of others (which must be shown to be of equal pollen fertility). Female choice has also been suggested to occur in the process of selecting which fruits to mature and which to abort. (For reviews of female choice see Bookman, 1984; Devlin and Stephenson, 1985; Marshall and Ellstrand, 1986; Wilson and Burley, 1983.)

## Post-pollination events: female choice and genetic complementarity

In animals, female choice has been hard to demonstrate, partly because if one offers a female two or more males between which to choose, it is difficult to determine whether the outcome is due to a process acting between the males, with the female passively accepting the 'victorious' male, or whether the female can exercise a true choice (Bateson, 1983). In contrast, much of the discussion of sexual selection in plants has emphasized female choice. Several experiments to demonstrate it have been done using pollen from single donors applied to individual flowers. The data are usually analysed by asking whether female parents discriminate between different plants as pollen donors. However, if one wishes to detect complementarity between different plants as male and female parents, one should determine how much of the reproductive success among individuals is due to male parent, female parent, or the male x female interaction. A significant interaction term indicates that some male-female combinations are better than others, and is evidence for male-female complementarity of some sort. An effect of male parent indicates variation in male traits, but this need not have a genetic basis. To distinguish between sexual and natural selection on such traits, one must determine whether the trait affects the mating probability (sexual selection) or the viability or fecundity of the offspring (natural selection).

Bertin's (1982) data on variation on fruit set in the hermaphroditic vine, *Campsis radicans,* can be analysed in this way. Bertin found significant heterogeneity among pollen donors in ability to sire offspring, and concluded that genetic complementarity was of primary importance in determining fruit set because different plants had different 'preferred' donors. A reanalysis of the data using a log-linear model for categorical data confirms his conclusion: most of the variation is explained by the interaction, but male and female effects were also significant. We note that female choice is an unlikely explanation of an observed case of complementarity when each female appears to prefer a different pollen donor (as in the study of Bertin, 1982), because it is unlikely that sexual selection could maintain a large number of female types differing in preference as well as many male types differing in the qualities discriminated by the females.

There are at least four possible causes of male-female complementarity: female choice, complementarity between pollen and the stylar environment, zygotically expressed complementarity between genes in the male and female parents (including inbreeding depression, specific combining ability and outbreeding depression), and self-incompatibility. Therefore, before one can conclude that female choice is happening, one must eliminate the other possibilities. We next discuss these other phenomena.

*Inbreeding depression.* There is much evidence (Stephenson, 1981) that plants sometimes select among fruits (and seeds), retaining those of 'better' quality and aborting others; quality may be in terms of inbreeding coefficient, seed number in fruits, etc. Inbred progeny are likely to be detected as such, and aborted, because of slower growth rates or similar signs of lower viability. This means that abortion due to inbreeding depression can be classified as an example of abortion depending on progeny genotype, not as female choice based on male quality. This distinction is explained very clearly by Queller (1987). Inbreeding depression could be one possible cause of different complementarity between different female plants and possible pollen parents, because matings between related plants would be expected to be less successful than those between non-relatives, due to lowered viability or growth of the resultant embryos. Such early mortality after selfing has been shown to occur in conifers (Bingham and Squillace, 1955; Sorenson, 1970). In doing this type of experiment with species that have no self-incompatibility, one should therefore be careful to ensure that the different pollen donors used all have the same degree of relatedness with the maternal plants. This has not yet been done.

*Self-incompatibility.* In self-incompatible species (with a genetically based self-incompatibility system controlled by one or a few incompatibility loci) rejection of incompatible pollen is based on the genotype of the pollen donor (or on that of its pollen – De Nettancourt, 1977). Although this is clearly a form of female choice, it is such a distinct phenomenon from choice based on the phenotype of the pollen donor that we think that it should be explicitly separated from the phenomenon of sexual selection. It is likely that a factor in the original evolution of self-incompatibility may have been avoidance of inbreeding depression (Charlesworth and Charlesworth, 1979), but self-incompatible plants today accept or reject pollen on the basis of its incompatibility type without regard to the inbreeding coefficient of the progeny that would result.

In studying the phenomenon of mate choice in plants, one must therefore be very careful if one is using a self-incompatible species. Bertin's (1982, 1985) studies of *Campsis radicans* show clear evidence for differential success of different pollen donors on different plants used as females, but this could reflect their incompatibility types as this species is known to be self-incompatible (Bertin, 1982), though the precise genetic basis is not known. If there is a gametophytic system, pollen donors would be heterozygous and would produce pollen of two incompatibility 'types', and if the maternal plant being tested carries one of these alleles, then half of the pollen will be incompatible with this plant (half-compatibility). The extremely high repeatability of the differences (Bertin, 1985) is consistent with this being the cause. Bertin tried to avoid this problem of half-compatibility lowering fertility of certain crosses by applying excess pollen to all stigmas (Bertin, 1982), but this is not necessarily an adequate safeguard. If there is some degree of clogging of stigma or style by incompatible pollen, a half-compatible cross might have lower fertility than a wholly compatible one, however much pollen was applied. This must be taken seriously, as a pollen clogging effect has been found in *Campsis radicans* (Sullivan and Bertin, 1986). The complicating effects of self-incompatibility are also clearly evident in the study by Marshall and Ellstrand (1986). In that study, complementarity between different females and males is not clearly demonstrated because their study plants included several incompatible pairs.

*Effects due to wide crossing.* Obtaining useful estimates of selection pressures and intensities for either natural or sexual selection requires that experiments or observations be carried out at the appropriate level of population structure, i. e. within demes. This criterion is particularly important for plants, where effective population sizes may often be small and one must be very careful to distinguish between selection for characters associated with the avoidance of inbreeding and those which are the result of sexual selection. Furthermore, if evidence is obtained for differential male success, or female choice, between pollen donors and maternal plants taken from different populations, this may be due to partial reproductive isolation, which has been extensively studied in plants (Stebbins, 1950). To be convincing evidence for sexual selection, such data should be obtained using plants from within a naturally interbreeding population. Thus the data of Bookman (1984), which involve pollinations between plants from different populations, are not adequate as a demonstration of female choice in the context of sexual selection.

Results obtained from crosses between cultivars may also be hard to interpret, as these will usually differ at many loci, so that it would not be surprising, or relevant to the situation in natural populations, if one found evidence for differential male success, or discrimination by females of certain genotypes against pollen from other sources. This criticism applies to several studies, e. g. Mulcahy (1971).

*Genetic complementarity: specific combining ability and pollen-pistil complementarity.* The term 'specific combining ability' is used by quantitative geneticists to refer to dominance and epistatic components of variance in a character (Kempthorne, 1957). It has been widely used in plant breeding to identify crosses between lines that produce particularly good yield, and therefore is exactly analogous to what is seen in tests for complementarity between different male and female parents in a natural population, where the results of the crosses are assessed in terms of the progeny quality.

Physiological complementarity between pollen genotypes and the stylar environment may be a similar phenomenon, except that the interactions occur between a haploid pollen nucleus and the diploid stylar tissue, rather than within the zygote nucleus after fertilisation. At present, there is no definite evidence that such interactions occur, though it is plausible that they might.

In our view, it is not helpful to consider any of the above phenomena as instances of sexual selection. It would certainly be helpful to separate them explicitly from other causes of complementarity between male and female plants as potential parents. Perhaps studies of male-female complementarity should be done with dioecious species, in which there can be no problem with self-incompatibility, and inbreeding depression can be avoided by crossing non-relatives. However, the high fruit-set of females of dioecious species (Sutherland, 1986) suggests that little female choice occurs in these species, unless multiple pollinations are occurring.

## Gametophytic selection

"Clearly a higher plant is at the mercy of its pollen grains. A gene which greatly accelerates pollen tube growth will spread through a species even if it causes moderately disadvantageous changes in the adult plant." (Haldane, 1932)

As Haldane (1930) realised, competition among pollen grains for access to ovules can result in selection for faster pollen tube growth rates, provided there is genetic variation for this character in a population. At present, we have very little information about the existence of such genetic variation. Evidence for differences in pollen tube growth rates between different genotypes exists in several species (Baker, 1975; Snow, 1986), but many of the data are from cultivars, which probably differ at many loci. Schemske and Fenster (1983) showed an effect of male parent on pollen tube growth rate *in vitro*, using plants recently descended from a natural population. If genetic variation exists for pollen tube growth rates, we might expect that there would be a negative genetic correlation between performance of the pollen and that of the zygotes produced by different pollen genotypes, because variation can be maintained at a locus if selection affects that locus in opposite ways in the haploid and diploid phases (Wright, 1969).

Studies of differential fertilisation have given conflicting results. Some workers have found positive associations between gametophytic performance and offspring quality. The experiments have been done with multiple pollen donors (Mulcahy, 1974) and also with single donors (Lee and Hartgerinck, 1986; McKenna and Mulcahy, 1983; Mulcahy and Mulcahy, 1975). In most of these experiments, seed size was standardised so that differences between treatments cannot be solely due to one of them producing larger seed than the other. Earlier competition experiments using pollen from two strains of maize or cotton gave the opposite result: pollen from the same strain produced a disproportionate share of the seeds, even though the progeny of inter-strain crosses had higher fitness due to heterosis (Jones, 1928; see also Hornby and Li, 1975, and Currah, 1986 for similar data from other species).

There are some problems with the interpretation of the recent data and one must be careful to rule out other possible explanations for the differences in offspring quality. In most of the recent experiments cited above, the experimental design was to apply either little or excess pollen and to measure fitness components of the resulting offspring. The rationale for this is that when excess pollen is given there will be opportunity for competition between pollen of different quality. However, the differences are found even within inbred strains, in which there will be little variation among pollen genotypes. A substantial effect of pollen load size was found (Stephenson et al., 1986) on four different measures of progeny quality in an inbred cultivar of zucchini which must contain relatively little genetic variation compared with outbred populations. In that study, reductions of 13–20% were observed in the estimated quality of progeny of the low pollen treatment, compared with those grown from seeds of the same size produced by high pollen loads. We may therefore suspect that genetic differences in the populations may not be necessary for this type of effect to occur. Furthermore, as mentioned above, if male gametophytic performance were strongly correlated with the performance of the resulting sporophyte due to genetic factors operating in similar ways in both stages, we would not expect populations under selection for these traits to maintain much genetic variation for either gametophytic performance or sporophyte fitness. It is thus surprising that several studies have demonstrated large effects of pollen number on sporophyte fitness (Lee and Hartgerinck, 1986; Marshall and Ellstrand, 1986; McKenna and Mulcahy, 1983; Mulcahy, 1974; Mulcahy and Mulcahy, 1975), and the results would be less surprising if there were a contribution to these effects from non-genetic causes. No study has yet attempted to quantify the relative contributions of genetic variation and non-genetic factors to effects of this kind. In *Cassia fasciculata* (the same species as used by Lee and Hartgerinck (1986), see above), Lee and Bazzaz (1982) found no difference in fruit set when self or cross pollen was used, and Sork (unpubl.) found that mixed loads of self and outcrossed pollen from other plants from the same natural sub-population result in random fertilisation (as judged from seed weight and progeny performance, which differ between the two types of pollination). These findings indicate that there is no strong pollen competition among individuals within the normal range of gene flow for this species, and suggests that the effect of pollen dilution in reducing progeny quality found by Lee and Hartgerinck (1986) is most likely due to non-genetic causes.

An alternative explanation for the differences in progeny quality found in these experiments is that plants respond to the quality of pollen a flower receives (perhaps because application of high pollen loads leads to high numbers of pollen tubes, see Cruzan, 1986; Schemske and Fenster, 1983). Flowers that get large amounts of pollen may trigger a response in the maternal parent plant that leads to their being favoured in terms of nourishment, compared with other flowers on the same plant, so that such flowers would have low abortion rates and could perhaps have better quality progeny than the average for seeds of the same size. It certainly seems likely that plants have such a response at the time of flowering, and that nourishment of fruits by the maternal plant is not just a function of the seeds that fruits contain, because parthenocarpic fruits, with no seeds, can sometimes be very large. It is also clear that a fruit developed after the application of a very small pollen load may produce poor quality progeny despite the fact that the nutrients available to the fruit are not shared between many progeny. Indeed Darwin (1876) was careful to avoid low pollen loads in doing his inbreeding depression experiments, because: "Naudin also found in the case of *Mirabilis* that if only one or two of its very large pollen grains were placed on the stigma, the plants raised from such seeds were dwarfed". It seems unlikely that this would occur regularly if its cause is fertilisation by non-competitive pollen genotypes: surely 'normal' genotypes of pollen would be present in sufficient frequency for production of some healthy progeny.

There may be another reason for better seed set (and perhaps also better offspring quality) of flowers given pollen from more than one source in mixed pollination experiments, if the pollen sources differ in quality – for example in the frequency of viable grains. Whether the differences are genetic or environmental in origin, e. g. due to age of the pollen, there will be greater variance between flowers in the quality of the pollen received in pure compared with mixed treatments, though one would expect the same mean quality in both cases. This could lead to a difference in the mean number of seeds produced, or their quality, with the pure treatment producing the lower value, for the following reason. Seed set must be a concave (diminishing returns) function of amount of viable pollen, because it is limited by ovule number. Thus pollen loads with low numbers of viable grains will result in low seed set because there will not be enough good pollen for the maximum seed set to be achieved; when loads containing excess amounts of pollen are used, there will be less dependence of seed set on pollen quality, and the treatment with the greater variance, i. e. the pure treatment, will yield the lower mean.

(B. Charlesworth pointed out to us that this can be proved as follows. Let seed quality or number of a flower be denoted by $y = f\{x\}$ where $x$ is the quantity of good pollen applied to the stigmas of the flower. Then we have $\bar{y} \sim f\{\bar{x}\} + E\{d\,x\}f'\{\bar{x}\} + V(x).f''\{\bar{x}\}/2$, where a bar indicates a mean. If $f''\{x\} < 0$, as assumed, then if $\bar{x}_1 = \bar{x}_2$ and $V_{x1} > V_{x2}$ it follows that $\bar{y}_1 < \bar{y}_2$.)

This property was pointed out by Fowler and Partridge (1986) in a discussion of factors determining yield of *Drosophila* vials. If the maternal plant gives more or better resources to fruits with high than low seed set, the progeny from these fruits could be of better quality. One could test whether this is happening by doing both pure and mixed pollinations with low total pollen loads, which should eliminate the treatment difference because then only the means will matter and these should be the same. Of course, it is well documented in several systems that the better results from mixed than pure pollinations are most marked when high pollen loads are used. The considerations above show that this finding does not necessarily imply a genetic basis for the results.

In the context of selection among microgametophytes, we note that there is a considerably body of work on segregation ratios in plants (Grant, 1975; Jones, 1928; Meincke, 1982; Muller, 1963; Snow, 1986) showing that these are often normal, so that mutants with deleterious phenotypic effects do not by any means always have pollen with lower competitive ability, though they sometimes do. There is probably a bias towards underestimating the frequency of mutants that cause serious effects in the microgametophytic stage, because pollen-lethals would not be included in many of these studies. In these experiments, segregation distortion could be another cause of non-random fertilisation, though this has only rarely been documented in microgametogenesis in plants (Zimmering et al., 1970).

In conclusion, it is not yet clear whether true male-male competition occurs during the process of pollen tube growth in the style, mainly because there is very little evidence for sporophytically (i. e. male parent) controlled genetic variation in pollen tube growth rates or competitive abilities in natural populations. It seems likely that such competition could occur, but it is still unclear whether this has led to fixation for the genes determining the most competitive pollen type, or whether good pollen competitive ability is often associated with lower fitness at other stages of the life cycle, in which case there may be variation in natural populations. The evidence for female choice in the style is less convincing, and in no case that we are aware of have alternative explanations for different pollen tube growth rates been excluded.

## Fruit and seed abortion. 'Love's Labour's Lost'

Evidence is available suggesting that the crop of seeds matured by an individual is a non-random sample of early zygote phenotypes, due to abortion of fruits with low seed numbers, insect predation, or other causes of poor quality (Stephenson, 1981; Stephenson and Bertin, 1983; Stephenson and Winsor, 1986). Fruit abortion has been proposed as a mechanism of post-zygotic mate selection which has evolved by sexual selection (Stephenson and Bertin, 1983; Willson and Burley, 1983), but is probably better viewed as choice among the offspring zygotes, as Queller (1986) makes clear. We suggest that the term sexual selection should be used in relation to post-zygotic events only when there is evidence that female choice is based on a male character that is associated with an increase in mating success.

We also note that plants incur a cost by being choosy; they may end up with fewer seeds than they could mature. One would therefore expect plants to forego choice in situations when fertilization is less well assured, and so one certainly would not expect choice to be found in all situations where variation in male 'quality' exists. One might expect choosiness at the start of the season, when there is a high chance of later pollination, but it might decrease later, especially if seed set were still not good (e. g. late-season selfing). Such effects have been found in several plant species (Lee and Bazzaz, 1982; Nakamura, 1986; Stephenson, 1981; also Fenster unpubl.).

To demonstrate female choice, one needs to show that the aborted seeds are of normal viability, i. e. that their abortion is conditional on the presence of other (better) progeny and is not due to dysfunction caused by genetic differences between the pollen source and the female parent (such as hybrid inviability or outbreeding depression). Seavey and Bawa (1986) suggest that if one observes a clearly-defined syndrome of abortion, this proves that the control of abortion is by the female parent plant, i. e. that the maternal plant is choosing which embryos to abort, whereas if control is by the progeny genotype, one would observe various times and types of degeneration. While it seems likely that a uniform syndrome of abortion indicates the occurrence of a maternal abortion process, it is not necessarily true that it rules out control by the progeny genotype. If early progeny inviability is detectable by the maternal plant, it may begin the abortion process before one can detect any morphological abnormality in the embryo, but the abortion or survival of the embryos depends largely on their own genotypes.

We also note here that the chance of demonstrating that a given female shows a preference for a given male parent is very small unless the preference is extremely strong or very large numbers of progeny are studied (Jacquard, 1974, p. 316). But we know that additive genetic variance in total fitness is unlikely to be large, though additive variance in individual fitness components can be large if the genes responsible have opposite effects on fitness at different stages of the life cycle (Charlesworth, 1986). It therefore seems improbable that female choice in plants will be easy to demonstrate, and we should therefore consider very carefully whether some of the other effects mentioned above may be more plausible explanations for the findings.

## Pre-pollination events: male floral display and allocation to attraction of pollinators

*Theory.* The evolution of traits which may increase fitness through male function, e. g. pollen amount, floral display, and the timing and duration of pollen production, have often been discussed as examples of an inter-male sexual selection process (Bateman, 1948; Bawa, 1980a; Bell 1985a; Charnov, 1979b; Couvet et al., 1985; Haldane, 1930; Lloyd and Yates, 1982; Queller, 1983; Willson, 1979). Although this is an appealing way of thinking, one must bear in mind that plants do not attract their mates directly, but via pollinators as intermediaries (except in wind-pollinated species, which also do not directly attract mates), and that this introduces complications that have no counterpart in the animal world.

There are two possible approaches to thinking about attraction of pollinators in cosexual plants. One approach is to assign gender to secondary structures that attract pollinators, such as nectar and petals (Bell, 1985a). The other, which is probably more fundamental, because it is explicitly based on a definition of fitness, is to treat this as a case of sex-allocation theory. In most of the work on sex-allocation so far, attraction has been alluded to (e. g. Charnov, 1979b), but not explicitly taken into account (Goldman and Willson, 1986), partly because, as will be seen, it is not easy to make general models. Instead of assuming that allocation to reproduction can be divided into allocation to male and female functions, such a model requires us to add allocation to attractiveness as a third category, and to make an explicit model of how this allocation affects both male and female fertilities (Charnov and Bull, 1986). This approach has the advantage that one is less likely to overlook the possibility that attraction of pollinators is also required for female fertility, which certainly must sometimes be the case, and is clearly shown by the phenomenon of female mimicry of male flowers in some dioecious plants (Baker, 1976; Bawa, 1980b; Charlesworth, 1984).

An example of the second approach to the theory of allocation to attractiveness can be developed for the case of partial self-fertilisation (Charlesworth and Charlesworth, manuscript). This is based on the formulation for total fitness (Charlesworth and Charlesworth, 1981; Lloyd, 1975), assuming that there are no survival differences, except that progeny of self-fertilization have survival rate $1-\delta$ relative to a value of 1 for non-inbred progessy. The model assumes that resources for reproduction are available at two periods in a reproductive season, at flowering and at the time when fruits are matured. The relative allocations to primary male and female structures and to attraction at flowering time are assumed to be $\mathbf{M}$, $\mathbf{F}$ and $\mathbf{A}$, with the restriction that $\mathbf{M}+\mathbf{F}+\mathbf{A}=1$. To model female fertility, ovule output is assumed to be proportional to $\mathbf{F}$, the allocation to primary female structures. A fraction $\mathbf{s}$ of the ovules is assumed to be self-fertilised. The probability that a non-selfed ovule is fertilised is assumed to be a function of attractiveness of the plant: $f_2\{\mathbf{A}\}$. It seems reasonable to assume that this will be a concave (i. e. diminishing returns) function, because the more attractive a plant is the closer it will get to having 100% of its non-selfed ovules fertilised, but it cannot of course exceed that value. The number of seeds successfully matured is assumed to be another function, $f_1$, of the number fertilised; a concave function is assumed for this process also, because resources at the time of fruit maturation will probably limit the total number of seeds matured. With these assumptions, one can derive an expression for the female component of fitness (Charlesworth and Charlesworth, manuscript).

For male fertility, pollen output is assumed to be proportional to **M**, and the chance that a pollen grain will be exported to an non-selfed ovule and be incorporated in a successful seed is given by a function $g\{A\}$ of the plant's attractiveness. This function could be convex or concave. The male fertility of a plant allocating **M** to male structures and **A** to attraction can be written as $\mathbf{M} \, x \, g\{A\}$. The male fitness can then be derived, and total fitness is given by

Fitness $\propto$ Female fertility + Male fitness

This gives us all the values needed to calculate fitness by the standard formulae for hermaphrodite partially selfing plants and to calculate the ESS allocations to the three functions (see e. g. Charlesworth and Charlesworth, 1981).

This type of model is useful to help one find out under what circumstances allocation to attraction is likely to have an important effect on female fertility, and in what circumstances it is reasonable to consider attraction as subserving male function only. The results of the ESS calculations were examined in the light of these questions, by calculating the derivatives of the male and female fitness components of the hermaphrodite ESS, relative to the ESS values of the components. This gives a way of asking how great an effect changing allocation to attractive structures would have on the fitness through each sex function. The results show that the derivative of male fitness with respect to **A** is larger than that of female fitness, but the magnitude of the difference varies depending on the parameter values assumed. The effect on female fitness is negligible compared with that on male fitness when seed output is strongly limited by resources available for fruiting, but if increased seed initiation increases seed output the effect can be of comparable magnitude (Charlesworth and Charlesworth, manuscript). Therefore both male and female fertility can be affected by allocation to attraction of pollinators. Thus one cannot in practice separate this allocation into a male and a female part, nor can one consider it as male allocation, without definite evidence that this is justified.

Clearly this model is just one possibility, and other factors might be taken into account in addition to or instead of those listed above. We have ignored the possibility that inbreeding depression might act before seeds are matured, and the possibility that pollen is also an attractant for pollinators. We have also neglected the fact that pollen and ovules are not single entities, independent of other pollen grains or ovules produced by the same plant, but that they are produced by flowers so that their production is correlated, as is the fertilisation of different ovules in the same flower.

Another possibility, which might be important in some species, is that increasing attractiveness might reduce the chance that pollinators leave a plant, as well as increasing the chance that they will visit it. It is rarely mentioned that attractiveness may have a cost, and that plants are selected to be somewhat, but not too, attractive. The case of male *Catasetum* plants, in which deposition of pollinia is aversive to pollinators and has the effect of making them visit the visibly different female flowers, may be relevant here (Romero and Nelson, 1986). In *Combretum farinosum,* Schemske (1983) showed that pollinator species changed with changes in the number of inflorescences on a plant, and that although total seed production per plant increased as inflorescence number inceased, mean seed set per inflorescence decreased. In *Catalpa speciosa,* which is self-incompatible, Stephenson (1982) found that seeds tended to be produced after the time of maximal attractiveness, not at its peak. This type of effect could reduce male fitness when **A** becomes very large, and would also, in a self-fertile plant, tend to increase the selfing rate, whereas in the above model we have assumed that **s** is independent of **A**. There is some evidence that within a self-compatible species larger plants tend to have higher selfing rates (Crawford, 1984). On the other hand, increasing attraction will often be likely to increase the average distance from which pollen is received, so that seeds produced by plants with high allocation to attraction might tend to have lower than average inbreeding coefficients. Nevertheless, the model shows explicitly that female fertility could be affected by allocation to attractive structures, and that we are not forced to assume that these are purely male. The fact that Charnov and Bull (1986) do make this assumption is simply a choice that they have made, not a necessary feature.

The model also allows one to study the difference in allocation that would be expected with differences in the shape of the male fitness function $g\{\mathbf{A}\}$, for example, between wind-pollinated plants, where there is no need to allocate resources to attraction, and animal pollinated species, and thus to generate predictions about how much attraction is being favoured in animal pollinated species. For example, if it were the case that males of wind-pollinated dioecious species have more flowers than females, one could not attribute this to sexual selection for attractiveness, and this could help us to distinguish how much flower number is determined by selection for pollen output, and how much by selection for attraction of pollinators. Bell (1985a) applied similar reasoning to data on weights of male and female flower parts and suggested that the ratio of male to female weights is bigger for animal-pollinated than wind-pollinated species, which suggests a rôle of pollinator attraction in increasing the weight of male flower parts. However the data are at present too few to base a firm conclusion on them. There are of course many other differences between wind- and animal-pollinated species, and it will be important to consider how these affect the ratio in question.

*Empirical data on the effect of floral display on male and female fertility.* The existence of large, conspicuous floral displays and nectar-rich flowers in animal-pollinated species may be due to sexual selection on floral display (Gilbert, 1975; Willson, 1979), with gains accruing largely to the male component of fitness. Evidence from field studies in support of this view is not conclusive because it is difficult to distinguish sexual selection from fertility selection, for traits associated with increased pollen output, and to measure the male component of fitness in natural populations.

The effect of floral display on female fertility was studied in three species by Bell (1985 a) who also reviews some earlier data. In *Asclepias syriaca* the number of fruits per inflorescence, the number of inflorescences per stem, and the number of pods per inflorescence decreased as the number of flowers per inflorescence was reduced, though there were no effects on the numbers of seeds per pod in any of the experiments. In *Impatiens capensis,* artificially reducing flower size did not affect the number of seeds in fruits, but the number of flowers altered per plant is not stated, and it is likely that most flowers would have been intact. Since the pollinators are probably attracted by the whole plant or inflorescence, the lack of an effect on single flowers is not surprising. In *Viburnum alnifolium* removal of the atttractive sterile flowers reduced the numbers of seeds in inflorescences. Thus in two of the three species studied, reducing the floral display reduced the female fertility.

In species with at least some outcrossing, selection for any trait related to male fertility could be said to qualify as sexual selection. The distinction between sexual and natural selection for those characters is unclear, because any increase in pollen production or in pollinator attraction will increase the ability to compete for mates, and so would qualify as sexual selection, following Darwin's (1871) original definition. The intensity of selection on male characters will often depend on the shape of the male fitness function. A linear relation between male fitness and some character implies that the increase in fitness is directly proportional to the representation in the pollen pool. This might sometimes be expected to occur when pollen is dispersed passively, e. g. by wind, so that there is little opportunity for different rates of removal and deposition of pollen of different plants, though there must always be an upper limit to male fertility, at least in theory, because the number of available ovules is finite. It is unknown whether the limit is ever reached in natural populations. In contrast, a convex (accelerating gains) fitness function, with male fertility increasing disproportionately with the character, seems most likely in animal-pollinated plants because pollinators may exert preferences for particular phenotypes.

Very few studies include estimates of fitness through male function. It has been suggested that male function can be estimated in terms of pollen removal. Several studies have been done in species of milkweeds (*Asclepias* spp.) and orchids, in which pollen is packaged in pollinia whose removal from flowers is technically relatively easy to study. Using natural variation in inflorescence size, or artificially reduced flower heads, several studies (Bell, 1985 a; Queller, 1985; Willson and Price, 1977; Willson and Rathke, 1974) have found linear or concave relationships between pollinium removal and inflorescence size, but not convex ones. Schemske (1980 a) found a convex increase in pollinium removal with increasing inflorescence size in the orchid *Brassavola nodosa,* but fruit set was pollen-limited, which would reduce the strength of sexual selection on floral display. One must be careful to remember that pollinium removal may sometimes not be a good guide to number of fertilisations effected, as is required if we are to use it as an estimate of male fertility. In milkweeds, there is a very large excess of pollinia insertions into flowers over the number of fruit produced (e. g. Bell, 1985a), so the relation between pollinium removal and fruits sired is tenuous. Piper and Waite (manuscript) have data from an English population of broad-leaved helleborine, which appears suitable for this type of study. The number of pollinia inserted was more than half of the number removed from flowers, so pollinium removal may be a good indicator of male fertility in this species, though they present no data on fruit production. In this study, pollinium removal was positively correlated with inflorescence size, with a suggestion of an accelerating gain which, however, was not statistically significant. Another study of the role of attractiveness in determining amount of pollen removed is that of Bell (1985a) on *Impatiens*. In this study, individual flowers were made smaller by removing flower parts, and the effect of weight of pollen removed by pollinators was studied. This study has the same problem as those using pollinia, but also suffers from the difficulty that the pollinators are attracted by the inflorescence as a whole, so that a single altered flower will be affected differently from what would occur in the case of a genetic difference in flower size, which would probably affect all flowers on the plant similarly. Thus the empirical data using indirect measures of male fertility do not strongly support the idea that sexual selection, rather than fertility selection, has determined flower number.

In a recent study of siring of seed in *Raphanus raphanistrum,* Stanton et al. (1986) conclude that male fertility is affected by colour discrimination by pollinators, because yellow-flowered individuals were more successful as pollen donors than white-flowered ones (which were also visited less). This conclusion was based on data for the progeny of yellow-flowered maternal plants only, so there remains the possibility that the result is not due to absolute pollinator preference, but to pollinator constancy to the previously visited colour. If this were so, one might find that pollen from white-flowered plants was disproportionately successful in siring progeny of white-flowered maternal parents as observed by Kay (1978) in the Sketty B population of the same species. However Stanton (pers. comm.) states that a predominance of yellow plants as sires is also found in the progeny of white-flowered maternal parents. This shows the importance of taking pollinator behaviour into account when one thinks about floral display in plants. One might also expect to find that preferences for a colour morph are weak when it is rare, because pollinators are unlikely to develop a search image for a rare morph, but become stronger as its frequency increases, so that more pollinators concentrate on this morph. Evidence for a search-image resulting in under-visitation of a rare morph, which thus had a higher selfing rate, has been found in an artificial population of two cultivars of *Phlox drummondii* that differed in flower shape (Levin, 1972). Kay (1978) found evidence of pollinator preferences that seem to fit this pattern in natural populations of *Raphanus raphanistrum*, with a decrease in the preference of *Pieris* butterflies for the yellow colour, in populations in which plants with this flower colour were rare. In *Ipomoea*, colour morph visitation also differs with the morph frequencies: the attractiveness of white-flowered plants was low (as judged by their high selfing rate) in a population in which they were rare, but higher (as indicated by a higher outcrossing rate) in an artificial population consisting of 50 % white (Epperson and Clegg, manuscript; Schoen and Clegg, 1985).

While exaggerated characters related to male function are frequent in animals, and have presumably evolved by sexual selection, they are not obvious in plants, even dioecious ones (Queller, 1986). Indeed, at the time of flowering, dioecious species frequently seem to show less sign of adaptations for pollinator attraction than hermaphrodites; dioecious species often have small, greenish flowers (Bawa, 1980a) though there are complex multiple correlations with other factors such as woodiness and mode of pollination (Muenchow, 1987) and also with geographic range. It is not obvious which came first. It is quite possible that dioecy has a greater tendency to evolve in species with inconspicuous flowers. This might be the case if inconspicuous flowers are a correlate of reliable pollination and this is a necessary pre-condition for the evolution of dioecy, as Darwin (1877) suggested. The alternative possibility, that this flower type evolves after dioecy has evolved, seems less likely, but certainly cannot be ruled out as one can readily imagine possible reasons why this might occur. Comparative studies may enable us to answer this question.

The greater number of flowers on male than female plants in dioecious species has been attributed to sexual selection, on the view that floral display and attraction are of value largely in relation to male fertility; a similar argument has been put forward for the evolution of the larger size of male than female flowers in dioecious and monoecious species (Bell, 1985a). The finding that male flowers are smaller than female ones, and have less nectar, in a high proportion of tropical dioecious species (Bawa and Opler, 1975) may contradict this, though the amount of nectar will be affected by removal as well as production so that large nectar amounts in females could be due to their receiving fewer pollinator visits.

It is also possible that the females of dioecious species have been modified from the original cosexual state, and now have fewer flowers. This could be selected for if the original cosexual species was partially selfing, so that plants would tend to allocate resources to female reproduction sufficient to allow for some abortion of inbred progeny, and could reduce this allocation once an effective outbreeding system had evolved. Reduction in the flower number of females is especially likely to be selected for in perennial species, in which conflict between allocation of resources to ripening fruit and to survival will have most effect, and so could have been a factor in the evolution of dioecious species, which are mostly perennial. This explanation is consistent with the finding that females of dioecious species mature fruits in a higher proportion of their flowers than do hermaphrodites (Sutherland, 1986).

Despite the inadequate empirical support, it still seems likely that male function has played an important rôle in the evolution of floral traits in many plant species. Perhaps the best current evidence comes from studies which disprove alternative hypotheses (Horvitz and Schemske, manuscript; Sutherland, 1986), although there is often evidence for substantial effects of floral display on the total female component of fitness (e. g. Bell, 1985 a). Ideally, one should measure male fertility directly, using genetic markers to determine paternity, and thus to estimate transmission of genes to surviving progeny. Schoen and Stewart (1986) obtained data relating the numbers of male cones in white spruce plants growing in an experimental population, in which they could estimate the proportion of seed sired by each multi-locus genotype. The results in 1984 showed a concave male gain curve, but data from 1985 may show an accelerating gain (Schoen, pers. comm.). Using analysis of paternity of established seedlings, Meagher (1987) found that male reproductive success in the dioecious lily, *Chamaelirium luteum,* was not positively correlated with inflorescence size. In these data, variance of male reproductive success was not greater than that of female reproductive success, which suggests that there is little current opportunity for sexual selection in that species (Wade and Arnold, 1980). This result does not demonstrate that male function was unimportant, especially as there was larger variance for males than females for fertilisation success, based on paternity analysis of seeds (Meagher, 1986). Clearly, more studies of these sorts are needed for other species. Unfortunately, the complex statistics and large sample sizes required make these techniques difficult, and the practicability of the paternity analysis method in partially self-fertilising natural populations has yet to be explored.

## Concluding remarks

Our objective in this paper was to focus attention on types of selective forces that act on plant reproductive traits, and to try to discriminate between the forces of natural and sexual selection. Because plants do not choose their mates, but rely on the actions of abiotic or biotic pollen vectors, events after pollination, e. g. gametophytic competition and abortion by female parents, could be very important in determining the genetic composition of offspring. But male and female characters expressed in these ways may not reflect evolution by sexual selection, as has been frequently claimed in the literature. In our opinion, a narrower view of sexual selection that does not include post-fertilisation events may be preferable (see also Queller, 1986). The most likely type of post-pollination sexual selection is pollen tube competition, and this has not yet been adequately demonstrated to occur as a phenomenon separate from self-incompatibility and genetic complementarity between pollen and pistil. To assess the importance of this kind of selection in contemporary natural plant populations, we need much more data on genetic variation in pollen characters, if possible in conjunction with estimates of genetic correlations with other components of fitness. Of course, the absence of substantial genetic variance for pollen characters does not mean that competition between different types of pollen has not occurred in the past history of the species; present-day properties of pollen may well have evolved in response to this kind of selection pressure. But if there is little genetic variation for pollen characters such as growth rates, it will be difficult to study this kind of selection, as pollen characters other than quantity cannot be experimentally manipulated.

Although the empirical data do not at present offer strong support, we believe that intrasexual selection for fitness gains through male function has promoted the evolution of many flowering traits, and that plants may often produce more flowers or pollen than required for achieving a level of seed set compatible with the resources they have available. Production of excess flowers whose cost is justified in terms of increased male fitness (Couvet et al., 1985; Queller, 1983) has sometimes been referred to as sexual selection (Willson, 1979). But it is not clear to what extent flower production is simply due to selection for pollen output, and to what extent to pollinator attraction.

It may be helpful to distinguish the following influences of flower numbers on plant reproductive success:

(1) flower numbers may affect male fertility through pollen quantity produced;

(2) flower numbers, flower size and nectar production may affect male and female fertility through attraction of pollinators, or by altering pollinator behaviour;

(3) high flower number may be favoured when full pollination of every flower is not certain (because high flower number might increase the chance that seeds are fertilised in sufficient numbers that the resources available for fruit ripening can be used up);

(4) high flower number may be favoured through female fitness because it ensures that an excess of fertilised ovules exists to compensate for mortality of fruits due damage, or to poor quality of the seeds in them (often because of inbreeding);

and (5) high flower number may be favoured through female fitness because it gives female plants an opportunity to exert mate choice via selective abortion of seeds sired by certain pollen donors.

Of these, categories (1) and (3) are not sexual selection, whereas (2) can be viewed either as fertility selection or sexual selection. The process listed as category (5) is one of sexual selection acting via female choice, though as already mentioned it is not direct choice of males but only of the zygotes they have sired. The evidence at present is strong for selective forces (1), (2) and perhaps (3) but, as discussed above, is less convincing for (5). The fact that female plants of dioecious species have high fruit/flower ratios (Sutherland and Delph, 1984) is consistent with (4), because females produce only outcrossed progeny so that this selective pressure for high flower number is absent and lower flower numbers in relation to fruit numbers are expected. However other explanations of this finding are possible.

This way of subdividing the selective processes acting at the time of flowering helps clarify the processes that can be classified as sexual selection, though it is sometimes a matter of taste whether one wishes to use this term in preference to fertility selection. The important point is that our knowledge of these selection processes in plants is still very imperfect. We need much more evidence on variation in male reproductive success, measured as contribution to progeny, and on the relation between flowering-time traits and the male component of fitness.

## Summary

The rôle of sexual selection in the evolution of plant reproductive characters has become a widely discussed topic among plant population biologists. For many reproductive characters, it is difficult to distinguish whether sexual selection or natural selection has been the more important force. We propose a definition of sexual selection for cosexual species and critically review the topic of sexual selection in plants. The emphasis of the chapter is on the evolution of plant reproductive characters, rather than on the issue of whether certain phenomena should be classified as sexual or natural selection. We do, however, argue that the category of sexual selection should not include the phenomena of inbreeding depression, genetically based self-incompatibility systems controlled by one or a few loci, or the effects of wide crossing and specific combining ability. We then discuss the available information on pollen competition and progeny quality, fruit and seed abortion, and the rôle of pollinator attraction in male and female fertility.

Acknowledgments. We thank B. Charlesworth for many helpful discussions, R. I. Bertin and N. C. Ellstrand for comments on an earlier version of this paper, and the following people for allowing us to see their unpublished manuscripts: J. Piper, S. Waite, T. R. Meagher, E. Thompson, D. J. Schoen, S. C. Stewart and M. F. Willson.

D. Charlesworth was supported by grant numbers BSR-8516629 and BSR-8516617 from the National Science Foundation, D. Schemske was supported by grant number BSR-8415666 from the National Science Foundation, and V. Sork was supported by an award through the NSF-VPW program (#RII-8503512).

# The selection-arena hypothesis

S. C. Stearns

Selection for juvenile survival is always strong, whether fecundity in the species is high or low. It is therefore at first sight puzzling that some organisms produce many offspring, then neglect, discard, resorb, or eat some of them, or allow them to eat each other.

The overproduction of offspring includes polyovulation (the production in mammals of more ova and zygotes than can be carried to term), superparasitism (the laying by insect parasitoids of more eggs into a single host than can survive to pupation), surplus flower production (the production of flowers that do not set seed), fruit abortion (the production of fruit that are abscised before finishing development), recurrent spontaneous abortions in humans, adjustment of sex ratios at birth in mammals, and some types of cannibalism, particularly kronism (the consumption of offspring by parents), sibling cannibalism, and adelphophagy – intrauterine embryonic cannibalism.

Because reproductive success is the product of survival and fecundity, overproduction of zygotes could have the following explanations: the excess offspring are converted into improved survival or fecundity (1) of the other offspring or (2) of the parent(s). While one of these consequences is necessarily true, none of them explain why the parents bothered to produce the excess offspring in the first place. Most explanations in the literature invoke improved survival of the other offspring and ignore the possible effects on sibling survival, parental fecundity, and parental survival.

The botanical literature is a rich source of examples of the overproduction of zygotes (e. g. Janzen, 1977; Lee and Bazzaz, 1982; Bawa and Webb, 1984; Lee, 1984; Bookman, 1984; Nakamura, 1986; Stephenson and Winsor, 1986; Sutherland, 1986; reviews in Charlesworth et al., 1987; Queller, 1987; Seavey and Bawa, 1986; Stephenson and Bertin, 1983; Willson and Burley, 1983; Stephenson, 1981). Botanical leadership in this areas results from early discovery, the development of a tradition of investigation, and energetic and imaginative experimental work. Darwin (1876) suggested that plants could improve offspring quality by selectively aborting fruit on the basis of paternity, and in 1922 Buchholz argued 'that intraovarian competition and seed abortion improves offspring quality' (Stephenson, pers. comm.). The botanists have had the advantage that it is easier to observe surplus flower production and fruit abortion than polyovulation or superparasitism.

Observations on animals, although often indirect, are not lacking, and interest in the subject has recently increased (e. g. Wourms, 1977; 1981; Charnov, 1979; Bemds and Barash, 1979; Tait, 1980; Wickler and Seibt, 1983; Gwynne, 1983; Taborsky, 1985; Birney and Baird, 1985; Labeyrie and Rojas-Rousse, 1985; Gosling, 1986; reviews in Fox, 1975; Polis, 1981; Eberhard, 1985; Clutton-Brock and Iason, 1986).

One general explanation may link these seemingly dissimilar and taxonomically unrelated events – the selection-arena hypothesis. For plants, Bertin (1982) and Stephenson and Bertin (1983) have already developed this hypothesis in all of its essential features: "We suggest that incompatibility systems, pollen competition, and selective seed or fruit abortion represent an integrated genetic sieve, favoring those pollen and zygote genotypes likely to make the greatest contributions to the fitness of the maternal sporophyte and simultaneously allowing a plant to adjust its fruit and seed crop to match the resources available for reproduction" (Stephenson and Bertin, p. 135). For priority on the general hypothesis, Eberhard (1985) should also be mentioned. He develops the idea primarily for female choice through sperm selection. Selection arenas are one of a series of mechanisms in which parental choices influence offspring quality.

## Choices that influence offspring quality

We can regard the life cycle as a series of parental choices that influence offspring quality: (1) Should reproduction be sexual or asexual? (2) If sexual, then with which mate(s)? (3) If multiple matings are possible and fertilization is internal in the female, then which sperm should be used to fertilize the eggs? In plants, which pollen should be used to fertilize the ova? If males care for the fertilized eggs, then which eggs should be preferred and which exposed to risk or eaten? (4) If fertilization is internal and females care for the fertilized zygotes, then which embryos, seeds, flowers, or fruit should be nourished, and which ignored, resorbed or discarded? (5) If offspring remain in a situation controlled by the parent after birth, then which offspring should be killed, if any, and for what reason? Under what conditions should that choice be left to a competition among the offspring?

These questions are normally dealt with in isolation from one another. Question (1) concerns the evolution of sex, questions (2–4) are often dealt with in discussions of sexual selection, and question (5) is normally discussed in the context of cannibalism, parental care or clutch size. All represent choices that affect offspring quality. Ideas on the evolution of sex have recently been reviewed in Bell (1982), in this book, and in Michod and Levin (1987). The fascinating diversity of mechanisms for sperm selection has been elegantly described by Eberhard (1985). Ideas on sexual selection are reviewed in Bradbury and Andersson (1987). Here the emphasis is on an idea that cuts across all these areas.

## The selection-arena hypothesis

The selection arena hypothesis depends on selection at an early stage of the life cycle when offspring have not yet cost very much. It makes four basic assumptions, (1) that zygotes are cheap, (2) that after conception there is a continuing investment of parental time, energy, or risk into offspring, (3) that offspring vary in fitness, and (4) that this variation in fitness can be identified at an early stage of the life cycle. If these four assumptions are met, then it should be advantageous to overproduce zygotes, identify those with lower expectations of lifetime fitness, and either kill them outright or at least stop investing in them so that investment goes only into offspring with higher fitness expectations.

The implicit argument here is that a gene causing overproduction of cheap zygotes will be favored because it will be born by adults that carry high-fitness genes or combinations of genes at other loci. The crucial supporting observation would be genotypic correlation of fitness between individuals of different ages (zygotes and adults). The cleanest system in which to make such observations is a flowering plant, where the relevant correlation is between gametophytic and sporophytic vigor. Such correlations have been observed (Mulcahy et al., 1975; McKenna and Mulcahy, 1983; McKenna, 1986; Schlichting et al., 1987 – but see the previous chapter). Further tests of this assumption, particularly on animals now that the trend has been demonstrated in plants, appear to be well justified. (I thank Graham Bell for most of the content of this paragraph.)

### Selection agents: Developmental defects

Probably the most widespread genetical problem encountered by developing zygotes is the expression of mutations or deleterious recessive alleles in homozygotes. Although mutation rates per locus are generally low, on the order of 0.0001 or 0.00001, mutation rates per genome are appreciably higher, on the order of 1.0 to 0.01, and the expression of a new mutant or deleterious homozygote should not be construed as a rare event (Dobzhansky, 1970). Thus developmental problems of genetic origin in early zygote stages should be widespread and recurrent. We should expect adaptations to have evolved that deal with the problem, the simplest of which would be overproduction of zygotes followed by viability selection.

In the best-studied mammal, man, there is increasing evidence that the frequency of spontaneous abortions is higher than previously believed (Beer and Quebbeman, 1982). In any large collection of fertilized ova, a certain percentage do not complete development, whether the organism is a fruitfly, a chicken, or a wheat plant. If deleterious mutants could be screened and discarded at early stages or used as a source of nutrition for sibs, the remaining offspring would have higher average fitness.

Stephenson (1981, p. 262) notes that "some undamaged juvenile fruits abscise because of genetic or developmental abnormalities". The dominant *proximate* influence on fruit abortion seems, however, to be food limitation (Stephenson, 1981).

## Selection agents: Parasites and disease

The second reason to place zygotes in a selection arena has to do with parasites, disease organisms, and the Red Queen hypothesis for the maintenance of sex (Bell, 1982). Let us assume, as suggested by Levin (1975), Jaenike (1978), Hamilton (1980), Rice (1983a), and Bremermann (1985), that genetically diverse offspring are less susceptible to diseases and parasites. Where parent-offspring contagion is likely, sexual reproduction would reduce the chances that a disease organism with a short generation time that had already 'evolved onto' the parental genotype would be able to flourish in the offspring.

If this hypothesis holds, then placing an overproduction of zygotes in an arena where they could be exposed to selection for resisting disease organisms living in the mother's tissues would improve on the advantages of sexual reproduction. This advantage would exist if disease organisms transmitted from the parents were capable of causing a significant portion of offspring mortality, as should be the case for all organisms with parental care, especially hole-nesters and den-builders, and for all organisms in which offspring grow up closer to their parents than to a genotype selected at random from the deme, as do most plants.

Two mechanisms might be involved in the selection. Either the embryos could be exposed directly to the disease organisms, or selection could be mediated, in organisms that have such reactions, by the female's immune system. The latter mechanism would involve a molecular inspection of the surface of the zygote resulting in the identification of genotypes that differ most from the mother immunologically or of genotypes that specifically indicated high survival probability in the presence of disease organisms for which antibodies were already circulating in the mother's body. This may be the evolutionary basis of the phenomenon perceived by medical doctors as 'immunological incompatibility' (cf. Beer and Quebbeman, 1982).

Given the ubiquity of both disease and sex, one would expect all viviparous organisms to polyovulate and practice selection for resistant offspring. That some may do so has been indicated by provocative studies on both rats and humans. Beer and Quebbeman (1982, p. 304) summarize studies on rats as follows [MHC refers to the Major Histocompatibility Complex]:

"In summary, (1) there are intense selective pressures operating during gestation against individuals that are homozygous with their mother with regard to the MHC antigens (Michie and Anderson, 1966); (2) there is decreased reproductive efficiency when MHC homozygous young are gestated in homozygous mothers; (3) there is selective elimination of MHC homozygotes either prenatally or early postnatally as a result of immunological responses to fetal antigens not associated with the major histocompatibility complex (Palm, 1974)."

These results from white rats indicate that the immune system can act to screen embryos, but do not establish definitively whether the criteria used are the elimination of homozygotes, the preferential production of recombinants, or both, or whether the screening takes place only with respect to MHC loci or in general.

In humans, Beer et al. (1981) found that women with recurrent consecutive spontaneous abortions of unknown cause were significantly more likely to share major histocompatibility genes with their husbands than were women belonging to a control group. In other words, human females tend to reject zygotes that are homozygous at the loci that produce antigens. It was precisely at these loci that Bremermann (1985, and this volume) suggested disease organisms would select for recombination and heterozygosity.

It is clear that the immune system is active during pregnancy in mammals and can act in a sophisticated manner to screen embryos. All regularities so far noted in screening are consistent with evolutionary hypotheses that see such screening as amplifying the advantages of sex, either by accelerating the elimination of deleterious mutants, or by conferring adaptive advantages in a coevolutionary race with disease organisms.

Nevertheless, polyovulation followed by immunological screening has apparently not become widespread. Either the selection arena hypothesis does not hold, or many more organisms practice polyovulation, superparasitism, and indeterminate flowering than previously recorded, or other factors are at work. Without rejecting the first two possibilities, I would like to suggest what two of those factors might be.

Two complications would make a selection arena either unnecessary or ineffective. First, many mammals produce antibodies in their milk that endow their offspring with immediate resistance to certain disease organisms (see Lascelles, 1977, for a review). To the extent that such antibodies solve the same problems that the selection arena is designed to solve, the selection pressure for polyovulation should decrease. The mechanism envisaged by Bremermann, however, is quite general and does not require parent-offspring disease transmission. It functions at the level of the population, not just at the level of the family. Thus transmission of antibodies in milk would not significantly reduce the advantage to the mother of selectively eliminating zygotes homozygous for MHC genes.

Secondly, it is clearly not in the offspring's interests to be killed by the parent, no matter how high the probability of later sickening and dying as a juvenile. If screening for disease resistance is going on, then there is a direct conflict between zygote and parent. The parent should create conditions under which a large number of genetically and immunologically different offspring can be screened for viability, and the offspring should try to trick the parent into thinking that they are viable even if they are not. If they could do this, they could escape parental surveillance and enter into normal competition with the other embryos for the few places available. Once this had occurred, polyovulation would no longer be advantageous to the parent, and the ovulation level should sink. Human embryos are clearly able to do this at a later stage of development, following implantation, as many studies of the placenta as an immunological barrier demonstrate (see Beer and Sio, 1982, for a review). It may be that the embryos mask themselves as early in life as possible, but that females take advantage of the single-cell or few-cell stage prior to implantation to carry out screening.

If the evolution of sex finds an important part of its general explanation in disease and parasites, and if in fact polyovulation is not common (we should look harder), then at least one plausible explanation of why polyovulation is not more frequent is that the offspring have won the conflict and disguised themselves immunologically to avoid parental surveillance and rejection.

## Selection agents: Sex-allocation

In addition to the elimination of developmental defects and selection for disease resistance, one could claim that polyovulation is one mechanism used to achieve an adaptive adjustment of the sex ratio. The overproduction of fertilized zygotes would be followed by the selective elimination of one sex or the other. While in certain lineages it is not likely that this could be the only explanation – in many families of flowering plants, for example, the evolution of all male flowers from hermaphrodites seems to pose no problem – it may well be that after polyovulation has evolved for other reasons, it would be exploited to this end.

A particularly appropriate example has recently been published by Clutton-Brock et al. (1986). As part of their continuing study of red deer on Rhum, they discovered (a) that sons born to mothers above median rank were more successful than daughters, (b) that daughters born to mothers below median rank were more successful than sons, and (c) that dominant mothers produced significantly male-biased sex ratios at birth, while subordinate mothers produced either 1:1 or female-biased sex ratios.

Gosling (1986) has documented another telling example in the coypu, in which young females in good condition abort small litters of mostly female offspring and retain large litters and small litters that are predominantly male.

Only three mechanisms could be used for such adaptive adjustments of sex ratio in mammals: (1) selection of sperm of the appropriate sex, which might be achieved by changing the chemical environment of the genital tract during and following copulation (see articles in Smith, 1984) or through the timing of insemination (Clutton-Brock and Iason, 1986), (b) intrauterine selection of zygotes of the appropriate sex, or (c) abortion of entire litters. Clutton-Brock and Iason (1986) review the evidence for adaptive variation in sex-ratios at birth in mammals, and conclude that while much of it must be discarded, some recent studies, including those two just cited, document preferential investment in the sex with higher expectation of lifetime reproductive success.

### Selection agents: Post-copulatory and post-pollination sexual selection

Eberhard (1985), for animals, and Queller (1987) and Charlesworth et al. (1987), for plants, have reviewed the possibilities that females have for choosing among potential fathers of their offspring after having been inseminated or pollinated. Eberhard (Chapter 7) points out that females can control whether intromission will follow genetical contact, whether insemination will follow intromission, and whether sperm transport to appropriate sites will follow insemination. Sperm may be stored and selectively activated, or destroyed. The morphological and physiological mechanisms that have evolved to accomplish control over sperm are diverse and complex. Eberhard suggests (p. 115) that females could also control implantation "of the fetus in the uterus in placental mammals and (probably) . . . in a number of other groups, such as scorpions, sharks and rays, onychophorans, flies, mites, and cockroaches, in which the young are nourished inside the female's body." He clearly envisioned a type of selection arena.

In plants female choice of pollen type has been well documented (see Stephenson and Bertin, 1983; Queller, 1987; Charlesworth et al., this book; Mulcahy and Mulcahy, 1987). Here the opportunities for selection are at least as great as those in the complex reproductive tracts of some insects, for the pollen must grow a considerable distance through female tissues to reach the ova. The selection of father may occur either before or after nuclear fusion. Mulcahy (1979) showed that pollen competition and interactions between pollen and pistil result in non-random fertilization and improve seed quality.

Because they cluster their offspring together into groups (fruit), plants have the opportunity of superimposing another stage of selection that resembles the abortion of entire litters by coypus (Gosling, 1986). Bookman's (1984) study of *Asclepias* provides convincing evidence of female choice of pollen type carried out after fusion through fruit abortion. Stephenson and Winsor (1986) have shown that *Lotus corniculatus* selectively aborts fruit with the fewest seeds. They argue that these differences among fruit in seed number are associated with differences in pollen quality. The most vigorously growing pollen (microgametophytes) produce the most seed, and the vigor in pollen growth is thought to be associated with vigor in the resulting offspring (sporophytes) because many genes are expressed in both stages of the life cycle (Mulcahy, 1979). Whether plants also select among individual seeds within fruits or pods remains an open question.

Winsor, Davis and Stephenson (1987) and Davis, Stephenson, and Winsor (1987) have examined the impact of pollen competition on a whole series of fitness components. Seeds produced under conditions of high pollen competition emerged more frequently and more rapidly and grew into plants that produced more flowers and fruit than did seeds produced under conditions of low pollen competition (Davis et al., 1987). Winsor et al. conclude that "progeny resulting from intense pollen tube competition (high pollen loads) are more vigorous than progeny produced under conditions of little or no pollen competition (low pollen loads) and that zucchini can, as Lee (1984) hypothesized, improve the average quality of its seed crop by selectively aborting fruits on the basis of seed number."

While sperm and pollen selection are cheaper than zygote selection because eggs and fruit do not have to be produced and nourished (Stephenson and Bertin, 1983), they are also less reliable than selection carried out after nuclear fusion for three reasons. First, the fitness of the offspring phenotype depends on the complementarity of gene combinations in the diploid state. This complementarity cannot be assessed until nuclear fusion has occurred and gene products produced by the diploid nucleus have started to have impact on the phenotype, usually only after several cell cleavages have taken place. Secondly, the quality of genetic information carried in sperm and pollen is not directly observable in many cases because their phenotypes are at least strongly influenced by paternal effects, and in some cases may be completely determined by paternal cytoplasm. Thirdly, in groups such as birds in which the male is homogametic and the female is heterogametic (see chapter by Bull), sperm selection cannot be used to adjust sex ratios (Wickler, pers. comm.), although it could be used to select for other traits. For these reasons we can expect sperm and pollen selection to be followed by more reliable zygote selection.

*Borderline cases and alternative explanations*

A number of cases in both plants and animals resemble selection arenas but may actually be by-products of adaptations to other problems, primarily flexible adjustments in clutch size and dealing with interspecific competitors.

One mechanism widespread in plants could be called bet-hedging in the face of an unpredictable nutrient supply. Stephenson's (1981) review documents numerous cases in which the proportion of fruits that abort increases with the number of fruits initiated, with the amount of herbivory, with defoliation, and with leaf shading. Addition of macronutrients, carbohydrates and water often increases the proportion of fruit that mature. Stephenson concludes (p. 264), "the evidence that flower and fruit abortions are a response to limited resources is prodigious." He clearly meant this statement to be limited to the proximate mechanisms involved, for his article concludes with a review of hypotheses on the adaptive value of surplus flower and fruit production. Sutherland's (1986) comparative study suggests that nutrient limitation is primarily important in determining fruit-set values in self-compatible plants.

The bet-hedging mechanism also probably explains facultative resorption of embryos in kangaroos and facultative parental cannibalism of nursing pups in mice (Koenig, in press), wood rats (McClure, 1981), and hamsters (Day and Galef, 1977 – see also Smart, 1983 for a review). Many raptors lay more eggs than they could rear in all but the best years; this has been interpreted as a bet-hedging mechanism for the flexible adjustment of clutch sizes to an unpredictable nutritional regime (Stearns, 1976). The analogy to fruit abortion under stress is direct.

It is important to stress that in these cases, the selection is not in principle among offspring that differ genetically (although they certainly may and probably do), but among offspring that differ in their expectation of getting resources. Such selection would still be advantageous among genetically identical offspring because the decision hinges on the expectations of resources, not on the implications for fitness of a particular combination of genes.

That an organism might overproduce zygotes to create the possibility of direct sib-competition and thereby selection of a few, high-quality offspring is most plausibly exemplified in superparasitism and adelphophagy. In superparasitism the parasite lays many more eggs into the host than will hatch. The ichneumonid wasp *Diadromus pulchellus,* for example, normally lays 2–3 eggs into each pupa of its bean-weevil host, *Acrolepiopsis assectella,* but only a single imago survives and emerges (Labeyrie and Rojas-Rousse, 1985). The first instar larvae have sickle-shaped mandibles useful for fighting but not for feeding on the soft tissues of the host (Hagen, 1964; Salt, 1961). Larval fights eliminate all but one survivor.

That such adaptations are primarily responses to interspecific competitors rather than maternally controlled selection among siblings is quite plausible. Support for this position is provided by the bizarre case of *Copidosomopsis tanytmemus,* an encytrid hymenopteran egg-larval parasite of the Mediterranean flour moth (Cruz, 1986). *Copidosomopsis,* like many other hymenopteran parasites, is polyembryonic: the developing embryo splits into a large number of clonally derived siblings early in development. In this case, however, there is a special twist. A specialized larval morph, the precocious larva, clonally derived from the same cells that produce the normal larvae, has well-developed head and mouth parts and is extremely mobile. Precocious larvae injured or killed interspecific competitors of three other species in 74, 98 and 97 % of the observed cases of multiple parasitism. Then the most remarkable event occurs: precocious larvae always die in the host hemolymph without pupating. They are evidently an altruist soldier caste engaged in defending a clone of genetically identical siblings (Cruz, 1986).

Adelphophagy, eating one's brother, is intrauterine embryonic cannibalism (Wourms, 1977). It is probably a derivative of oophagy, one of the mechanisms by which mothers may nourish viviparous offspring, in which unfertilized eggs are continuously ovulated during gestation and eaten *in utero* by the developing embryos (Wourms, 1981). It is difficult to document adelphophagy because after the egg or embryo has been eaten, it is hard to decide whether it was an unfertilized trophic egg or a fertilized developing sibling. The best-documented case appears to be the sand tiger shark, *Eugomphodus taurus*. Springer (1948) found a single embryo in each oviduct. At term, these embryos were more than 1 m long and weighed at least 6.4 kg. Living 26-cm embryos were very active in the mother's reproductive tract. They moved rapidly and bit freely. Stribling et al. (1980) found that each oviduct held more than 30 eggcases, each with 12–15 small eggs or embryos, at the start of gestation. Only one embryo per oviduct survives to term.

Oophagy also occurs in many other sharks (Wourms, 1981), buccinid and muricid gastropods, and in some amphibians (Wake, 1980; Duellmann and Trueb, 1986), but no other well-documented cases of adelphophagy are known to me.

Tree-hole mosquitos normally lay many eggs into a single pool of water. Some species are predatory, and cases of sibling cannibalism have been documented (cf. Jenkins and Carpenter, 1946; Corbet and Griffiths, 1963). In addition, some species (in the genera *Toxorhynchites* and *Culex*) have larvae that will kill all other larvae that they encounter, without eating them, just prior to pupating. This is thought to be an adaptation that eliminates the danger of being eaten by another, younger larva during the relatively defenseless pupal period (Haddow, 1942; Corbet and Griffiths, 1963).

The larvae of both tiger salamanders (Collins and Cheek, 1983) and spadefoot toads (Bragg, 1964) are capable of developing special morphological and behavioral adaptations for cannibalism. In brief, it appears that those individuals in a cohort that grow most rapidly are capable of changing their head shape, dentition, and feeding behavior to make cannibalism possible. Whether the mechanisms are genetic, developmental, or both has not yet been adequately investigated. The pools of water into which amphibians lay their eggs are certainly potential selection arenas. However, in the case of amphibian larvae, as with superparasitism and the case of tree-hole mosquitos, sibling cannibalism can be plausibly explained as a by-product of adaptations that evolved primarily in the context of unrelated competitors. In all these cases, larvae of several species and of several families within a species may occupy the same arena. More careful work needs to be done on the relatedness of the larvae that are eaten, or simply killed, before any conclusion can be drawn on the applicability of the selection-arena hypothesis. In the toads and salamanders, it would be quite interesting to know if we are dealing with a developmental switch, a genetic polymorphism, or a continuous reaction norm with a polygenic basis.

## Hypotheses pertaining only to certain lineages

Within any given lineage particular traits can suggest plausible alternative explanations. For example, Birney and Baird (1985) suggested a 4-step process to explain the adaptive advantage of polyovulation in mammals, including some bats, elephant shrews, tenrecs, the plains viscacha, and the pronghorn antelope. When (a) large precocial young are selected, (b) a litter size of two is favored, and (c) balanced implantation of one embryo in each horn of a bifurcate uterus is necessary for mobility in avoiding predation, then (d) ovulation rates should increase to the point where at least one viable embryo is produced in each horn of the uterus. Birney and Baird stated that they could not, through this mechanism, explain the high rates of ovulation observed in one elephant shrew and the plains viscacha, where up to 50–60 times as many ova are produced as can be carried to term. Their hypothesis is clearly local, lineage specific, not exclusive of selection hypotheses, but plausible.

It has also been suggested that some plants may produce many flowers in which they have 'no intention' of ever forming seeds, simply to create an impressive visual display that attracts pollinators that would then home in on the functional flowers through non-visual cues at short range (Schemske, 1980; Willson and Schemske, 1980). These flowers might produce the appearance of selection-arenas, but if they were actually functioning purely according to the pollinator-attraction hypothesis, they should never produce fertilized embryos. Sutherland's (1986) review indicates that pollinator limitation and pollinator attraction are hypotheses not widely applicable in plants, although perhaps functioning in specific cases.

Similar additional explanations for apparent or real selection arenas could in principle be erected and based on the purely local taxonomic features of almost any group. They should not be considered as logical alternatives only one of which could be correct, but as local members of a series of hypotheses that could all potentially hold.

## How the selection-arena hypothesis could be tested

The selection-arena hypothesis has already survived experimental tests in which the selection criterion was either expectation of resources (especially well done for flowering plants and reviewed by Stephenson, 1981, and for mice by Koenig, in press), microgametophyte vigor (Mulcahy et al., 1975; McKenna, 1986; Schlichting et al., 1987), or sex ratio (Clutton-Brock et al., 1986; Gosling 1986). In haplodiploid parasitic Hymenoptera where the arena is a host insect egg, larva, or pupa, the female can control sex ratio simply by deciding whether or not to fertilize the egg (Charnov, 1982), and thus is able to avoid having to select among sperm or zygotes to achieve this end. The arena is there, and the selection pressure is there, but the mechanism is not necessary because a more efficient one is available.

I therefore consider the application of the selection-arena hypothesis to decisions based on expectation of resources, microgametophyte vigor, or fitness advantages of one sex over the other to be fairly well established. Further experimental tests are well justified, and I think their design is clear. Less well-established is the idea that a female mammal might select among her offspring on the basis of some molecular indicator of expectation of survival. Here the selection criterion is a much more finely graded genetic difference than the simple distinction between male and female or rate of pollen tube growth.

Experimental manipulations of mice and rats seem within the reach of available technology, and could be aimed at testing the critical assumptions of the selection arena hypothesis. I take it as self-evident (1) that zygotes are cheap in mammals (a human female is born with over a million ova but uses only about 300–400 in 30 years of menstruation) and (2) that pregnancy and parental care through weaning cost the female time, energy and risk. One can now assert quite generally (3) that offspring produced by outcrossed matings vary in fitness (cf. Clutton-Brock, 1987). The key assumption is therefore (4): that variation in fitness among offspring can be identified at an early stage of the life cycle and manipulated to the mother's advantage.

The selection of a model system seems clear: either rats or mice, where methods and background are well known. The general statement of experimental goals also seems clear: to manipulate the genetic and immunological characteristics of zygotes, study rejection rates by mothers, then see if the rejected class of zygotes have lower expectations of lifetime reproductive success through viability, fecundity, or both. More concretely, one would want to carry out at least the following manipulations. (1) Vary the degree of homozygosity at the MHC loci and measure the rate of rejection by the mother or foster-mother. (2) Suppress the rejection mechanisms *in utero*, allowing all offspring to reach birth, in both diseased and healthy mothers. Then see if the variance in fitness among offspring is higher in the suppressed mothers, and yet higher in those suppressed mothers that were also diseased.

A methodologically more straightforward test is nevertheless indirect: measure the gene and genotype frequencies of the MHC loci. If they are not in linkage disequilibrium, then selection against homozygotes *in utero* is not occurring (Koella, pers. comm.).

Two other species, an elephant shrew and the plains viscacha, in which 50–60 times as many ova are produced as can be carried to term, also clearly deserve further study (cf. Birney and Baird, 1985).

Such work would combine the latest in reproductive biotechnology with some of the most difficult problems in evolutionary ecology (measurement of lifetime fitnesses in natural situations). It should certainly not be undertaken lightly. That it should be undertaken at all is convincingly argued by the fact that it would exemplify in female mammals a process also involved in the maintenance of sex, kronism, surplus flower production, sibling cannibalism, adaptive adjustment of the sex-ratio, polyovulation, the selection of zygotes, and the rate of spontaneous abortions in humans.

The notion that natural selection would have designed adaptations that themselves use natural selection to be effective is especially appealing. They resemble recursive algorithms that invoke themselves with a compact computational elegance.

## Summary

The selection arena hypothesis offers one answer to a puzzling question. Why do some organisms produce many more fertilized zygotes than are actually reared to hatching, birth, or release – then neglect, discard, resorb, or eat some of them, or allow them to eat each other? It makes four assumptions: (1) zygotes are cheap; (2) after conception the investment of parental time, energy, or risk into offspring continues; (3) offspring vary in fitness; (4) variation in offspring fitness can be identified by the mother at an early stage of the life cycle. If these assumptions hold, then one general prediction follows: the parent should overproduce zygotes, identify those with lower expected fitness, then either kill and reabsorb them, let them be eaten by sibs, or simply stop feeding them in order to invest in more promising offspring.

The explanation appears to apply to a wide range of phenomena whose common cause had not previously been appreciated. These include: (1) polyovulation in some bats, tenrecs, the plains viscacha, and the pronghorn antelope; (2) cases of recurrent, consecutive, spontaneous abortions in humans; (3) some cases of surplus flower production and fruit abortion; (4) sex-ratio adjustment in red deer, mice, and coypus; (5) some types of cannibalism, including possible cases in mice, sharks, and wasps.

Some cases that might be explained by the selection arena hypothesis are also plausibly explained by other causes, including bet-hedging reproductive investment in the face of unpredictable food supplies, and inter-specific or inter-familial aggression as an alternative to parent-offspring or sib-sib cannibalism.

Acknowledgments. Comments by Jacob Koella, Arie van Noordwijk, Barbara Koenig, Pierre-Henri Gouyon, Wolfgang Wickler, Paul Schmid-Hempel, Hugh Rowell, Graham Bell and Andy Stephenson improved the manuscript. A talk by Hanne Ostergard stimulated these ideas. The preparation of this chapter was supported by the Swiss National Science Foundation, Grant 3.642-0.84.

# References

Abplanalp, H. 1956. Selection procedures for poultry stocks with many hatches. Poultry Sci. 35: 1285–1304

Adams, M. W., A. H. Ellinghoe, and E. C. Rossmann. 1971. Biological uniformity disease epidemics. Bioscience 21: 1067–1070.

Ahokas, H. 1979. Cytoplasmic male-sterility in Barley. III. Maintenance of sterility and restoration in the msml cytoplasm. Euphytica 28: 409–420.

Aida, T. 1936. Sex reversal in *Aplocheilus latipes* and a new explanation of sex differentiation. Genetics 21: 136–153.

Alberts, B., D. Bray, J. Lewis, M. Raff, K. Roberts, and J. D. Watson. 1983. Molecular Biology of the Cell. Garland, New York, London.

Alcaino, H. A., N. V. Baker, and R. A. Fisk. 1976. Enzyme polymorphism in *Fasciola hepatica* L.: Esterases. Am. J. vet. Res. 37: 1153–1157.

Alexander, R. D., and P. Sherman. 1977. Local mate competition and parental investment in social insects. Science 196: 494–500.

Allard, R. W. 1961. Relationship between genetic diversity and consistency of performance in different evironments. Crop Sci. 1: 127–133.

Allard, R. W., and J. Adams. 1969. Population studies in predominantly self-fertilizing species. XIII. Intergenotypic competition and population structure in barley and wheat. Am. Nat. 103: 621–645.

Altmann, S. A. 1979. Altruistic behaviour: the fallacy of kin deployment. Anim. Behav. 27: 958–959.

Anderson, R. M. 1979. The persistence of direct life cycle infectious diseases within populations of hosts. In: Some Mathematical Questions in Biology. Lectures on Mathematics in the Life Sciences, Vol. 12. (ed. S. A. Levin), pp. 1–67. Am. Math. Soc. Providence, R. I.

Anderson, R. M., and R. M. May (eds). 1982. Population Biology of Infectious Diseases. Dahlem Konferenzen. Springer Verlag, Berlin, Heidelberg, New York.

Anderson, R. M., and R. M. May. 1979. Population biology of infectious diseases I. Nature 280: 361–367.

Anderson, R. M., and R. M. May. 1981. The population dynamics of micro-parasites and their invertebrate hosts. Phil. Trans. R. Soc. London B 291: 451–524.

Andersson, M. 1982. Sexual selection, natural selection and quality advertisement. Biol. J. Linn. Soc. 17: 375–393.

Andersson, M. 1986a. Evolution of condition-dependent sex ornaments and mating preferences: sexual selection based on viability differences. Evolution 40: 816.

Andersson, M. 1986b. Sexual selection and the importance of viability differences: a reply. J. theor. Biol. 120: 251–254.

Antonovics, J., and N. C. Ellstrand. 1984. Experimental studies of the evolutionary significance of sex. I. A test of the frequency-dependent selection hypothesis. Evolution 38: 103–115.

Antonovics, J., and N. C. Ellstrand. 1985. The fitness of dispersed progeny: experimental studies with *Anthoxanthum odoratum*. In: Genetic Differentiation and Dispersal in Plants (eds P. Jacquard et al.), pp. 369–381. Springer Verlag, Berlin.

Antonovics, J., N. C. Ellstrand, and R. N. Brandon. 1987. Genetic variation and environmental variation: expectations and experiments. In: Plant Evolutionary Biology (eds L. Gottlieb, and S. Jain): in press.

Antonovics, J., and R. Primack. 1982. Experimental ecological genetics in *Plantago*. VI. The demography of seedling transplants of *P. lanceolata*. J. Ecol. 70: 55–75.

Apple, J. L. 1977. In: Plant Disease, Vol. 1 (eds J. G. Horsfall and E. B. Cowling), pp. 79–101. Academic Press, New York.

Arber, W., and D. Dussoix. 1962. Host specificity of DNA produced by *Escherichia coli* I. Host controlled modification of bacteriophage II. Control over acceptance of DNA from infecting phage. J. molec. Biol. 5: 18–49.

Arnold, S. J. 1976. Sexual behavior, sexual interference and sexual defense in the salamanders *Ambystoma maculatum, Ambystoma tigrinum* and *Plethodon jordani.* Z. Tierpsychol. 42: 247–300.

Arnold, S. J. 1983. Sexual selection: the interface of theory and empiricism. In: Mate Choice (ed. P. Bateson), pp. 67–107. Cambridge University Press, Cambridge.

Arnold, S. J. 1985. Quantitative genetic models of sexual selection. Experientia 41: 1296–1310.

Arnold, S. J., and M. J. Wade. 1984. On the measurement of natural and sexual selection: theory. Evolution 38: 709–719.

Austad, S. N. 1984. A classification of alternative reproductive behaviors and methods for field-testing ESS models. Am. Zool. 24: 309–319.

Aviles, L. 1986. Sex ratio bias and possible group selection in the social spider *Anelosimus eximius.* Am. Nat. 128: 1–12.

Avtalion, R. R., and I. S. Hammerman. 1978. Sex determination in *Sarotherodon* (Tilapia) I. Introduction to a theory of autosomal influence. Bamidgeh 30: 110–115.

Babcock E. B., and G. L. Stebbins. 1938. The American species of *Crepis.* Their interrelationships and distribution as affected by polyploidy and apomixis. Carnegie Inst. Washington Publ. No. 504.

Bacci, C. 1965. Sex Determination. Pergamon Press. Oxford.

Bacci, G. C., and A. M. Vaccari. 1961. Endomeiosis and sex determination in *Daphnia pulex.* Experientia 17: 505–506.

Baker, A. M. J., and D. H. Dalby. 1981. Morphological variation between some isolated populations of *Silene maritima* within the British isles with particular reference to inland populations of metalliferous soils. New Phytol. 84: 123–138.

Baker, B. S., and K. A. Ridge. 1980. Sex and the single cell. I. On the action of major loci affecting sex determination in *Drosophila melanogaster.* Genetics 94: 383–423.

Baker, H. G. 1959. Reproductive methods as factors in speciation in flowering plants. Cold Spring Harb. Symp. quant. Biol. 24: 177–191.

Baker, H. G. 1974. The evolution of weeds. A. Rev. Ecol. Syst. 5: 1–24.

Baker, H. G. 1975. Sporophyte-gametophyte interactions in *Linum* and other genera with heteromorphic self-incompatability. In: Gamete Competition in Plants and Animals (ed. D. L. Mulcahy). North-Holland Publ. Co., Amsterdam.

Baker, H. G. 1976. 'Mistake' pollination as a reproductive system with special reference to the *Caricaceae.* In: Tropical Trees, Breeding and Conservation (eds J. Burley and B. T. Styles). Academic Press, London.

Baker, J. P. 1979. Electrophoretic studies on populations of *Myzus persicae* in Scotland from October to December 1976. Ann. appl. Biol. 91: 159–164.

Baltzer, F. 1912. Ueber die Entwicklungsgeschichte von *Bonellia.* Verh. dt. zool. Ges. 22: 252–261.

Banga, S. S., and K. S. Labana. 1985. Male-sterility in Indian Mustard (*Brassica junceae* L) IV. Genetics of MS-4. Can. J. gen. Cytol. 27: 487–490.

Barber, W. S. 1913. The remarkable life-history of a new family (Micromaltidae) of beetles. Proc. Biol. Soc. Wash. 26: 185–190.

Barker, D. M., and P. D. N. Hebert. 1986. Secondary sex ratio of the cyclic parthenogen *Daphnia magna* (Crustacea: Cladocera) in the Canadian arctic. Can. J. Zool. 64: 1137–1143.

Barker, W. W. 1966. Apomixis in the genus *Arnica* (Compositae). PhD thesis, University of Washington, Seattle.

352

Barrett, J. A. 1981. The evolutionary consequences of monoculture, In: Genetic Consequences of Man-made Change (eds J. A. Bishop and L. M. Cook), pp. 209–249. Academic Press, London.

Bateman, A. J. 1948. Intra-sexual selection in *Drosophila*. Heredity 2: 349–368.

Bateson, P. G. 1983. Mate Choice. Cambridge University Press, Cambridge.

Battaglia, B., and H. Smith. 1961. The Darwinian fitness of polymorphic and monomorphic population of *Drosophila pseudoobscura* at 16C. Heredity 16: 475–484.

Bawa, K. S. 1980a. Evolution of dioecy in flowering plants. A. Rev. Ecol. Syst. 11: 15–39.

Bawa, K. S. 1980b. Mimicry of male by female flowers and intrasexual competition for pollinators in *Jacaratia dolichaula* (D. Smith) Woodson (Caricaceae). Evolution 34: 467–474.

Bawa, K. S., and C. J. Webb. 1984. Flower, fruit, and seed abortion in tropical forest trees: implications for the evolution of paternal and maternal reproductive strategies. Am. J. Bot. 71: 736–751.

Bawa, K. S., and P. A. Opler. 1975. Dioecism in tropical forest trees. Evolution 29: 167–179.

Bayer, R. J., and G. L. Stebbins. 1983. Distribution of sexual and apomictic populations of *Antennaria parlinii*. Evolution 37: 555–561.

Beaman, H. G. 1957. The systematics and evolution of *Townsendia* (Compositae). Contrib. Gray Herbarium (Harvard) No. 183.

Beardmore, J. A., Th. Dobzhansky, and O. Pavlovsky. 1960. An attempt to compare the fitness of polymorphic and monomorphic experimental populations of *Drosophila pseudoobscura*. Heredity 14: 19–33.

Beer, A. E., J. F. Quebbeman, J. W. T. Ayers, and R. F. Haines. 1981. Major histocompatibility complex antigens, maternal and paternal immune responses, and chronic habitual abortions in humans. Am. J. Obstet. Gynec. 141: 987–999.

Beer, A. E., and J. F. Quebbeman. 1982. The immunobiology and immunopathology of the maternal-fetal relationship. In: Physiopathology of Hypophysial Disturbances and Diseases of Reproduction. pp. 289–326. A. R. Liss, Inc.

Beer, A. E., and J. O. Sio. 1982. Placenta as an immunological barrier. Biol. Reprod. 26: 15–27.

Beermann, W. 1955. Geschlechtszellenbestimmung und Evolution der genetischen Y-Chromosomen bei *Chironomus*. Biol. Zbl. 74: 525–544.

Bell, G. 1978a. The evolution of anisogamy. J. theor. Biol. 73: 247–270.

Bell, G. 1978b. The handicap principle in sexual selection. Evolution 32: 872–885.

Bell, G. 1982. The Masterpiece of Nature – The Evolution and Genetics of Sexuality. University of California Press, Berkeley, California.

Bell, G. 1985a. On the function of flowers. Proc. R. Soc. London B 224: 223–265.

Bell, G. 1985b. Two theories of sex and variation. Experientia 41: 1235–1244.

Bell, G., and M. Praiss. 1986. Optimality and constraint in a self-fetilized alga. Evolution 40: 194–198.

Bemds, W. P., and D. P. Barash. 1979. Early termination of parental investment in mammals, including humans. In: Evolutionary Biology and Human Social Behavior: An Anthropological Perspective (eds N. A. Chagnon and W. Irons), pp. 487–506. Duxbury, North Scituate, Mass.

Benazzi, M., and G. Benazzi Lentati. 1976. Animal Cytogenetics. Vol 1. Platyhelminthes. (ed. B. John) Gebrüder Bornträger, Berlin.

Bennett, M. D. 1972. Nuclear DNA content and minimum generation time in herbaceous plants. Proc. R. Soc. Lond. B 181: 109–135.

Bennett, M. D. 1973. The duration of meiosis. In: The Cell Cycle in Development and Differentiation (eds M. Balls and F. S. Billett), p. 111–131. Cambridge University Press, Cambridge.

Beranek, A. P. 1974. Esterase variation and organophosphate resistance in populations of *Aphis fabae* and *Myzus persicae*. Ent. exp. app. 17: 129–142.

Bergelson, J., and P. Bierzychudek. MS. Phenotypic variation among sexual and apomictic progeny of *Antennaria parvifolia*.

Berger, E. 1976. Heterosis and the maintenance of enzyme polymorphism. Am. Nat. 110: 823–839.

Berger, R. D. 1973. Infection rates of *Cercospora apii* in mixed populations of susceptible and tolerant celery. Phytopathology 63: 535–537.

Bernstein, H. 1977. Germ line recombination may be primarily a manifestation of DNA repair processes. J. theor. Biol. 69: 371–380.

Bernstein, H. 1983. Recombinational repair may be an important function of sexual reproduction. Bioscience 33: 326–331.

Bernstein, H., H. C. Byerly, F. A. Hopf, and R. E. Michod. 1985. Genetic damage, mutation, and the evolution of sex. Science 229: 1277–1281.

Bernstein, H., G. S. Byers, and R. E. Michod. 1981. Evolution of sexual reproduction: Importance of DNA repair, complementation, and variation. Am. Nat. 117: 537–549.

Bertin, R. I. 1982. Paternity and fruit production in trumpet creeper *(Campsis radicans)*. Am. Nat. 119: 694–709.

Bertin, R. I. 1985. Non-random fruit production in *Campsis radicans:* between year consistency and effect of prior pollination. Am. Nat. 126: 750–759.

Bertin, R. I., and A. G. Stephenson. 1983. Toward a definition of sexual selection. Evol. Theor. 6: 293–295.

Beverly, S. M., and A. C. Wilson. 1985. Ancient origin for Hawaiian Drosphilinae inferred from protein comparisons. Proc. natl Acad. Sci. USA 82: 4753–4757.

Bhalla, S. C., and R. R. Sokal. 1964. Competition among genotypes in the housefly at varying densities and proportions (the green strain). Evolution 18: 213–230.

Bierzychudek: P. 1981. Pollinator limitation of plant reproductive effort. Am. Nat. 117: 838–840.

Bierzychudek: P. 1985. Patterns in plant parthenogenesis. Experientia 41: 1255–1264.

Bierzychudek: P. 1987. Pollinators increase the cost of sex by avoiding female flowers. Ecology: in press.

Bingham, R. T., and A. R. Squillace. 1955. Self-compatibility and effects of self-fertility in western white pine. For. Sci. 1: 121–129.

Birky, C. W. 1978. Transmission genetics of mitochondria and chloroplasts. A. Rev. Genet. 12: 471–512.

Birky, C. W. 1983. Relaxed cellular controls and organelle heredity. Science 222: 468–475.

Birley, A. J., and J. A. Beardmore. 1977. Genetical composition, temperature, density and selection in an enzyme polymorphism. Heredity 39: 133–144.

Birney, E. C., and D. D. Baird. 1985. Why do some mammals polyovulate to produce a litter of two? Am. Nat. 126: 136–140.

Blackman, R. L. 1972. The inheritance of life-cycle differences in *Myzus persicae* (Sulzer). Bull. ent. Res. 62: 281–294.

Blackman, R. L. 1979. Stability and variation in aphid clonal lineages. Biol. J. Linn Soc. 11: 259–277.

Blackman, R. L. 1980. Chromosomes and parthenogenesis in aphids. In: Insect Cytogenetics (eds R. L. Blackman, G. M. Hewitt and M. Ashburner). Symp. R. ent. Soc. Lond. 10. Blackwell Scientific, Oxford.

Blackman, R. L. 1981. Species, sex and parthenogenesis in aphids. In: The Evolving Biosphere (ed. P. L. Forey). Cambridge University Press, Cambridge.

Blackman, R. L. 1985. Aphid cytology and genetics. In: Evolution and Biosystematics of Aphids. Proc. Int. Aphid. Symp. Polska. Akad. Nauk. Inst. Zool.

Blakeslee, A. F. 1904. Sexual reproduction in the Mucorineae. Proc. Am. Acad. Sci. 40: 205–319.

Blank, J. L., and V. Nolan Jr. 1980. Offspring sex ratio in Red-Winged Blackbirds is dependent on maternal age. Proc. natl Acad. Sci. USA 80: 6141–6145.

Bloom, B. R. 1979. Games parasites play: How parasites evade immune surveillance. Nature 279: 21–26.

Blum, M. S., and N. A. Blum (eds). 1979. Sexual Selection and Reproductive Competition in Insects. Academic Press, New York, San Francisco, London.

354

Blyth, D. E., R. L. Eisemann, and I. H. de Lacy. 1976. Two-way pattern analysis of a large data set to evaluate genotypic adaptation. Heredity 37: 215–230.

Boake, C. R. B. 1984. Male display and female preferences in the courtship of a gregarious cricket. Anim. Behav. 32: 690–697.

Boake, C. R. B. 1985. Genetic consequences of mate choice: a quantitative method for testing sexual selection theory. Science 277: 1061–1063.

Boake, C. R. B. 1986. A method for testing adaptive hypotheses of mate choice. Am. Nat. 127: 654–666.

Bolla, R. J., and L. S. Roberts. 1968. Gametogenesis and chromosomal complement in *Strongyloides ratti* (Nematoda: Rhabdiasoidea). J. Parasit. 54: 849–855.

Bookman, S. S. 1983. Costs and benefits of flower abscission and fruit abortion in *Asclepias speciosa*. Ecology 64: 264–273.

Bookman, S. S. 1984. Evidence for selective fruit production in *Asclepias*. Evolution 38: 72–86.

Borgia, G. 1979. Sexual selection and the evolution of mating systems. In: Sexual Selection and Reproductive Competition in Insects (eds M. S. Blum and N. A. Blum), pp. 19–80. Academic Press, New York.

Borlaug, N. E. 1959. The use of multilineal or composite varieties to control airborne epidemic deseases of self-pollinated crop plants. Proc. 1st int. Wheat Genetics Symp.: 12–26.

Borror, D. J., D. M. Delong, and C. A. Triplehorn. 1976. An Introduction to the Study of Insects. Holt, Reinhart and Winston, New York.

Bowman, T. E., and L. G. Abele. 1982. Classification of the Recent Crustacea. In: The Biology of the Crustacea, vol. 1 (ed. L. G. Abele). Academic Press, New York.

Bradbury, J. W., and M. B. Andersson (eds.). 1987. Sexual Selection: Testing The Alternatives. Dahlem Konferenzen. Springer Verlag, Berlin, Heidelberg, New York.

Bragg, A. N. 1964. Further study of predation and cannibalism in spadefoot tadpoles. Herpetologica 20: 17–24.

Breese, E. L. 1969. The measurement and significance of genotype-environment interactions in grasses. Heredity 24: 27–44.

Bremermann, H. J. 1979. Theory of spontaneous cell fusion. Sexuality in cell populations as an evolutionary stable strategy. Applications to immunology and cancer. J. theor. Biol. 76: 311–334.

Bremermann, H. J. 1980a. Sex and polymorphism and strategies of host-pathogen interactions. J. theor. Biol. 87: 641–702.

Bremermann, H. J. 1980b. Disease and epidemics from the pathogen's perspective. Applications to cervical cancer, ecological invasion, and bacteriophages. Technical Report PAM-17, Center for Pure and Applied Mathemathics, University of California, Berkeley.

Bremermann, H. J. 1983a. Game-theoretical models of parasite virulence. J. theor. Biol. 100: 255–274.

Bremermann, H. J. 1983b. Parasites at the origin of life. J. math. Biol. 16: 165–180.

Bremermann, H. J. 1985. The adaptive significance of sexuality. Experientia 41: 1245–1254.

Bremermann, H. J., and B. Fiedler. 1985. On the stability of polymorphic host-pathogen populations. J. theor. Biol. 117: 621–631.

Bremermann, H. J., and J. Pickering. 1983. A game theoretical model of parasite virulence. J. theor. Biol. 100: 411–426.

Bremermann, H. J., and H. Thieme. 1987. A competitive exclusion principle for pathogen virulence, manuscript.

Bridges, C. B. 1916. Non-disjunction as proof of the chromosome theory of heridity. Genetics 1: 1–52, 107–163.

Bridges, C. B. 1925. Sex in relation to genes and chromosomes. Am. Nat. 59: 127–137.

Britten, J. R. 1986. Rates of DNA sequence evolution differ between taxonomic groups. Science 231: 1393–1398.

Brockman, H. J., A. Grafen, and R. Dawkins. 1979. Evolutionarily stable strategy in a digger wasp. J. theor. Biol. 77: 473–496.

Brook, A. J. 1981. The Biology of Desmids. University of California Press, Berkeley, California.

Browne, R. A. 1980. Competition experiments between parthenogenetic and sexual strains of the brine shrimp, *Artemia salina*. Ecology 61: 471–474.

Browning, J. A. 1974. Relevance of knowledge about natural ecosystems to development of pest management programs for ecosystems. Am. phytopath. Soc. Proc. 1: 191–199.

Bütschli, O. 1887–1889. Protozoa III Ciliata. Bronn's Klassen und Ordnungen des Tierreichs. Winter Leibzig.

Bull, J. J. 1979. An advantage for the evolution of male haploidy and systems with similar genetic transmission. Heredity 43: 361–381.

Bull, J. J. 1980. Sex determination in reptiles. Q. Rev. Biol. 55: 3–21.

Bull, J. J. 1981. Sex ratio evolution when fitness varies. Heredity 46: 9–26.

Bull, J. J. 1983. Evolution of Sex Determining Mechanisms. Benjamin/Cummings Publishing.

Bull, J. J., and E. L. Charnov. 1977. Changes in the heterogametic mechanism of sex determination. Heredity 39: 1–14.

Bull, J. J., and E. L. Charnov. 1985. On irreversible evolution. Evolution 39: 1149–1155.

Bull, J. J., R. C. Vogt, and C. J. McCoy. 1982. Sex determining temperatures in turtles: a geographic comparison. Evolution 36: 326–332.

Bulmer, M. G. 1980a. The sib competition model for the maintenance of sex and recombination. J. theor. Biol. 82: 335–345.

Bulmer, M. G. 1980b. The Mathematical Theory of Quantitative Genetics. Oxford University Press, Oxford.

Bulmer, M. G. 1986. Sex ratio theory. Science 233: 1436–1437.

Bulmer, M. G. 1987. Sex ratios in geographically structured populations. T. R. E. E.: in press.

Bulmer, M. G., and J. J. Bull. 1982. Models of polygenic sex determination and sex ratio control. Evolution 36: 13–26.

Bulmer, M. G., and P. D. Taylor. 1980. Dispersal and the sex ratio. Nature 284: 448–449.

Bulnheim, H.-P. 1978. Interaction between genetic, external and parasitic factors in sex determination of the crustacean amphipod *Gammarus duebeni*. Helgoländer wiss. Meeresunters. 31: 1–33.

Burdon, J. J., and D. R. Marshall. 1981. Biological control and the reproductive mode of weeds. J. appl. Ecol. 18: 649–658.

Burdon, J. J., and G. A. Chilvers. 1976. Epidemiology of *Pythium*-induced damping-off in mixed-species seedling stands. Ann. appl. Biol. 82: 233–240.

Burdon, J. J., and R. Whitbread. 1979. Rates of increase of barley mildew in mixed stands of barley and wheat. J. appl. Ecol. 16: 253–258.

Burger, M. M., and R. S. Turner, W. J. Kuhns, and G. Weinbaum. 1975. A possible model for cell-cell recognition via surface macromolecules. Phil. Trans. R. Soc. Lond. B. 271: 379–393.

Burnet, F. M. 1971. 'Self-recognition' in colonial marine forms and flowering plants in relation to the evolution of immunity. Nature 232: 230–235.

Burton, J. A. (ed.). 1973. Owls of the World: Their Evolution, Structure, and Ecology. E. P. Dutton, New York.

Butcher, A. C., J. Croft, and M. Grindle. 1972. Use of genotype-environment interaction analysis in the study of natural populations of *Aspergillus nidulans*. Heredity 29: 263–283.

Cahn, M. G., and J. L. Harper. 1976. The biology of the leaf-mark polymorphism in *Trifolium repens* L. I. Distribution of phenotypes at a local scale. Heredity 37: 309–325.

Campell, B. (ed.) 1972. Sexual Selection and the Descent of Man. Aldine, Chicago.

Cannon, L. R. G. 1971. The life-cycle of *Bunodera sacculata* and *B. luciopercae* in Algoquin Park, Ontario. Can. J. Zool. 49: 1417–1429.

Caraco, T. 1981. Risk-sensitivity and foraging groups. Ecology 62: 527–531.

Carter, C. I. 1971. Conifer woolly aphids (Adelgidae) in Britain. Forestry Commission Bulletin 42. Her Majesty's Stationery Office, London.

Carvalho, G. R. 1985. Studies on the ecological genetics of *Daphnia magna* Straus. PhD. Thesis. University of Wales.

Case, T. J., and M. L. Taper. 1986. On the coexistence and coevolution of asexual and sexual competitors. Evolution 40: 366–387.

Catling, P. M. 1982. Breeding systems of Northeastern North American *Spiranthes* (Orchidaceae). Can. J. Bot. 60: 3017–3039.

Caullery, M., and M. Comas. 1928. Le déterminisme du sexe chez un nematode *(Paramermis contorta),* parasite des larves des *Chironomus.* C. r. Acad. Sci. Paris 186: 646–648.

Cavalier-Smith, T. 1978. Nuclear volume control by nucleosceletal DNA, selection for cell volume and cell growth rate, and the solution of the DNA C-value paradox. J. Cell Sci. 34: 247–278.

Cavalier-Smith, T. 1985. Cell volume and the evolution of eukaryotic genome size. In: The Evolution of Genome Size (ed. T. Cavalier-Smith), pp. 105–184. Wiley, New York.

Cavalier-Smith, T. 1985. The Evolution of Genome Size. John Wiley & Sons, Chichester.

Celarier, R. P., K. L. Mehra, and M. L. Wulf. 1958. Cytogeography of the *Dichanthium annulatum* complex. Brittonia 10: 59–72.

Chanock, R. M., and R. A. Lerner, 1984. Modern Approaches to Vaccines. Molecular and Chemical Basis of Virus Virulence Immunogenicity. Cold Spring Harbor Laboratory. Cold Spring Harbor, New York.

Charles, A. H. 1964. Differential survival of plant types in swards. J. Brit. Grassld Soc. 19: 198–204.

Charlesworth, B. 1977. Population genetics, demography and the sex ratio. In: Measuring Selection in Natural Populations (eds F. B. Christiansen and T. M. Fenchel), pp. 345–363. Springer-Verlag, Berlin.

Charlesworth, B. 1978. The population genetics of anisogamy. J. theor. Biol. 73: 347–357.

Charlesworth, B. 1980. Evolution in age-structured populations. Cambridge University Press, Cambridge.

Charlesworth, B. 1983. Mating types and uniparental transmission of chloroplast genes. Nature 304: 211.

Charlesworth, B. 1987. The heritability of fitness. In: Sexual Selection: Testing the Alternatives (eds J. Bradbury and M. Andersson). Dahlem Conference, Berlin.

Charlesworth, D. 1981. A further study of the problems of the maintenance of females in gynodioecious species. Heredity 46: 27–39.

Charlesworth, D. 1984. Androdioecy and the evolution of dioecy. Biol. J. Linn. Soc. 23: 333–348.

Charlesworth, D., and B. Charlesworth. 1978. A model for the evolution of dioecy and gynodioecy. Am. Nat. 112: 975–997.

Charlesworth, D., and B. Charlesworth. 1979. The evolution and breakdown of self-incompatibility systems. Heredity 43: 41–55.

Charlesworth, D., and B. Charlesworth. 1980. Sex differences in fitness and selection for centric fusions between sex-chromosomes and autosomes. Genet. Res. 35: 205–214.

Charlesworth, D., and B. Charlesworth. 1981. Allocation of resources to male and female function in hermaphrodites. Biol. J. Linn. Soc. 15: 57–74.

Charlesworth, D., and B. Charlesworth. MS. The effect of investment in attractive structures on allocation to male and female functions in plants.

Charnier, M. 1966. Action de la température sur la sex-ratio chez l'embryon d'*Agama agama* (Agamidae, Lacertilien). Soc. biol. Ouest Afr. 160: 620–622.

Charnov, E. L. 1978. Sex ratio selection in eusocial hymenoptera. Am. Nat. 112: 317–326.

Charnov, E. L. 1979a. Natural selection and sex change in pandalid shrimp: a test of a life history theory. Am. Nat. 113: 715–734.

Charnov, E. L. 1979b. Simultaneous hermaphroditism and sexual selection. Proc. natl Acad. Sci. USA 76: 2480–2484.

Charnov, E. L. 1982. The Theory of Sex Allocation. Princeton University Press, Princeton.

Charnov, E. L., and J. J. Bull. 1977. When is sex environmentally determined? Nature 266: 828–830.

Charnov, E. L., and J. J. Bull. 1986. Sex allocation, pollinator attraction, and fruit dispersal in cosexual plants. J. theor. Biol. 118: 321–326.

Charnov, E. L., R. L. Los-den Hartogh, W. T. Jones and J. van den Assem. 1981. Sex ratio evolution in a variable environment. Nature 289: 27–33.

Charnov, E. L., J. Maynard Smith, and J. J. Bull. 1976a. Why be a hermaphrodite? Nature 263: 125–126.

Charnov, E. L., G. H. Orians, and K. Hyatt. 1976b. The ecological implications of resource depression. Am. Nat. 110: 247–259.

Cherfas, J., and J. Gribbin. 1984. The Redundant Male. Pantheon, New York.

Cheverud, J. M. 1984. Evolution by kin selection: a quantitative genetic model illustrated by maternal performance in mice. Evolution 38: 766-777.

Christie, J. R. 1929. Some observations on sex in the Mermithidae. J. exp. Zool. 53: 59–76.

Clark, A. B. 1981. Sex ratio and local resource competition in a prosimian primate. Science 201: 163–165.

Clark, W. C. 1974. Interpretation of life history pattern in the Digenea. Int. J. Parasit. 4: 115–123.

Clarke, B. C. 1976. The ecological genetics of host-parasite relationships. In: Genetic Aspects of Host-parasite Relationships (eds A. E. R. Taylor and R. Muller). Symp. Soc. Parasit. 14: 87–103.

Clarke, B. C. 1979. The evolution of genetic diversity. Proc. R. Soc. London (B) 205: 453–474.

Clausen, J. 1954. Partial apomixis as an equilibrium system in evolution. Caryologia, 6 suppl.: 469–479.

Clausen, J., and W. M. Hiesey. 1958. Experimental studies on the nature of species. IV. Genetic structure of ecological races. Carnegie Inst. of Washington, Publ. 615.

Clermont, Y. 1977. Spermatogenesis. In: Frontiers in Reproduction and Fertility Control (eds R. D. Greep and M. A. Koblinsky), pp. 293–301. M. I. T. Press, Cambridge, Massachusetts.

Cline, T. W. 1979. A male-specific lethal mutation in *Drosophila melanogaster* that transforms sex. Devl. Biol. 72: 266–275.

Clutton-Brock, T. H. 1986. Sex ratio variation in birds. Ibis 128: 317–329.

Clutton-Brock, T. H. (ed.). 1987. Lifetime Reproductive Success. Univ. of Chicago Press: in press.

Clutton-Brock, T. H., F. E. Guinness, and S. D. Albon. 1982. Red Deer; Behavior and Ecology of Two Sexes. University of Chicago Press, Chicago.

Clutton-Brock, T. H., S. D. Albon and F. E. Guinness. 1986. Great expectations: dominance, breeding success and offspring sex ratio in red deer. Anim. Behav. 34: 460–471.

Clutton-Brock, T. H., and G. R. Iason. 1986. Sex ratio variation in mammals. Q. Rev. Biol. 61: 339–374.

Cognetti, G. 1961. Endomeiosis in parthenogenetic lines of aphids. Experientia 17: 168–169.

Cohen, D. 1966. Optimizing reproduction in a randomly varying environment. J. theor. Biol. 12: 119–129.

Cohen, D. 1968. A general model of optimal reproduction in a randomly varying environment. J. Ecol. 56: 219–228.

Collins, J. P., and J. E. Cheek. 1983. Effect of food and density on development of typical and cannibalistic salamander larvae in *Ambystoma tigrinum nebulosum*. Am. Zool. 23: 77–84.

Colwell, R. K. 1981. Group selection is implicated in the evolution of female-biased sex ratios. Nature 290: 401–404.

Conover, D. O. 1983. Adaptive significance of temperature-dependent sex determination in a fish. Am. Nat. 123: 297–313.

Conover, D. O., and B. E. Kynard. 1981. Environmental sex determination: Interaction of temperature and genotype in a fish. Science 213: 577–579.

Cook, L. M. 1984. The problem. In: Evolutionary Dynamics of Genetic Diversity. (ed. G. S. Mani). Springer-Verlag, Berlin, Heidelberg.

Coombe, D. R., P. L. Ey, and C. R. Jenkin. 1984. Self/non-self recognition in invertebrates. Q. Rev. Biol. 59: 231–255.

358

Cooper, J. 1984. Expected time to extinction and the concept of fundamental fitness. J. theor. Biol. 107: 603–629.

Corbet, P. S., and A. Griffiths. 1963. Observations on the aquatic stages of two species of *Toxorhynchites* (Diptera: Culicidae) in Uganda. Proc. R. ent. Soc. London A 38: 125–135.

Correns. 1908. Die Rolle der männlichen Keimzellen bei der Geschlechtsbestimmung der gynodioecischen Pflanzen. Ber. dt. bot. Ges. 26a: 686–701.

Cort, W. W. 1921. Sex in the trematode family Schistosomidae. Science 53: 226–228.

Cosmides, L. M., and J. Tooby. 1981. Cytoplasmic inheritance and intragenomic conflict. J. theor. Biol. 89: 83–129.

Cotterman, C. W. 1953. Regular two-allele and three-allele phenotype systems. Am. J. Hum. Genet. 5: 193–235.

Couvet, D., F. Bonnemaison, and P. H. Gouyon. 1987. The maintenance of females among hermaphrodites: the importance of nuclear-cytoplasmic interactions. Heredity: in press.

Couvet, D., J. Henry, and P. Gouyon. 1985. Sexual selection in hermaphroditic plants: the case of gynodioecy. Am. Nat. 126: 294–299.

Cox, P. A., and J. A. Sethian. 1985. Gamete motion, search, and the evolution of anisogamy, oogamy, and chemotaxis. Am. Nat. 125: 74–101.

Craig, G. B. 1965. Genetic control of thermally-induced sex reversal in *Aedes aegypti*. In: Proc. XIIth Int. Congr. Ent., London (ed. P. Freemann), p. 263. R. ent. Soc., London.

Craik, C. S., W. J. Rutter, and R. Fletterick. 1983. Associated variation in protein structure. Science 220: 1125–1129.

Crandall, M. 1977. Mating-type interactions in micro-organisms. In: Receptors and Recognition, A, Vol 3 (eds. P. Cuatrecasas and M. F. Greaves). Chapman and Hall, London.

Crawford, T. J. 1984. What is a population? In: Evolutionary Ecology (ed. B. Shorrocks), pp. 135–173. Blackwell, Oxford.

Crease, T. J. 1981. Genetic variation in natural populations of *Daphnia*. MSc. thesis. University of Windsor.

Crease, T. J. 1986. Mitochondrial DNA variation in populations of *Daphnia pulex* Leydig reproducing by obligate and cyclic parthenogenesis. PhD thesis. Washington University, St. Louis.

Crew, F. A. E. 1965. Sex Determination. 4th edn. Methuen & Co. Ltd. Reprinted: Dover, New York.

Crow, J. F. 1979. Genes that violate Mendel's rules. Sci. Am. 240: 104–113.

Crow, J. F., and M. Kimura. 1965. Evolution in sexual and asexual populations. Am. Nat. 99: 439–450.

Crow, J. F., and M. Kimura. 1970. An Introduction to Population Genetics Theory. Harper and Row, New York.

Crowne, D. R., and W. H. Parker. 1981. Hybridization and agamospermy of *Bidens* in northwestern Ontario. Taxon 30: 749–760.

Crozier, R. H. 1977. Evolutionary genetics of the Hymenoptera. A. Rev. Ent. 22: 263–288.

Cruz, Y. P. 1986. The defender role of the precocious larvae of *Copidosomopsis tanytmemus* (Encyrtidae, Hymenoptera). J. exp. Zool. 237: 309–318.

Cruzan, M. B. 1986. Pollen tube distributions in *Nicotiana glauca:* evidence for density dependent growth. Am. J. Bot. 73: 902–907.

Cuellar, O. 1977. Animal parthenogenesis. Science 197: 837–843.

Currah, L. 1986. Pollen competition and the breeding system in the onion (*Allium cepa* L.) In: Pollen: Biology and implications for Plant Breeding (eds·D. L. Mulcahy et al.), pp. 375–379. Elsevier, New York.

Czurda, V. 1933. Ueber einige Grundbegriffe der Sexualitätstheorie. Beih. Z. b. ZentBibl. I 50: 196–210.

Dalton, D. C. 1967. Selection of growth in mice on two diets. Anim. Prod. 9: 425–434.

Darnell, J., H. Lodish, and D. Baltimore. 1986. Molecular Cell Biology. W. H. Freeman, New York.

Darwin, C. 1859. The Origin of Species by Means of Natural Selection, or the Preservation of Favoured Races in the Struggle of Life. John Murray, London.

Darwin, C. 1862. On the two forms, or dimorphic condition, in the species of *Primula,* and on their remarkable sexual relations. In: The Collected Papers of Charles Darwin (ed. P. H. Barrett). University of Chicago Press, 1977.

Darwin, C. 1871. The Descent of Man and Selection in Relation to Sex. John Murray, London.

Darwin, C. 1876. The Effect of Cross and Self Fertilisation in the Vegetable Kingdom. John Murray, London.

Darwin, C. 1877. The Different Forms of Flowers on Plants of the Same Species. John Murray, London.

Davies, M. S., and R. W. Snaydon. 1973a. Physiological differences among populations of *Anthoxanthum odoratum* L., collected from the Park Grass Experiment, Rothamsted. I. Response to calcium. J. appl. Ecol. 10: 33–45.

Davies, M. S., and R. W. Snaydon. 1973b. Ibid. II. Response to aluminium. J. appl. Ecol. 10: 47–55.

Davies, M. S., and R. W. Snaydon. 1974. Ibid. III. Response to phosphate. J. appl. Ecol. 11: 699–708.

Davies, M. S. 1975. Ibid. IV. Response to potassium and magnesium. J. appl. Ecol. 12: 953–964.

Davies, M. S., and R. W. Snaydon. 1976. Rapid population differentiation in a mosaic environment. III. Measures of selection pressures. Heredity 36: 59–66.

Davis, L. E., A. G. Stephenson, and J. A. Winsor. 1987. Pollen competition improves performance and reproductive outpout of the common zucchini squash under field conditions. Am. Soc. hort. Sci.: in press.

Day, C. S. D., and B. G. Galef. 1977. Pup cannibalism: one aspect of maternal behavior in golden hamsters. J. comp. Physiol. Psychol. 91: 1179–1189.

De Nettancourt, D. 1977. Incompatability in Angiosperms. Springer-Verlag, Berlin, Heidelberg.

Delannay, X. 1979. Etude cytologique et mathématique de l'évolution des mécanismes de reproduction dans le genre *Cirsium* Miller. Doct. Sci. Université Catholique de Louvain, Belgique.

Delannay, X., P. H. Gouyon, and G. Valdeyron. 1981. Mathematical study of the evolution of gynodioecy with cytoplasmic inheritance under the effect of a nuclear restorer gene. Genetics 99: 169–181.

Devlin, B., and A. G. Stephenson. 1985. Sex differential floral longevity, nectar secretion, and pollinator foraging in a protandrous species. Am. J. Bot. 72: 303–310.

Dingle, H., and J. P. Hegmann (eds). 1982. Evolution and Genetics of Life Histories. Springer Verlag, New York, Heidelberg, Berlin.

Dinoor, A. 1974. The role of the alternate host in amplifying the pathogenic variability of oat crown rust. Res. Rep. Sci. Agric. Hebrew University Jerusalem 1: 734–735.

Dixon, A. F. G. 1985. Structure of aphid populations. A. Rev. Ent. 30: 155–174.

Dobzhansky, Th., R. C. Lewontin, and O. Pavlovsky. 1964. The capacity for increase in chromosomally polymorphic and monomorphic populations of *Drosophila pseudoobscura.* Heredity 19: 597–614.

Dobzhansky, Th., and O. Pavlovsky. 1961. A further study of fitness of chromosomally polymorphic and monomorphic populations of *Drosophila pseudoobscura.* Heredity 16: 169–179.

Dobzhansky, Th. 1970. Genetics of the Evolutionary Process. Columbia University Press, 505 p.

Dodson, C. H. 1975. Coevolution of orchids and bees. In: Coevolution of Animals and Plants (eds L. Gilbert and P. Raven), pp. 91–99. University of Texas Press, Austin.

Dominey, W. J. 1983. Sexual selection, additive genetic variance and the 'phenotypic handicap'. J. theor. Biol. 101: 495–502.

Dommée, B., M. W. Assoud, and G. Valdeyron. 1978. Natural selection and gynodioecy in *Thymus vulgaris* L. Bot. J. Linnean Soc. 77: 17–28.

360

Dommée, B., J. L. Guillerm, and G. Valdeyron. 1983. Régime de production et hétérozygotie des populations dans une succession post-culturale. C. r. Acad. Sci. Paris, Serie III 296: 111–114.

Donaldson, J. S., and G. H. Walter. 1984. Sex ratios of *Spalangia endius.* (Hymenoptera: Pteromalidae) in relation to current theory. Ecol. Ent. 9: 395–402.

Doncaster, L. 1914. The Determination of Sex. Cambridge University Press, Cambridge.

Doncaster, L. 1916. Gametogenesis and sex-determination of the gall fly *Neuroterus lenticularis (Spathegaster baccarum).* III. Proc. R. Soc. Lond. B. 89: 183–200.

Doolittle, W. F., and C. Sapienza. 1980. Selfish genes, the phenotype paradigm and genome evolution. Nature 284: 601–603.

Dournon, C., and C. Houillon. 1983. Déterminisme génétique du sexe: démonstration à partir d'animaux à phenotype sexuel inverse sous l'action de la température chez l'Amphibian Urodele *Pleurodeles waltlii* Michah. C. r. Acad. Sci. Paris 296: 770–782.

Downhower, J. F., and K. B. Armitage. 1971. The yellow-bellied marmot and the evolution of polygyny. Am. Nat. 105: 355–370.

Drebes, G. 1977. Sexuality. In: The Biology of Diatoms (ed. D. Werner), pp. 250–283. University of California Press, Berkeley.

Dressler, R. L. 1968. Pollination by euglossine bees. Evolution 22: 202–210.

Dressler, R. L. 1981. The Orchids. Harvard University Press, Cambridge MA.

Dressler, R. L. 1982. Biology of the Orchid Bees (Euglossini). Ann. Rev. Ecol. Syst. 13: 373–394.

Duellmann, W. E., and L. Trueb. 1986. Biology of Amphibians. McGraw-Hill, New York, St. Louis and San Francisco.

Dumont, H. J. 1983. Biogeography of rotifers. Hydrobiologia 104: 19–30.

Eastop, V. F., and D. Hille Ris Lambers. 1976. Survey of World Aphids. W. Junk, The Hague.

Eberhard, W. G. 1982. Beetle horn dimorphism: making the best of a bad lot. Am. Nat. 119: 420–426.

Eberhard, W. G. 1985. Sexual Selection and Animal Genitalia. Harvard University Press, Cambridge MA, 244 p.

Ehrendorfer, F. 1980. Polyploidy and distribution. In: Polyploidy: biological relevance (ed. W. H. Lewis), pp. 45–60. Plenum Press, New York.

Eicher, E. M. 1982. Primary sex determining genes in mice: a brief review. In: Prospects for Sexing Mammalian Sperm (eds R. P. Amann and G. E. Seidel), pp. 121–135. Colorado Assoc. University Press, Colorado.

Eickwort, K. R. 1973. Cannibalism and kin selection in *Labidomera clivicollis* (Coleoptera: Chrysomelidae). Am. Nat. 107: 452–453.

Elena-Rossello, J. A., A. Kheyr-Pour, and G. Valdeyron. 1976. La structure génétique et le régime de la fécondation chez *Origanum vulgare* L. Répartition d'un marqueur enzymatique dans deux populations naturelles. C. r. Acad. Sci. Paris Ser. D, 283: 1587–1589.

Ellstrand, N. C., and J. Antonovics. 1985. Experimental studies of the evolutionary significance of sexual reproduction. II. A test of the density-dependent selection hypothesis. Evolution 39: 657–666.

Ellstrand, N. C., and D. L. Marshall. 1986. Patterns of multiple paternity of *Raphanus sativus.* Evolution 40: 837–842.

Emlen, S. T., J. M. Emlen, and S. A. Levin. 1986. Sex ratio selection in species with helpers at the nest. Am. Nat. 127: 1–8.

Engelke, H. 1935. Versuche mit Weizenmischsaaten. Z. Landwirt. 83: 63–83.

England, F. 1968. Competition in mixtures of herbage grasses. J. appl. Ecol. 5: 227–242.

Epperson, B. K., and M. T. Clegg. MS. Frequency dependent variation for outcrossing rate among flower color morphs of *Ipomoea purpurea.*

Eshel, I. 1978. On the handicap principle – a critical defense. J. theor. Biol. 70: 245–250.

Ettl, H., D. G. Mueller, K. Neumann, H. A. von Stosch, and W. Weber. 1967. Vegetative Fortpflanzung, Parthenogenese und Apogamie bei Algen. Handbuch der Pflanzenphysiologie, Band XVIII. Springer Verlag, Berlin.

Fahraeus, G. 1980. *Sorbus teodori* and its distribution in Gotland, Sweden. Svensk Bot. Tidskr. 74: 377–382.

Falconer, D. S. 1960. Selection of mice for growth on high and low planes of nutrition. Genet. Res. 1: 91–113.

Falconer, D. S. 1981. Introduction to Quantitative Genetics, 2nd. ed. Longmans, London.

Falconer, D. S., and M. Latyszewski. 1952. The environment in relation to selection. J. Genet. 51: 67–80.

Felsenstein, J. 1974. The evolutionary advantage of recombination. Genetics 78: 737–756.

Felsenstein, J. 1985. Recombination and sex: is Maynard Smith necessary? In: Evolution: Essays in Honour of John Maynard Smith (eds P. J. Greenwood, P. H. Harvey, and M. Slatkin), pp. 209–220. Cambridge University Press, Cambridge.

Felsenstein, J., and S. Yokoyama. 1976. The evolutionary advantage of recombination. II. Individual selection for recombination. Genetics 83: 845–859.

Ferguson, M. W. J. 1982. Temperature of egg incubation determines sex in *Alligator mississippiensis*. Nature 296: 850–853.

Ferguson, M. W. J., and T. Joanen. 1983. Temperature-dependent sex determination in *Alligator mississippiensis*. J. Zool. London 200: 143–177.

Fischer, E. A. 1980. The relationship between mating system and simultaneous hermaphroditism in the coral reef fish *Hypoplectrus nigricans* (Serranidae). Anim. Behav. 28: 620–634.

Fischer, E. A. 1981. Sexual allocation in a simultaneously hermaphroditic coral fish. Am. Nat. 117: 64–82.

Fischer, E. A. 1984. Egg trading in the Chalk Bass, *Serranus tortugarum*, a simultaneous hermaphrodite. Z. Tierpsychol. 66: 143–151.

Fischer, E. A., R. L. Foster, and M. Kipersztok. 1987. Local mate competition, body size and sex allocation in a simultaneous hermaphrodite. Evolution: in press.

Fischer, E. A., and A. Harper, 1986. Local mate competition in finite groups. Evolution 40: 862–863.

Fisher, R. A. 1915. The evolution of sexual preference. Eug. Rev. 1: 184–192.

Fisher, R. A. 1930. The Genetical Theory of Natural Selection. Clarendon Press, Oxford.

Fisher, R. A. 1941. Average excess and average effect of a gene substitution. Ann. Eug. 11: 53–63.

Fisher, R. A. 1958. The Genetical Theory of Natural Selection. 2nd edn. Dover, New York.

Fitch, W. M., and R. Atchley. 1985. Evolution in inbred strains of mice appears rapid. Science 228: 1169–1175.

Fletcher, M., P. T. LoVerde, and D. S. Woodruff. 1981. Genetic variation in *Schistosoma mansoni*: enzyme polymorphism in populations from Africa, Southwest Asia, South America, and the West Indies. Am. J. trop. Med. Hyg. 30: 406–421.

Flint, R. F. 1957. Glacial and Pleistocene Geology. John Wiley & Sons, Inc., New York.

Flor, H. H. 1955. Host-parasite interaction in flax rust – its genetics and other implications. Phytopathology 45: 680–685.

Flor, H. H. 1956. The complementary genic systems in flax and flax rust. In: Advances in Genetics, pp. 29–54. Academic Press, New York.

Ford, H. 1981. Competitive relationships among apomictic dandelions *(Taraxacum)*. Biol. J. Linn. Soc. 15: 355–368.

Forest, C. L., and R. K. Togasaki. 1975. Selection for conditional gametogenesis in *Chlamydomonas reinhardi*. Proc. natl Acad. Sci. USA 72: 3652–3655.

Fowler, K., and L. Partridge. 1986. Variation in male fertility explains an apparent effect of genotypic diversity on success in larval competition in *Drosophila melanogaster*. Heredity 57: 31–36.

Fox, L. R. 1975. Cannibalism in natural populations. A. Rev. Ecol. Syst. 6: 87–106.

Fraenkel, G. S., and D. L. Gunn. 1961. The Orientation of Animals. Dover, New York.

Frank, S. A. 1985. Hierarchical selection theory and sex ratios. II. On applying the theory, and a test with fig wasps. Evolution 39: 949–964.

Frankel, O. H. 1958. The dynamics of plant breeding. J. Aust. Inst. agric. Sci. 24: 112–123.

Frankham, R. 1968. Sex and selection for a quantitative character in *Drosophila*. I. Single-sex selection. Aust. J. biol. Sci. 21: 1215–1223.

Fredga, K., A. Gropp, H. Wiinking, and F. Frank. 1977. A hypothesis explaining the exceptional sex ratio in the wood lemming *(Myopus schisticolor)*. Hereditas 85: 101–104.

Fretwell, S. D. 1972. Populations in Seasonal Environment. Princeton University Press, Princeton.

Fripp, Y. J., and C. E. Caten. 1971. Genotype-environmental interaction in *Schizophyllum commune*. Heredity 27: 393–407.

Fritsch, F. E. 1965. The Structure and Reproduction of the Algae. Cambridge University Press. Cambridge.

Frohlich, D. R., and V. J. Tepedino. 1986. Sex ratio, parental investment, and interparental variability in nesting success in a solitary bee. Evolution 40: 142–151.

Gagne, R. J. 1973. Cecidomyiidae from Mexican tertiary amber. Proc ent. Soc. Wash. 75: 169–171.

Gagne, R. J. 1979. A new species of *Neostenoptera* from the Congo (Diptera: Cecidomyiidae). Annls Soc. ent. Fr. (N. S.) 15: 345–347.

Galloway, R. E., and U. W. Goodenough. 1985. Genetic analysis of mating locus linked mutations in *Chlamydomonas reinhardii*. Genetics 111: 447–461.

Geitler, L. 1932. Der Formwechsel der pennaten Diatomen (Kieselalgen). Arch. Protistenk. 78: 1–226.

Ghiselin, M. T. 1969. The evolution in hermaphroditism among animals. Q. Rev. Biol. 44: 189–208.

Ghiselin, M. T. 1974. The Economy of Nature and Evolution of Sex. University of California Press, Berkeley, California.

Ghiselin, M. T. 1987. The evolution of sex: a history of competing points of view. In: The Evolution of Sex (eds R. E. Michod and B. Levin). Sinauer, Sutherland, MA, in press.

Gilbert, L. E. 1975. Ecological consequences of coevolution between butterflies and plants. In: Coevolution of Animals and Plants, pp. 211–240 (eds L. E. Gilbert and P. H. Raven). University of Texas Press, Austin, Texas.

Gileva, E. A. 1980. Chromosomal diversity and an aberrant genetic system of sex determination in the Arctic lemming, *Dicrostonyx torquatus,* Pallas (1779). Genetica 52/53: 99–103.

Gillespie, J. H. 1974. Natural selection for within generation variance in offspring number II. Discrete haploid models. Genetics 81: 403–413.

Gillespie, J. H. 1977. Natural selection for variance in offspring numbers. Am. Nat. 111: 1010–1014.

Glesener, R. R. 1979. Recombination in a simulated predator-prey environment. Am. Zool. 19: 763–771.

Glesener, R. R., and D. Tilman. 1978. Sexuality and the components of environmental uncertainty: clues from geographic parthenogenesis in terrestrial animals. Am. Nat. 112: 659–673.

Gliddon, C., E. Belhassen, and P. H. Gouyon. 1987. Genetic neighborhoods in plants with diverse systems of mating and different patterns of growth. Heredity: in press.

Goldman, D. A., and M. F. Willson. 1986. Sex allocation in functionally hermaphroditic plants: a review and critique. Bot. Rev. 52: 157–194.

Goodenough, U. W., C. J. Hwang, and A. J. Warren. 1978. Sex-limited expression of gene loci controlling flagellar membrane agglutination in the *Chlamydomonas* mating reaction. Genetics 89: 235–243.

Gooding, H. J., D. L. Jennings, and P. B. Topham. 1975. A genotype-environment experiment on strawberries in Scotland. Heredity 34: 105–115.

Goodman, D. 1984. Risk spreading as an adaptive strategy in iteroparous life histories. Theor. Pop. Biol. 25: 1–20.

Gosling, L. M. 1986. Selective abortion of entire litters in the coypu: Adaptive control of offspring production in relation to quality and sex. Am. Nat. 127: 772–795.

Gould, F. W. 1959. Notes on apomixis in sideoats grama. J. Range Mgmt. 12: 25–28.

Gouyon, P. H. 1978. Contribution à l'étude du régime de réproduction et du maintien de la variabilité dans les populations naturelles d'une éspece gynodioique *Thymus vulgaris*. Thes. Doct. Ing. INAPG, Paris.

Gouyon, P. H., and D. Couvet. 1985. Selfish cytoplasm and adaptation: variations in the reproductive system of thyme. In: Structure and Functioning of Plant Populations/2 (eds J. Haeck and J. W. Woldendorp); North Holland Publ. Co., New York.

Gouyon, P. H., and P. Vernet. 1980. Etude de la variabilité génétique dans une population naturelle de *Thymus vulgaris*. Oec. Plant. 1: 165–178.

Gouyon, P. H., and P. Vernet. 1982. The consequences of gynodioecy in natural populations of *Thymus vulgaris*. Theor. appl. Genet. 61: 315–320.

Gowans, C. S. 1976. Publications by Franz Moewus on the genetics of algae. In: The Genetics of Algae (ed. R. A. Lewin). Blackwell, Oxford.

Gowaty, P. A., and M. R. Lennartz. 1985. Sex ratios of nestling and fledgling Red-Cockaded Woodpeckers *(Picoides borealis)* favor males. Am. Nat. 126: 347–353.

Gowe, R. S. 1956. Environment and poultry-breeding problems. II. A comparison of seven SC White Leghorn strains housed in laying batteries and floor pens. Poultry Sci. 35: 430–435.

Gowe, R. S., and W. S. Wakely. 1954. Environment and poultry-breeding problems. I. The influence and viability of different genotypes. Poultry Sci. 33: 691–703.

Gracen, V. 1983. Role of genetics in etiological phytopathology, A. Rev. Phytopath. 20: 219–233.

Grant, V. 1975. Genetics of Flowering Plants. Columbia University Press, New York.

Green, R. F., G. Gordh, and B. A. Hawkins. 1982. Precise sex ratios in highly inbred parasitic wasps. Am. Nat. 120: 653–665.

Greene, C. W. 1980. The systematics of *Calamacrostis* (Gramineae) in eastern North America. PhD Diss., Harvard University, Cambridge, MA.

Greene, C. W. 1984. Sexual and apomictic reproduction in *Calamacrostis* (Gramineae) from eastern North America. Am. J. Bot. 71: 285–293.

Gregorius, H. R., M. D. Ross, and E. Gillet. 1982. Selection in plant populations of effectively infinite size. III. The maintenance of females among hermaphrodites for a biallelic model. Heredity 48: 329–343.

Griffiths, D. J. 1950–1956. Cereal, bean and Brassicae breeding. Rep. Welsh Plant Breeding Sta., pp. 66–93.

Gross, L. 1983. Oncogenic Viruses, 3rd edn. Pergamon Press, Elmsford, New York.

Grossman, A. I., R. B. Short, and G. D. Cain. 1981. Karyotypic evolution and sex chromosome differentiation in schistosomes (Trematoda, Schistosomatidae). Chromosoma 84: 413–430.

Grun, P. 1976. Cytoplasmic Genetics and Evolution. Columbia University Press, New York.

Guppy, G. A. 1978. Species relationship of *Hieracium* (Asteraceae) in British Columbia. Can. J. Bot. 56: 3008–3019.

Gustafsson, A. 1946a. Apomixis in higher plants. I. The mechanism of apomixis. Acta univ. Lund. 42: 1–66.

Gustafsson, A. 1946b. Apomixis in higher plants. II. The causal aspect of apomixis. Acta univ. Lund. 43: 71–178.

Gustafsson, A. 1947. Apomixis in higher plants. III. Biotype and species formation. Acta univ. Lund. 43: 183–370.

Gustafsson, A. 1953a. Mutations, viability and population structure. Acta agric. scand. 4: 6601–6032.

Gustafsson, A. 1953b. The cooperation of genotypes in barley. Hereditas 39: 1–18.

Gutteridge, H. S., and O'Neil. 1942. The relative effects of environment and heredity upon body measurements and production characteristics in poultry. II. Period of egg production. Sci. Agric. 22: 482–491.

Gutzke, W. H. M., G. L. Paukstis, and L. L. McDaniel 1985. Skewed sex ratios for adult and hatchling bullsnakes, *Piuophis melanoleucus,* in Nebraska. Copeia: 649–652.

364

Gutzke, W. H. N., and G. L. Paukstis. 1983. Influence of the hydric environment on sexual differentiation of turtles. J. exp. Zool. 226: 467–469.

Gwynne, D. T. 1983. Male nutritional investment and the evolution of sexual differences in Tettigoniidae and other Orthoptera. In: Orthopteran Mating Systems (eds D. T. Gwynne and G. K. Morris); pp. 337–366. Westview Press, Boulder, Colorado.

Haas, W. 1976. Die Anheftung (Fixation) der Cercarie von Schistosoma mansoni. Z. Parasit. 49: 63.

Haddow, A. J. 1942. A note on the predatory larva of the mosquito Culex (Lutzia) tigripes G. and D. (Diptera). Proc. R. ent. Soc. Lond. (A) 17: 73–74.

Hagen, K. S. 1964. Developmental stages of parasites. In: Biological Control of Insect Pests and Weeds (ed. P. de Bach), pp. 844–866. Chapman and Hall, London.

Haldane, J. B. S. 1930. The Causes of Evolution. Longmans and Green, London.

Hamilton, W. D. 1964. The genetical evolution of social behavior. J. theor. Biol. 7: 1–52.

Hamilton, W. D. 1967. Extraordinary sex ratios. Science 156: 477–488.

Hamilton, W. D. 1979. Wingless and fighting males in fig wasps and other insects. In: Reproductive Competition and Sexual Selection in Insects (eds M. S. Blum and N. A. Blum), pp. 167–220. Academic Press, New York.

Hamilton, W. D. 1980. Sex versus non-sex versus parasite. Oikos 35: 282–290.

Hamilton, W. D. 1983. Pathogens as causes of genetic diversity in their host populations. In: Population Biology of Infectious Diseases (eds R. M. Anderson and R. M. May), pp. 269–296. Springer Verlag, Berlin.

Hamilton, W. D., P. A. Henderson, and N. Moran. 1981. Fluctuation of environment and coevolved antagonist polymorphism as factors in the maintenance of sex. In: Natural Selection and Social Behavior (eds R. D. Alexander and D. W. Tinkle), pp. 363–381. Chiron Press, New York.

Hancock, J. F. Jr, and R. E. Wilson. 1976. Biotype selection in Erigeron annuus during old fields succession. Bull. Torrey bot. Club 103: 122–125.

Hann, B. J., and P. D. N. Hebert. 1982. Re-interpretation of genetic variation in Simocephalus (Cladocera, Daphniidae). Genetics 102: 101–107.

Hann, B. J., and P. D. N. Hebert. 1986. Genetic variation and population differentiation in species of Simocephalus (Cladocera, Daphniidae). Can. J. Zool.: in press.

Hanson, A. A., R. J. Garber, and W. M. Myers. 1952. Yields of individual and combined apomictic strains of Kentucky bluegrass (Poa pratensis). Agron. J. 44: 125–128.

Harberd, D. J. 1962. Some observations on natural clones of Festuca ovina. New Phytol. 61: 85–100.

Harberd, D. J. 1967. Observation on natural clones in Holcus mollis. New Phytol. 66: 401–408.

Harberd, D. J., and M. Owen. 1969. Some experimental observations on the clone structure of a natural population of Festuca rubra. New Phytol. 68: 93–108.

Hardin, R., and A. E. Bell. 1967. Two-way selection for body weight in Tribolium on two levels of nutrition. Genet. Res. 9: 309–330.

Harley, C. B. 1981. Learning the evolutionarily stable strategy. J. theor. Biol. 89: 611–633.

Harris, R. G. 1925. Reversal of function in a species of Oligarces. Biol. Bull. 48: 139–144.

Harrison, R. G. 1980. Dispersal polymorphisms in insects. A. Rev. Ecol. Syst. 11: 95–118.

Hartl, D. L., and S. W. Brown. 1970. The origin of male haploid genetic systems and their expected sex ratio. Theor. Pop. Biol. 1: 165–190.

Hartmann, M. 1909. Autogamie bei Protisten und ihre Bedeutung für das Befruchtungsproblem. Arch. Protistenk. 14: 264–334.

Hartmann, M. 1925. Über relative Sexualität bei Ectocarpus siliculosus. Ein experimenteller Beitrag zur Sexualitätshypothese der Befruchtung. Naturwissenschaft 13: 975–980.

Hartmann, M. 1956. Die Sexualität. 2nd edn. Fischer Verlag, Stuttgart.

Harvey, P. H., and S. J. Arnold. 1982. Female mate choice and runaway sexual selection. Nature 297: 533–534.

Haskell, G. 1966. The history, taxonomy, and breeding system of apomictic British Rubi. In: Reproduction Biology and Taxonomy of Vascular Plants (ed. J. G. Hawkes), pp. 141–151. Pergamon Press, Oxford.

Herbert, P. D. N. 1974. Enzyme variability in natural populations of *Daphnia magna*. I. Population structure in East Anglia. Evolution 28: 546–556.

Hebert, P. D. N. 1981. Obligate asexuality in *Daphnia*. Am. Nat. 117: 784–789.

Hebert, P. D. N. 1986. Genotypic characteristics of the Cladocera. Hydrobiologia: in press.

Hebert, P. D. N. 1987. Genetics of *Daphnia*. In: The Biology of *Daphnia* (eds R. Peters and R. de Bernardi). Occasional Pupl. ital. Inst. Idrobiol.

Hebert, P. D. N., R. D. Ward, and L. J. Weider. MS. Clonal diversity patterns and breeding system variation in *Daphnia pulex*: an asexual-sexual complex.

Hebert, P. D. N., and C. Moran. 1980. Enzyme variability in natural populations of *Daphnia carinata* King. Heredity 45: 313–312.

Hebert, P. D. N., and J. M. Loaring. Systematics of the *Daphnia pulex* complex: morphological variation in an agamic complex and description of a species new to North America. Biochem. Syst. Ecol. 14: 333–340.

Hebert, P. D. N., and T. J. Crease. 1983. Clonal diversity in populations of *Daphnia pulex* reproducing by obligate parthenogenesis. Heredity 51: 353–369.

Heimcke, J. W., and R. C. Starr. 1979. The sexual process in several heterogamous *Chlamydomonas* strains in the subgenus *Pleichloris*. Arch. Protistenk. 133: 20–42.

Heisler, I. L. 1981. Offspring quality and the polygyny threshold: a new model for the 'sexy-son' hypothesis. Am. Nat. 117: 316–328.

Heisler, I. L. 1984. A quantitative genetic model for the origin of mating preferences. Evolution 38: 1283–1295.

Heisler, I. L. 1985. Quantitative genetic model of female choice based on 'arbitrary' male characters. Heredity 55: 187–198.

Herre, E. A. 1985. Sex ratio adjustment in fig wasps. Science 228: 896–898.

Herrnstein, R. J., and W. Vaughan. 1980. Melioration and behavioral allocation. In: Limits to Action (ed. J. E. R. Staddon), pp. 143–156. Academic Press, New York.

Hertwig, O. 1906. Allgemeine Biologie. 8th edn. Fischer Verlag, Jena.

Heslop-Harrison, J., and H. F. Linskens. 1984. Cellular interaction: a brief conspectus. In: Cellular Interactions. Encyclopedia of Plant Physiology, New Series, Vol. 17. Eds H. F. Linskens and J. Heslop-Harrison. Springer Verlag, Berlin.

Hickey, D. H. 1982. Selfish DNA: a sexually-transmitted nuclear parasite. Genetics 101: 519–531.

Hiesey, W. M. 1953. Growth and development of species and hybrids of *Poa* under controlled temperatures. Am. J. Bot. 40: 205–221.

Hiesey, W. M., and M. A. Nobs. 1952. Experimental studies on the nature of species. VI. Interspecific hybrid derivates between facultatively apomictic species of bluegrass and their responses to contrasting environments. Carnegie Inst. of Washington Publication 636.

Hiesey, W. M., and M. A. Nobs. 1970. Genetic and transplant studies on contrasting species and ecological races of the *Achillea millefolium* complex. Bot. Gaz. 131: 245–259.

Hill, J. F., and A. W. Nordstrog. 1956. Efficiency of performance testing in poultry. Poultry Sci. 35: 256–265.

Hiraizumi, Y., L. Sandler, and J. Crow. 1960. Meiotic drive in natural populations of *Drosophila melanogaster*. III. Populational implications of the segregation distorter laws. Evolution 14: 433–444.

Hiroyoshi, T. 1964. Sex limited inheritance and abnormal sex ratio in strains of the housefly. Genetics 50: 373–385.

Hodgkin, J. 1980. More sex determination mutants of *Caenorhabditis elegans*. Genetics 96: 649–664.

Hoekstra, R. F. 1980. Why do organisms produce gametes of only two different sizes? Some theoretical aspects of the evolution of anisogamy. J. theor. Biol. 87: 785–793.

Hoekstra, R. F. 1982. On the asymmetry of sex: evolution of mating types in isogamous populations. J. theor. Biol. 98: 427–451.

Hoekstra, R. F. 1984. Evolution of gamete motility differences II. Interaction with the evolution of anisogamy. J. theor. Biol. 107: 71–83.

Hoekstra, R. F. 1985. Evolution of asymmetry in sexual reproduction. In: Dynamics of Macrosystems (eds J.-P. Aubin, D. Saari and K. Sigmund), Springer Verlag, Berlin.

Hoekstra, R. F., R. F. Janz, and A. J. Schilstra. 1984. Evolution of gamete motility differences I. Relations between swimming speed and pheromonal attraction. J. theor. Biol. 107: 57–70.

Hoekstra, R. F., and N. van der Hoeven. 1984. The evolution of sexual reproduction. Some population genetic models. Nieuw Archf. Wisk. 4: 5–24.

Hoffmann, S. G., M. P. Schildhauer, and R. R. Warner. The cost of changing sex and the ontogeny of males under contest competition for mates. Evolution 39: 915–927.

Hood, L., M. Kronenberg, and T. Hunkapiller. 1985. T cell antigen receptors and the immunoglobulin supergene family. Cell 40: 225–230.

Hoogland, J. L. 1981. Sex ratio and local resource competition. Am. Nat. 117: 796–797.

Hornby, C. A., and S.-C. Li. 1975. Some effects of multiparental pollination in tomato plants. Can. J. Plant Sci. 55: 127–132.

Horovitz, A., and A. Beiles. 1980. Gynodioecy as a possible population strategy for increasing reproductive output. Theor. appl. Genet. 57: 11–15.

Horvitz, C. C., and D. W. Schemske. MS. A test of the pollinator-limitation hypothesis.

Houck, L. D., S. J. Arnold, and R. A. Thisted. 1985. A statistical study of mate choice: sexual selection in a plethodonid salamander *(Desmognathus ochrophaeus)*. Evolution 39: 370–386.

Howe, H. F. 1977. Sex ratio adjustment in the Common Grackle. Science 198: 744–746.

Hull, P., and R. S. Gowe. 1962. The importance of interactions detected between genotype and environmental factors for characters of economic significance in poultry. Genetics 47: 143–159.

Hull, V. J., and R. H. Groves. 1973. Variation in *Chrondrilla juncea* L. in southeastern Australia. Aust. J. Bot. 21: 113–135.

Hutchinson, J. B., and R. L. M. Ghose. 1937. Studies on crop ecology. I. Indian J. agric. Sci. 7: 1–34.

Hutson, V., and R. Law. 1981. Evolution of recombination in a population experiencing frequency-dependent selection with time delay. Proc. R. Soc. London. Ser. B. 213: 345–359.

Huxley, J. S. 1914. The courtship-habits of the great-crested grebe *(Podiceps cristatus)* with an addition to the theory of sexual selection. Proc. Zool. Soc. London 35: 491–562.

Huxley, J. S. 1938. Darwin's theory of seuxal selection and the data subsumed by it, in the light of recent research. Am. Nat. 72: 416–433.

Iestwaart, J. H., R. A. Barel, and M. E. Ikelaar. 1984. Male sterility in *Organum vulgare* populations. Acta bot. neerl. 33: 335–345.

Imam, A. G., and R. W. Allard. 1965. Population studies in predominantly self-pollinated species. VI. Genetic variability between and within natural populations of wild oats, *Avena fatua* L., from differing habitats in California. Genetics 51: 49–62.

Innes, D. J., and P. D. N. Hebert, MS. The origin of obligate parthenogenesis in *Daphnia pulex.*

Innes, D. J., S. S. Schwartz, and P. D. N. Hebert. 1986. Variation in mode of reproduction among populations in the *Daphnia pulex* group. Heredity: in press.

Jackson, J. B. C., L. W. Buss and R. E. Cook. 1985. Population Biology and Evolution of Clonal Organisms. Yale University Press, New Haven.

Jacquard, D. F. 1974. The Genetic Structure of Populations. Springer Verlag, Berlin.

Jaenike, J., and R. K. Selander. 1979. Evolution and ecology of parthenogenesis in earthworms. Am. Zool. 19: 729–737.

Jaenike, J. 1978. An hypothesis to account for the maintenance of sex within populations. Evol. Theor. 3: 191–194.

Jain, S. K. 1968. Gynodioecy in *Organum vulgare:* computer simulation of a model. Nature 217: 764–765.

Janzen, D. H. 1977. A note on optimal mate selection by plants. Am. Nat. 111: 365–371.

Javier, R. T., F. Sedarati, and J. G. Stevens. 1986. Two avirulent herpes simplex viruses generate lethal recombinants in vivo. Science 234: 746–747.

Jayakar, S. D. 1970. A mathematical model for interaction of frequencies in a parasite and its host. Theor. Pop. Biol. 1: 140–164.

Jelnes, J. E. 1983. Phosphoglucose isomerase: a sex-linked character in *Schistosoma mansoni*. J. Parasit. 69: 780–781.

Jenkins, D. W., and S. J. Carpenter. 1946. Ecology of the tree hole breeding mosquitos of nearctic North America. Ecol. Monogr. 16: 31–47.

Jensen, N. F. 1952. Intra-varietal diversification in oat breeding. Agronom. J. 44: 30–34.

Jinks, J. L., J. M. Perkins, and H. S. Pooni. 1973. The incidence of epistasis in normal and extreme environments. Heredity 31: 263–270.

Johnson, A. W., and J. G. Packer. 1965. Polyploidy and environment in arctic Alaska. Science 148: 237–239.

Jolly, A. 1985. The Evolution of Primate Behavior. Macmillan, New York.

Jones, D. F. 1928. Selective Fertilization. University of Chicago Press, Chicago.

Jones, J. S., and L. Partridge. 1983. Tissue rejection: the price of sexual acceptance. Nature 304: 484–485.

Kagel, J. H., R. C. Battalio, L. Green, and H. Rachlin. 1980. Consumer demand theory applied to choice behavior of rats. In: Limits to Action (ed. J. E. R. Staddon), pp. 237–268. Academic Press, New York.

Kallmann, K. D. 1968. Evidence for the existence of transformer genes for sex in the teleost *Xiphophorus maculatus*. Genetics 60: 811–821.

Kallmann, K. D. 1984. A new look at sex determination in poeciliid fishes. In: Evolutionary Genetics of Fishes (ed. B. J. Turner), pp. 95–171. Plenum, New York.

Kalmus, H. 1932. Ueber den Erhaltungswert der phänotypischen (morphologischen) Anisogamie und die Entstehung der ersten Geschlechtsunterschiede. Biol. Zbl. 52: 716–726.

Kalmus, H., and C. A. B. Smith. 1960. Evolutionary origin of sexual differentiation and the sex-ratio. Nature 186: 1004–1006.

Kamil, A. C. 1983. Optimal foraging theory and the psychology of learning. Am. Zool. 23: 291–302.

Kaplan, M. M., and R. G. Webster. 1977. The epidemiology of influenza. Sci. Am. 237: 88–106.

Karlin, S. A., and S. Lessard. 1986. Theoretical Studies on Sex Ratio Evolution. Princeton University Press, Princeton, NJ.

Kay, Q. O. N. 1978. The role of preferential and assortive pollination in the maintenance of flower colour polymorphisms. In: The Pollination of Flowers by Insects (ed. A. J. Richards), pp. 175–190. Linnean Soc. London.

Kays, S., and J. L. Harper. 1968. Components and regulation of a natural population of *Rumex acetosella* L. J. Ecol. 56: 421–431.

Kearsey, M. J. 1965. The interaction of food supply and competition in two lines of *Drosophila melanogaster*. Heredity 20: 169–181.

Keeney, R. L., and H. Raiffa. 1976. Decisions with Multiple Objectives: Preferences and Value Tradeoffs. Wiley, New York.

Kelley, S. E. 1985. The mechanism of sib competition for the maintenance of sex in *Anthoxanthum odoratum*. Ph. D. thesis Duke University, Durham, N. C.

Kempthorne, O. 1957. Introduction to Genetic Statistics. John Wiley. New York.

Kendal, A. P., N. J. Cox, S. Nakajima, L. Raymond, A. Canton, G. Brownlee, and G. Webster. 1984. Structures in influenza A/USSr/90/77-haemagglutinin associated with epidemiologic and antigenic changes. In: Modern Approaches to Vaccines (eds R. M. Chanock and R. A. Lerner), pp. 151–157. Cold Spring Harbor Laboratory, Cold Spring Harbor, New York.

Kerr, R. W. 1970. Inheritance of DDT resistance in a laboratory colony of the housefly, *Musca domestica*. Aust. J. biol. Sci. 23: 377–400.

Kesavan, V., P. Crisp, A. R. Gray, and B. D. Dowker. 1976. Genotypic and environmental effects on the maturity time of autumn cauliflowers. Theor. appl. Genet. 47: 133–140.

Khan, M. A., P. D. Putwain, and A. D. Bradshaw. 1975. Population interrelationships. 2. Frequency-dependent fitness in *Linum*. Heredity 34: 145–163.

Kheyr-Pour, A. 1980. Nucleo-cytoplasmic polymorphism for male-sterility in *Organum vulgare*. J. Hered. 71: 253–260.

Khokhlov, S. S. 1976. Evolutionary-genetic problems of apomixis in angiosperms. In: Apomixis and Plant Breeding (ed. S. S. Khokhlov), pp. 3–17. Nauka Publisher, Moscow.

Kiester, A. R., R. Lande, and D. W. Schemske. 1984. Models of coevolution and speciation in plants and their pollinators. Am. Nat. 124: 220–243.

Killick, R. J., and N. W. Simmonds. 1974. Specific gravity of potato tubers as a character showing small genotype-environment interactions. Heredity 32: 109–112.

Kimura, M. 1979. Model of effectively neutral mutations in which selective constraint is incorporated. Proc. natl Acad. Sci. USA 76: 3440–3444.

Kimura, M. 1983a. The Neutral Theory of Molecular Evolution. Cambridge University Press, Cambridge.

Kimura, M. 1983b. The neutral theory of molecular evolution, in: Evolution of Genes and Proteins. (eds M. Nei and R. K. Koehn). Sinauer Associates, Sunderland, MA.

Kimura, M., and T. Ohta. 1971. Theoretical Aspects of Population Genetics. Princeton University Press, Princeton, NJ.

King, C. E. 1972. Adaptation of rotifers to seasonal variation. Ecology 53: 408–418.

King, C. E. 1977a. Genetics of reproduction, variation and adaptation in rotifers. Arch. Hydrobiol. Ergeb. Limnol. 8: 187–201.

King, C. E. 1977b. Effects of cyclical ameiotic parthenogenesis on gene frequencies and effective population size. Arch. Hydrobiol. Ergeb. Limnol. 8: 357–360.

King, C. E. 1980. The genetic structure of zooplankton populations. In: Evolution and Ecology of Zooplankton Communities (ed. W. C. Kerfoot). University Press of New England, Hanover, N. H.

King, C. E., and T. W. Snell. 1977. Sexual recombination in rotifers. Heredity 39: 357–360.

King, J. W. B., and G. B. Young. 1955. A study of three breeds of sheep wintered in four environments. J. agric. Sci. 45: 331–338.

King, M. 1977. The evolution of sex chromosomes in lizards. In: Evolution and Reproduction (eds J. Calaby and H. Tyndale-Briscoe), pp. 55–60. Aust. Acad. Sci., Canberra.

Kirkpatrick, M. 1982. Sexual selection and the evolution of female choice. Evolution 36: 1–12.

Kirkpatrick, M. 1985. Evolution of female choice and male parental investment in polygynous species: the demise of the 'sexy-son'. Am. Nat. 125: 788–810.

Kirkpatrick, M. 1986. The handicap mechanism of sexual selection does not work. Am. Nat. 127: 222–240.

Kirkwood, T. B. L., R. F. Rosenberger, and D. J. Galas. 1986. Accuracy in Molecular Processes: Its Control and Relevance to Living Systems. Chapman and Hall.

Kjellberg, F., P. H. Gouyon, M. Ibrahim, M. Raymond, and G. Valdeyron. 1987. The stability of the symbiosis between dioecious figs and their pollinators: a study of *Ficus carica* and *Blastophaga psenes* L. Evolution: in press.

Klinkowski, M. 1970. Catastrophic plant disease. A. Rev. Phytopath. 8: 37–60.

Klinman, N. R., N. H. Sigal, E. S. Metcalf, P. J. Gearhart, and S. K. Pierce 1977. Cold Spring Harbor Symp. quant. Biol. 41: 165.

Knossow, M., R. S. Daniels, A. R. Douglas, J. J. Skehel, and D. C. Wiley. 1984. Three-dimensional structure of an antigenetic mutant of the influenza virus haemagglutinin. Nature 311: 678–680.

Knowlton, N. 1974. A note on the evolution of gamete dimorphism. J. theor. Biol. 46: 283–285.

Knowlton, N. 1979. Reproductive synchrony, parental investment, and the evolutionary dynamics of sexual selection. Anim. Behav. 27: 1022–1033.

Kobori, J. A., E. Strauss, K. Minard, and L. Hood. 1986. Molecular analysis of the hotspot of recombination in the murine major histocompatibility complex. Science 234: 173–179.

Kochert, G. 1978. Sexual pheromones in algae and fungi. A. Rev. Pl. Physiol. 29: 461–486.

Kodric-Brown, A., and J. H. Brown. 1984. Truth in advertising: the kinds of traits favored by sexual selection. Am. Nat. 124: 309–323.

Kolman, W. 1960. The mechanism of natural selection for sex ratio. Am. Nat. 94: 373–377.

Korpelainen, H. 1984. Genic differentiation of *Daphnia magna* populations. Hereditas 101: 209–216.

Korpelainen, H. 1985. Ecological and evolutionary genetics of *Daphnia* (Crustacea: Cladocera). PhD Thesis. University of Helsinki.

Korpelainen, H. 1986. Temporal changes in the genetic structure of *Daphnia magna* populations. Heredity 57: 5–14.

Koste, W. 1978. Rotaria. Die Rädertiere Mitteleuropas. Gebrüder Bornträger, Berlin.

Krebs, J. R., and R. H. McCleery. 1984. Optimization in behavioural ecology. In: Behavioural Ecology, an Evolutionary Approach. 2nd edn. (eds J. R. Krebs and N. B. Davies), pp. 91–121. Sinauer, Sunderland, MA.

Kuhne, H. 1972. Entwicklungsablauf und -stadien von *Micromalthus debilis* Le Conte (Coleoptera, Micromalthidae) aus einer Laboratoriumspopulation. Z. angew. Ent. 72: 157–168.

Kulman, H. M. 1971. Effects of insect defoliation on tree growth and mortality of trees. A. Rev. Ent. 16: 286–324.

LaRue, G. R. 1957. The classification of digenetic *Trematoda:* a review and a new system. Exp. Parasit. 6: 306–349.

Labeyrie, V., and D. Rojas-Rousse. 1985. Superparasitism reconsidered: is it an adaptive competition? The example of *Diadromus pulchellus*. Experientia 41: 15–18.

Lacey, E. P., L. Real, A. Antonovics, and D. G. Heckel. 1983. Variance models in the study of life history. Am. Nat. 122: 114–131.

Lamb, R. Y., and R. B. Willey. 1979. Are parthenogenetic and related bisexual insects equal in fertility? Evolution 33: 774–775.

Lande, R. 1975. The maintenance of genetic variability by mutation in a polygenic character with linked loci. Genet. Res. 26: 221–235.

Lande, R. 1976. Natural selection and random genetic drift in phenotypic evolution. Evolution 30: 314–334.

Lande, R. 1977. The influence of the mating system on the maintenance of genetic variability in polygenic characters. Genetics 86: 485–498.

Lande, R. 1979. Quantitative genetic analysis of multivariate evolution, applied to brain: body size allometry. Evolution 33: 402–416.

Lande, R. 1980a. Sexual dimorphism, sexual selection, and adaptation in polygenic characters. Evolution 34: 292–305.

Lande, R. 1980b. The genetic covariance between characters maintained by pleiotropic mutation. Genetics 94: 203–215.

Lande, R. 1981. Models of speciation by sexual selection on polygenic traits. Proc. natl Acad. Sci. USA 78: 3721–3725.

Lande, R. 1982. Rapid origin of sexual isolation and character divergence in a cline. Evolution 36: 213–223.

Lande, R., and S. J. Arnold. 1983. The measurement of selection on correlated characters. Evolution 37: 1210–1226.

Lande, R., and S. J. Arnold. 1985. Evolution of mating preference and sexual dimorphism. J. theor. Biol. 117: 651–664.

Lascelles, A. K. 1977. Role of the mammary gland and milk in immunology. In: Comparative Aspects of Lactation. (ed. M. Peaker). Symp. zool. Soc. Lond. 41: 241–260.

Lee, T. D. 1984. Patterns of fruit maturation: a gametophyte competition hypothesis. Am. Nat. 123: 427–432.

Lee, T. D., and A. P. Hartgerinck. 1986. Pollination intensity, fruit maturation pattern and offspring quality in *Cassia fasciculata*. In: Biotechnology and Ecology of Pollen (eds D. L. Mulcahy, G. B. Mulcahy and E. Ottaviano), pp. 417–422. Springer Verlag, Berlin.

Lee, T. D., and F. A. Bazzaz, 1982. Regulation of fruit maturation pattern in an annual legume, *Cassia fasciculata*. Ecology 63: 1374–1388.

370

Lee, Y. M., D. J. Friedman, and F. Ayala. 1985. Superoxide dismutase: An evolutionary puzzle. Proc. natl Acad. Sci. USA 82: 824–828.

Leigh, E. G. Jr. 1970. Sex ratio and differential mortality between the sexes. Am. Nat. 104: 205–210.

Leigh, E. G. Jr. 1971. Adaptation and Diversity. Freeman, Cooper & Co., San Francisco, CA.

Leigh, E. G. Jr, E. L. Charnov, and R. R. Warner. 1976. Sex ratio, sex change, and natural selection. Proc. natl Acad. Sci. USA 73: 3656–3660.

Leigh, E. G. Jr. 1977. How does selection reconcile individual advantage with the good of the group? Proc. natl Acad. Sci. USA 74: 4542–4546.

Leigh, E. G. Jr. 1983. When does the good of the group override the advantage of the individual? Proc. natl Acad. Sci. USA 80: 2985–2989.

Leigh, E. G. Jr. 1987. Ronald Fisher and the development of evolutionary theory: 1. The role of selection. In: Oxford Surveys in Evolutionary Biology, no. 3 (ed. R. Dawkins): in press.

Leigh, E. G. Jr, E. A. Herre, and E. A. Fischer. 1985. Sex allocation in animals. Experientia 41: 1265–1276.

Leonard, K. J. 1969. Factors affecting rates of stem-rust increase in mixed plantings of susceptible and resistant oat varieties. Phytopathology 59: 1845–1850.

Lerche, W. 1937. Untersuchungen über Entwicklung und Fortpflanzung in der Gattung *Dunaliella*. Arch. Protistenk. 88: 236–268.

Leutert, R. 1975. Sex determination in *Bonellia*. In: Intersexuality in the Animal Kingdom (ed. E. Reinboth), pp. 84–90. Springer-Verlag, Berlin.

Levin, B. 1977. Gene Expression, vol. 3: Plasmids and Phages. Wiley, New York.

Levin, B. C. 1972. Low frequency disadvantage in the exploitation of pollinators by corolla variants of *Phlox* cultivars with different shapes. Am. Nat. 106: 453–460.

Levin, B. R., B. R. Stewart, and L. Chao. 1977. Resource limited growth, competition and predation: A model and some experimental studies with bacteria and bacteriophage. Am. Nat. 111: 3–24.

Levin, B. R., A. C. Allison, H. J. Bremermann, L. L. Cavalli-Sforza, B. C. Clarke, R. Frentzel-Beyme, W. D. Hamilton, S. A. Lewin, R. M. May, and H. R. Thieme. 1982. Evolution of parasites and hosts. Group report. In: Population Biology of Infectious Diseases (ed. R. M. May), pp. 213–243. Springer-Verlag, New York.

Levin, D. A. 1975. Pest pressure and recombination systems in plants. Am. Nat. 109: 437–451.

Levin, D. A. 1983. Polyploidy and novelty in flowering plants. Am. Nat. 122: 1–25.

Levin, S. A. 1976. Population dynamic models in heterogeneous environments. A. Rev. Ecol. Syst. 7: 287–310.

Levin, S. A. 1983. Some approaches to modeling coevolutionary phenomena. In: Coevolution. (ed. M. Nitecki). University of Chicago Press, Chicago, IL.

Levin, S. A., and J. D. Udovic. 1977. A mathematical model of coevolutionary populations. Am. Nat. 111: 657–675.

Levins, R. 1968. Evolution in Changing Environments. Princeton University Press, Princeton, NJ.

Lewin, B. 1975. Gene Expression, vol. 3: Plasmids and Phages. Wiley, New York.

Lewin, B. 1983. Genes. Wiley, New York.

Lewis, D. 1941. Male-sterility in a natural population of hermaphrodite plants. New Phytol. 40: 56–63.

Lewis, D., and L. K. Crowe. 1956. The genetics and evolution of gynodioecy. Evolution 10: 115–125.

Lewis, G. A., and D. S. Madge. 1984. Esterase activity and associated insecticide resistance in the damson-hope aphid, *Phorodon humuli* (Schrank) (Hemptera: Aphididae). Bull. Ent. Res. 74: 227–238.

Lewis: W. H. 1980. Polyploidy in species populations. In: Polyploidy: Biological Relevance (ed. W. H. Lewis), pp. 103–144. Plenum Press. New York.

Lewis, W. M. Jr. 1979. Zooplankton Community Analysis. Springer-Verlag, New York.

Lewis, W. M. Jr. 1983. Interruption of synthesis as a cost of sex in small organisms. Am. Nat. 121: 825–834.

Lewis, W. M. Jr. 1984. The diatom sex clock and its evolutionary significance. Am. Nat. 123: 73–80.

Lewontin, R. C. 1955. The effects of population density and composition on viability in *Drosophila melanogaster.* Evolution 9: 27–41.

Lewontin, R. C. 1974. The Genetic Basis of Evolutionary Change. Columbia University Press, New York.

Lewontin, R. C., and J. L. Hubby. 1966. A molecular approach to the study of genic heterozygosity in natural populations. II. Amount of variation and degree of heterozygosity in natural populations of *Drosophila pseudoobscura.* Genetics 54: 595–609.

Lloyd, D. G. 1974. Theoretical sex ratios of dioecious and gynodioecious angiosperms. Heredity 32: 11–34.

Lloyd, D. G. 1975. The maintenance of gynodioecy and androdioecy in angiosperms. Genetica 45: 325–339.

Lloyd, D. G. 1976. The transmission of genes via pollen and ovules in gynodioecious angiosperms. Theor. Pop. Bio. 9: 299–316.

Lloyd, D. G. 1983. Evolutionarily stable sex ratios and sex allocations. J. theor. Biol. 105: 525–539.

Lloyd, D. G. 1984a. Gender allocations in outcrossing cosexual plants. In: Perspectives on Plant Population Ecology (eds. R. Dirzo and J. Sarukhan), pp. 277–300. Sinauer, Sunderland, MA.

Lloyd, D. G. 1984b. Gene selection of Mendel's rules. Heredity 53: 613–624.

Lloyd, D. G. 1984c. Variation strategies of plants in heterogeneous environments. Biol. J. Linn. Soc. 21: 357–385.

Lloyd, D. G., and J. M. A. Yates. 1982. Intrasexual selection and segregation of pollen and stigmas in hermaphrodite plants, exemplified by *Wahlenbergia albomarginata* (Campanulaceae). Evolution 36: 903–913.

Lloyd, D. G., and K. S. Bawa. 1984. Modification of the gender of seed plants in varying conditions. Evol. Biol. 17: 255–338.

LoVerde, P. T., J. DeWals, and D. J. Minchella. 1985. Further studies of genetic variation in *Schistosoma mansoni.* J. Parasit. 71: 732–734.

Loaring, J. M. 1982. The ecological genetics of *Daphnia species.* MSc. thesis. University of Windsor.

Loaring, J. M., and P. D. N. Hebert. 1981. Ecological diffences among clones of *Daphnia pulex* (Leydig). Oecologia 51: 162–168.

Lokki, J., and A. Saura. 1980. Polyploidy in insect evolution. In: Polyploidy: Biological Relevance (ed. W. H. Lewis), pp. 277–312. Plenum Press, New York.

Lovett-Doust, L. 1981. Population dynamics and local specialization in a clonal perennial, *Ranunculus repens.* II. The dynamics of leaves and a reciprocal transplant-replant experiment. J. Ecol. 69: 757–768.

Lowry, D. C., I. M. Lerner, and L. W. Taylor. 1956. Intraflock genetic merit under floor and cage mangement. Poultry Sci. 35: 1034–1043.

Loxdale, H. D., I. J. Tarr, C. P. Weber, C. P. Brookes, P. G. N. Digby, and P. Costanera. 1985. Electrophoretic study of enzymes from cereal aphid populations of *Sitobion avenae* (Hemiptera: Homoptera: Aphididae). Bull. Ent. Res. 75: 121–142.

Lucas, G. B. 1980. The war against blue mold. Science 210: 147–153.

Lush, J. L. 1945. Animal Breeding Plans, 3rd edn. Iowa State College Press, Ames.

Lyman, J. C., and N. C. Ellstrand. 1984. Clonal diversity in *Taraxacum officinale* (Compositae), an apomict. Heredity 53: 1–10.

Lynch, M. 1983. Ecological genetics of *Daphnia pulex.* Evolution 37: 358–374.

Lynch, M. 1984. Destabilizing hybridization, general-purpose genotypes and geographic parthenogenesis. Q. Rev. Biol. 59: 257–290.

Lynch, M., and W. Gabriel. 1983. Phenotypic evolution and parthenogenesis. Am. Nat. 122: 745–764.

372

Lyttle, T. W. 1977. Experimental population genetics of meiotic drive systems. I. Pseudo-Y chromosomal drive as a means of eliminating cage populations of *Drosophila melanogaster.* Genetics 86: 413–445.

Lyttle, T. W. 1979. Experimental population genetics of meiotic drive systems. II. Accumulation of genetic modifiers of segregation distorter (SD) in laboratory populations. Genetics 91: 393–357.

Lyttle, T. W. 1981. Experimental population genetics of meiotic drive systems. III. Neutralization of sex-ratio distortion in *Drosophila* through sex-chromosome aneuploidy. Genetics 98: 317–334.

MacArthur, R. H. 1961. Population effects of natural selection. Am. Nat. 95: 195–199.

MacArthur, R. H. 1965. Ecological consequences of natural selection. In: Theoretical and Mathematical Biology (eds T. H. Waterman and H. J. Morowitz), pp. 388–397. Blaisdell, New York.

MacArthur, R. H., and E. R. Pianka. 1966. On the optimal use of a patchy environment. Am. Nat. 100: 603–609.

Machlis, L., and E. Rawitscher-Kunkel. 1963. Mechanism of gametic approach in plants. Int. Rev. Cytol. 15: 97–138.

Madson, J. D., and D. M. Waller. 1983. A note on the evolution of gamete dimorphism in algae. Am. Nat. 121: 443–447.

Malcolm, J. R., and K. Marten. 1982. Natural selection and the communal rearing of pups in the African Wild Dog *(Lycaon pictus).* Behav. Ecol. Sociobiol. 10: 1–13.

Mani, G. S. 1984. Evolutionary Dynamics of Genetic Diversity. Lecture Notes in Biomathematics, vol. 53. Springer-Verlag, Berlin, Heidelberg.

Manton, I. 1934. The problem of *Bicutella laevigata* L. Z. indukt. Abstamm.- u. Vererb.-Lehre 67: 41–57.

Margulis, L., and D. Sagan. 1986. Origins of Sex: Three Billion Years of Genetic Recombination. Yale University Press, New Haven, CT.

Marshall, D. L., and N. C. Ellstrand. 1986. Sexual selection in *Raphanus sativus:* experimental data on non-random fertilization, maternal choice, and consequences of multiple paternity. Am. Nat. 127: 446–461.

Martin, J., C. Kuvangkadilok, D. H. Peart, and B. T. O. Lee. 1980. Multiple sex determining regions in a group of related *Chironomus* species (Diptera: Chironomidae). Heredity 44: 367–382.

Matuszewski, B. 1982. Animal Cytogenetics, vol. 3. Insecta 3. Diptera I. Cecidommyiidae. Gebrüder Bornträger, Berlin.

May, B., and F. R. Holbrook. 1978. Absence of genetic variability in the green peach aphid, *Myzus persicae* (Hemiptera, Aphididae). Ann. ent. Soc. Am. 71: 809–812.

May, R. M., and R. M. Anderson. 1979. Population biology of infectious diseases II. Nature 280: 455–461.

May, R. M., and R. M. Anderson. 1983a. Epidemiology and genetics in the coevolution of parasites and hosts. Proc. R. Soc. London (B) 219: 281–313.

May, R. M., and R. M. Anderson. 1983b. Parasite-host coevolution. In: Coevolution. (eds D. J. Futuyma and M. Slatkin), pp. 186–206. Sinauer Associates, Sunderland, Massachusetts.

Maynard Smith, J. 1971a. What use is sex? J. theor. Biol. 30: 319–335.

Maynard Smith, J. 1971b. The origin and maintenance of sex. In: Group Selection (ed. G. C. Williams), pp. 163–175. Aldine-Atherton, Chicago.

Maynard Smith, J. 1976a. A short-term advantage for sex and recombination through sib-competition. J. theor. Biol. 63: 245–258.

Maynard Smith, J. 1976b. Group selection. Q. Rev. Biol. 51: 277–287.

Maynard Smith, J. 1976c. Sexual selection and the handicap principle. J. theor. Biol. 57: 239–242.

Maynard Smith, J. 1978. The Evolution of Sex. Cambridge University Press, Cambridge.

Maynard Smith, J. 1979. Game theory and the evolution of behaviour. Proc. R. Soc. Lond. (B) 205: 475–488.

Maynard Smith, J. 1980. A new theory of sexual investment. Behav. Ecol. Soziobiol. 7: 247–251.

Maynard Smith, J. 1982. Evolution and the Theory of Games. Cambridge University Press, Cambridge.

Maynard Smith, J. 1984. Game theory and the evolution of behaviour. Behav. Brain Sci. 7: 95–125.

Maynard Smith, J. 1985. Mini review: sexual selection, handicaps and true fitness. J. theor. Biol. 115: 1–8.

Maynard Smith, J. 1986. Contemplating life without sex. Nature, 324: 300–301.

Maynard Smith, J. 1987. Are we stuck with sex? Nature 325: 307–308.

Mayr, E. 1963. Animal Species and Evolution. Harvard University Press, Cambridge, MA.

McClure, P. A. 1981. Sex-biased litter reduction in food-restricted wood rats *(Neotoma floridana)*. Science 211: 1058–1060.

McDonald, I. C., P. Evenson, C. A. Nickel, and O. A. Johnson. 1978. House fly genetics: Isolation of a female determining factor on chromosome 4. Ann. ent. Soc. 71: 692–694.

McGraw, J. B., and Antonovics. 1983. Experimental ecology of *Dryas octopetala* ecotypes. I. Ecotypic differentiation and life-cycle stages of selection. J. Ecol. 71: 879–897.

McKenna, M., and D. Mulcahy. 1983. Ecological aspects of gametophytic competition in *Dianthus chinensis*. In: Pollen: Biology and Implications for Plant Breedings (ed. D. Mulcahy), pp. 419–424. Elsevier, New York.

McKenna, M. A. 1986. Heterostyly and microgametophytic selection: The effect of pollen tube competition on sporophyte vigor in two distylous species. In: Biotechnology and Ecology of Pollen (eds D. L. Mulcahy, G. B. Mulcahy and E. Ottavviano). Springer Verlag, Berlin.

McKenna, M. A., and D. L. Mulcahy. 1983. Gametophytic competition in *Dianthus chinensis:* effect on sporophytic competitive ability. In: Pollen: Biology and Applications to Plant Breeding (ed. D. L. Mulcahy). Elsevier, New York.

McKenzie, K. G. 1983. On the origin of Crustacea. In: The Biology and Evolution of Crustacea (ed. J. K. Lowrey). Australian Museum Memoir 18, Sydney.

Meagher, T. R. 1986. Analysis of paternity within a single natural population of *Chamaerilium luteum*. 1. Identification of most-likely male parents. Am. Nat. 128: 199–215.

Meagher, T. R. 1987. Analysis of parentage for naturally established seedlings within a population of *Chamaerilium luteum* (Liliaceae). Ecology: in press.

Meincke, D. W. 1982. Embryonic lethal mutants of *Arabidopsis thaliana:* evidence for gametophytic expression of the mutant genes. Theor. appl. Genet. 63: 381–386.

Michaels, H. J., and F. A. Bazzaz. 1986. Resource allocation and demography of sexual and apomictic *Antennaria parlinii*. Ecology 67: 27–36.

Michie, D., and M. F. Anderson. 1966. A strong selective effect associated with the histocompatibility gene in the rat. Ann. NY Acad. Sci. 192: 88.

Michod, R. E. and B. Levin (eds). 1987. The Evolution of Sex. Sinauer Assoc., Sunderland, MA: in press.

Milani, R. 1970. Genetics of factors affecting fertility and of sex ratio distortions in the housefly. Sterility principle for insect control or eradication. In: Proceedings of a Symposium, Athens, 14–18 September. International Atomic Energy Agency Proceedings Series, Vienna, pp. 381–397.

Milani, R. 1975. The house fly, *Musca domestica*. In: Handbook of Genetics, vol. 3: Invertebrates of Genetic Interest (ed. R. C. King), pp. 377–399. Plenum Press, New York.

Mittwoch, U. 1967. Sex Chromosomes. Academic Press, New York.

Mittwoch, U. 1973. Genetics of Sex Differentiation. Academic Press, New York.

Moav, R., and G. W. Wohlfarth. 1974. Magnification through competition of genetic differences in yield capacity of carp. Heredity 33: 181–202.

Mode, C. J. 1958. A mathematical model for the coevolution of obligate parasites and their hosts. Evolution 12: 158–165.

Mode, C. T. 1960. A model of a host-pathogen system with particular reference to the rusts of cereals. Biometrical Genetics, pp. 84–96. Pergamon Press, New York.

Moncol, D. J., and A. C. Triantophyllou. 1978. *Strongyloides ransomi:* factors influencing the in vitro development of the free living generation. J. Parasit. 64: 220–225.

Monroy, A., and F. Rosati. 1979. The evolution of the cell-cell recognition system. Nature 278: 165–166.

Montelaro, R., B. Parekh, C. Issel, and A. Orrego. 1984. Antigenic variation during persistent infection by equine Infectious Anemia virus, a retrovirus. J. Biol. Chem. 259: 105–139.

Moore, W. S. 1984. Evolutionary ecology of unisexual fishes. In: Evolutionary Genetics of Fish (ed. B. J. Turner). Plenum Press, New York.

Mort, M. A., and H. G. Wolf. 1985. Enzyme variability in large-lake *Daphnia* populations. Heredity 55: 27–36.

Mort, M. A., and H. G. Wolf. 1986. The genetic structure of large-lake *Daphnia* populations. Evolution 40: 756–766.

Mrosovsky, N. 1980. Thermal biology of sea turtles. Am. Zool. 20: 531–547.

Mrosovsky, N., P. H. Dutton, and C. P. Whitmore. 1984. Sex ratios of two species of sea turtles nesting in Surinam. Can. J. Zool. 62: 2227–2239.

Mueller, A. J. 1963. Embryonentest zum Nachweis rezessiver Letalfaktoren bei *Arabidopsis thaliana.* Biol Zbl. 82: 133–163.

Müller, D. G. 1976. Relative sexuality in *Entocarpus siliculosus.* Arch. Microbiol. 109: 89–94.

Muenchow, G. A. 1987. Is dioecy associated with fleshy fruit? Am. J. Bot.: in press.

Muesbeck, C. F. W., K. V. Krombein, and H. K. Townes. 1951. Hymenoptera of America North of Mexico. Synoptic Catalogue. United States Dept. of Agric. Agriculture Monograph 2, Washington.

Mulcahy, D. L. 1971. A correlation between gametophytic and sporophytic characteristics in *Zea mays* L. Science 171: 1155–1156.

Mulcahy, D. L. 1974. Adaptive significance of gamete competition. In: Fertilization in Higher Plants (ed. H. F. Linskens), pp. 27–30. North Holland, Amsterdam.

Mulcahy, D. L. 1979. The rise of the angiosperms: a genecological factor. Science 206: 20–23.

Mulcahy, D. L., and G. B. Mulcahy. 1975. The influence of gametophytic competition on sporophytic quality. Theor. Appl. Genet. 46: 277–280.

Mulcahy, D. L., and G. B. Mulcahy. 1983. Gametophytic incompatibility reexamined. Science 220: 1247–1251.

Mulcahy, D. L., G. B. Mulcahy, and E. Ottaviano. 1975. Sporophytic expression of gametophytic competition in *Petunia hybrida.* In: Gamete Competition in Plants and Animals (ed. D. L. Mulcahy). North-Holland, Amsterdam.

Mulcahy, D. L., and G. B. Mulcahy. 1987. The effects of pollen competition. Am. Sci. 75: 44–50.

Muller, H. J. 1932. Some genetic aspects of sex. Am. Nat. 66: 118–138.

Muntzing, A. 1966. Apomixis and sexuality in new material of *Poa alpina* from middle Sweden. Hereditas 54: 314–337.

Murphy, E. D. 1966. Biology of the Laboratory Mouse. McGraw-Hill, New York, 521 p.

Nakamura, D., S. S. Wachtel, and K. Kallmann. 1984. H-Y antigen and the evolution of heterogamety. J. Hered. 75: 353–358.

Nakamura, R. 1986. Maternal investment and fruit abortion in *Phaseolus vulgaris.* Am. J. Bot. 73: 1049–1057.

Nanney, D. L. 1980. Experimental Ciliatology. Wiley, New York.

National Academy of Sciences Committee Report. 1972. Genetic Vulnerability of Major Crops. National Academy of Sciences, Washington, D. C.

Nei, M. 1984. Genetic polymorphism and neomutationism. In: Evolutionary Dynamics of Genetic Diversity. Lecture Notes in Biomathematics, vol. 53 (ed. G. S. Mani). Springer Verlag, Berlin, Heidelberg, New York.

Nei, M., and R. K. Koehn (eds). 1983. Evolution of Genes and Proteins. Sinauer, Sunderland, MA.

Nevo, E., A. Beiles, and R. Ben-Shlomo. 1984. The evolutionary significance of genetic diversity: ecological, demographic, and life history correlates. In: Evolutionary Dynamics of Genetic Diversity (ed. G. S. Mani). Lecture Notes in Biomathematics 53: 13–21.

Newhook, F. J. and F. D. Podger. 1972. The role of *Phytophthora cinnamoni* in Australia and New Zealand forests. A. Rev. Phytopath. 10: 299–326.

Newton, I., and M. Marquiss. 1979. Sex ratio among nestlings of the European Sparrowhawk. Am. Nat. 113: 309–315.

Nicklas, R. B. 1960. The chromosome cycle of a primitive cecidomyiid *Mycophila speyeri*. Chromosoma 11: 402–418.

Nijs, J. M. C. den, and A. A. Sterk. 1980. Cytogeographical studies of *Taraxacum* (section *Taraxacum*) in Central Europe. Bot. Jb. 101: 527–554.

Noble, E. R., and G. A. Noble. 1982. Parasitology: The Biology of Animal Parasites. Lea & Febeger, Philadelphia, PA.

Nonaces, P. 1986. Ant reproductive strategies and sex allocation theory. Q. Rev. Biol. 61: 1–21.

Nordstrog, A. W., and O. Kempthorne. 1960. Importance of genotype-environment interactions in random sample poultry tests. In: Biometrical Genetics (ed. O. Kempthorne), pp. 159–186. Pergamon Press, London.

Novick, R., and F. C. Hoppenstaedt. 1978. On plasmid incompatibility. Plasmid 1: 421.

Nuding, J. 1936. Leistung und Ertragsstruktur von Winterweizensorten in Reinsaat und Mischung in verschiedenen deutschen Anbaugebieten. Pflanzenbau 12: 382–447.

Nunney, L. 1985. Female biased sex ratios: individual or group selection? Evolution 39: 349–361.

Nur, N., and O. Hasson. 1984. Phenotypic plasticity and the handicap principle. J. theor. Biol. 110: 275–297.

Nygren, A. 1954. Apomixis in the angiosperms. II. Bot. Rev. 20: 577–643.

O'Donald, P. 1980. Genetic Models of Sexual Selection. Cambridge University Press, Cambridge.

O'Donald, P. 1983. Sexual selection by female choice. In: Mate Choice (ed. P. Bateson), pp. 53–66. Cambridge University Press, Cambridge.

Ohno, S. 1967. Sex Chromosomes and Sex-Linked Genes. Springer Verlag, Berlin.

Ohno, S. 1979. Major Sex-Determining Genes. Springer Verlag, Berlin.

Ohta, S. 1967. Chromosomal polymorphism and capacity for increase under near optimal conditions. Heredity 22: 169–185.

Oliver, J. H. Jr. 1971. Parthenogenesis in mites and ticks. Am. Zool. 1: 283–289.

Oliver, J. H. Jr. 1977. Cytogenetics of mites and ticks. A. Rev. Ent. 22: 407–429.

Orgel, L. E., and F. H. C. Crick. 1980. Selfish DNA: the ultimate parasite. Nature 284: 604–607.

Orozco, F. 1976. A dynamic study of genotype-environment interaction with egg-laying of *Tribolium castaneum*. Heredity 37: 157–171.

Orzack, S. H. 1986. Sex ratio control in a parasitic wasp, *Nasonia vitripennis*. II. Experimental analysis of an optimal sex ratio model. Evolution 40: 341–356.

Orzack, S. H., J. J. Sohn, K. D. Kallmann, S. A. Levin, and R. Johnston. 1980. Maintenance of the three sex chromosome polymorphism in platyfish, *Xiphophorus maculatus*. Evolution 34: 663–672.

Orzack, S. H., and E. D. Parker. 1986. Sex ratio control in a parasitic wasp, *Nasonia vitripennis*. I. Genetic variation in facultative sex ratio adjustment. Evolution 40: 331–340.

Osborne, R. 1952. Sexual maturity in brown Leghorns. The interaction of genotype and environment. Proc. R. Soc. Edinburgh 64: 445–455.

Oster, G. F., and E. O. Wilson. 1978. Caste and Ecology in the Social Insects. Princeton University Press, Princeton, NJ.

Packard, G. C., M. J. Packard, T. J. Boardman, and M. D. Ashen. 1981. Possible adaptive value of water exchanges in flexible-shelled eggs of turtles. Science 213: 471–473.

Pagliai, A. M. B. 1983. Endomeiosis and parthenogenesis in two strains of *Megoura viciae* Buckt (Hom. Aphid.). In: Atti. XIII Congressa Nazionale Italiano di Entomologia, Turin.

Palm, J. 1974. Maternal-fetal histoincompatibility in rats: An escape from adversity. Cancer Res. 34: 2061.

376

Pandey, K. K. 1977a. Generation of multiple genetic specificities: origin of genetic polymorphism through gene regulation. Theor. Appl. Genet. 49: 85–93.

Pandey, K. K. 1977b. Evolution of incompatibility systems in plants: complementarity and the mating locus in flowering plants and fungi. Theor. Appl. Genet. 50: 89–101.

Panelius, S. 1971. Male germ line, spermatogenesis, and karyotypes of *Heteropeza pygmaea* Winnertz (Diptera: Cecidomyiidae). Chromosoma 32: 295–331.

Panopoulos, N. J., J. D. Walton, and D. K. Willis. 1984. Genetic and biochemical basis of virulence in plant pathogens. In: Genes Involved in Microbe-Plant Interactions. (eds D. P. S. Verma and T. H. Hohn). Springer-Verlag, New York.

Park, Y. I., C. T. Hansen, C. S. Chung, and A. B. Chapman. 1966. Influence of feeding regime on the effects of selection for postweaning gain in the rat. Genetics 54: 1315–1327.

Parker, E. D. Jr, and S. H. Orzack. 1985. Genetic variation for the sex ratio in *Nasiona vitripennis*. Genetics 110: 93–105.

Parker, G. A. 1978. Selection on non-random fusion of gametes during the evolution of anisogamy. J. theor. Biol. 73: 1–28.

Parker, G. A., R. R. Baker, and V. G. F. Smith. 1972. The origin and evolution of gamete dimorphism and the male-female phenomenon. J. theor. Biol. 36: 529–533.

Pederson, D. G. 1968. Environmental stress, heterozygote advantage and genotype-environment interaction in *Arabidopsis*. Heredity 23: 127–138.

Pederson, D. G. 1974. The stability of varietal performance over years. I. The distribution of seasonal effects for wheat grain yield. Heredity 32: 85–94.

Perelson, A. S., M. Mirmirani, and. G. F. Oster. 1976. Optimal strategies in immunology I. B-cell differentiation and proliferation. J. math. Biol. 3: 325–367.

Perelson, A. S., M. Mirmirani, and G. F. Oster. 1978. Optimal strategies in immunology II. B-memory cell production. J. math. Biol. 5: 213–256.

Perelson, A. S., and G. F. Oster. 1979. Theoretical studies of clonal selection: Minimal antibody repertoire size and reliability of self-non-self discrimination. J. theor. Biol. 81: 645–670.

Perkins, J. M., and J. L. Jinks. 1968. Environmental and genotype-environmental components of variability. III: Multiple lines and crosses. Heredity 23: 339–356.

Person, C. 1966. Genetic polymorphism in parasitic systems. Nature 212: 266–267.

Petersen, J. J. 1977. Effects of host size and parasite burden on sex ratio in the mosquito parasite *Octomymermis muspratti*. J. Nemat. 9: 343–346.

Phillip, M. 1980. Reproductive biology of *Stellari longipes* Goldie as revealed by a cultivation experiment. New Phytol. 85: 557–569.

Pickering, J. 1980. Larval competition and brood sex ratios in the gregarious parasitoid *Pachysomoides stupidus*. Nature 283: 291–292.

Pieau, C. 1972. Effects de la température sur le développement des glandes génitales chez les embryons de deux Cheloniens, *Emys orbicularis* et *Testudo graca*. C. r. Acad. Sci. Paris (D) 274: 719–722.

Pieau, C. 1974. Sur la différenciation sexuelle chez des embryons d'*Emys orbicularis* (Chelonien) issus d'œufs incubés dans le sol au cours de l'été 1973. Bull. Soc. zool. Fr. 99: 363–376.

Pieau, C. 1975a. Temperature and sex differentiation in embryos of two chelonians, *Emys orbicularis* and *Testudo graca*. In: Intersexuality in the Animal Kingdom (ed. R. Reinboth), pp. 332–339. Springer-Verlag, Berlin.

Pieau, C. 1975b. Effects des variations thermiques sur la différentation du sex chez les vertébrés. Bull. Soc. zool. Fr. 100: 67–76.

Pieau, C. 1982. Modalities of the action of temperature on sexual differentiation in field-developing embryos of the European pond turtle *Emys orbicularis* (Emydidae). J. exp. Zool. 220: 353–360.

Pieau, C. 1970. Sur la proportion sexuelle chez les embryons de deux Cheloniens (*Testudo graca* L. et *Emys orbicularis* L.) issus d'œufs incubés artificiellement. C. r. Acad. Sci. Paris (D) 272: 3071–3074.

Pimentel, D. MS. Genetic diversity and stability in parasite-host systems.

Piper, J., and S. Waite. MS. Pollen export and pollen import in broad-leaved helleborine.

Plack, A. 1958. Effect of giberellic acid on corolla size. Nature 182: 610.

Pollis, G. A. 1981. The evolution and dynamics of intraspecific predation. A. Rev. Ecol. Syst. 12: 225–251.

Pooni, H. S., J. L. Jinks, and N. E. M. Jayasekara. 1978. An investigation of gene action and genotype x environment interaction in two crosses of *Nicotiana rustica* by triple test cross and inbred line analysis. Heredity 41: 83–92.

Porsild, A. E. 1965. The genus *Antennaria* in eastern Arctic and subarctic America. Bot. Tidskr. 61: 22–55.

Price, M. V., and N. M. Waser. 1982. Population structure, frequency-dependent selection, and the maintenance of sexual reproduction. Evolution 36: 83–92.

Price, R. A. 1980. *Draba streptobrachia* (Brassicaceae), a new species from Colorado. Brittonia 32: 160–169.

Pringsheim, E. G., and K. Ondracek. 1939. Untersuchungen über die Geschlechtsvorgänge bei *Polytoma*. Beih. Bot. Zbl. I 59: 118–172.

Pringsheim, E. G. 1963. Farblose Algen. Gustav Fischer Verlag, Jena.

Pritchard, A. E. An new classification of the paedogenic gall midges formerly assigned to the subfamily *Heteropezinae* (Diptera: Ceidomyiidae). Ann. ent. Soc. Am. 53: 305–316.

Prout, T. 1953. Some effects of variations in the segregation ratio and of selection upon the freqency of alleles. Acta genet. statist. med. 4: 148–151.

Prout, T., J. Bungaard, and S. Bryant. 1973. Population genetics of modifiers of meiotic drive. I. The solution of a special case and some general implications. Pop. Biol. 4: 446-465.

Pulliam, H. R. 1981. Learning to forage optimally. In: Foraging Behaviour (eds A. C. Kamil and T. D. Sargent), pp. 379–388. Garland STPM Press, New York.

Pyke, G. H., H. R. Pulliam, and E. L. Charnov. 1977. Optimal foraging: a selective review of theory and tests. Q. Rev. Biol. 52: 137–154.

Queller, D. C. 1983. Sexual selection in hermaphroditic plants. Nature 305: 706–707.

Queller, D. C. 1985. Proximate and ultimate causes of low fruit production in *Asclepias exaltata*. Oikos 44: 373–381.

Queller, D. C. 1987. Sexual selection in flowering plants. In: Sexual selection: Testing the Alternatives (eds J. Bradbury and M. Andersson). Dahlem Conf. Reports, Springer Verlag: in press.

Rachlin, H. 1980. Economics and behavioral psychology. In: Limits to Action (ed. J. E. R. Staddon), pp. 205–236. Academic Press, New York.

Raper, J. R. 1966. Life cycles, basic patterns of sexuality and sexual mechanisms. In: The Fungi, an Advanced Treatise, vol. 2 (eds G. C. Ainsworth and A. S. Sussman), pp. 473–511. Academic Press, New York.

Ratner, L., R. C. Gallo, and F. Wong-Staal. 1985. HTLV-III, LAV, and ARV are variants of the same AIDS virus. Nature 313: 636–637.

Raven, P. H. 1979. A survey of the reproductive biology of *Onagracea*. New Zeal. J. Bot. 17: 575–593.

Real, L. 1980. Fitness, uncertainty and the role of diversification in evolution and behavior. Am. Nat. 115: 623–638.

Real, L. 1980. On uncertainty and the law of diminishing returns in evolution and behaviour. In: Limits to Action (ed. J. E. R. Staddon), pp. 37–64. Academic Press, New York.

Rechten, C., M. I. Avery, and T. A. Stevens. 1983. Optimal prey selection: Why do Great Tits show partial preference? Anim. Behav. 31: 576–584.

Renner, O. 1958. Auch etwas über F. Moewus, *Forsythia* und *Chlamydomonas*. Z. Naturforsch. 13b: 399–403.

Rhodes, I. 1969. The yield, canopy structure and light interception of two rye grass varieties in mixed and monoculture. J. Br. Grassld. Soc. 24: 123–127.

Rhomberg, L. R., S. Joseph, and R. S. Singh. 1985. Seasonal variation and clonal selection in cyclically parthenogenetic rose aphids *(Macrosiphum rosae)*. Can. J. Genet. Cytol. 27: 224–232.

Rice, W. R. 1983a. Parent-offspring pathogen transmission: a selective agent promoting sexual reproduction. Am. Nat. 121: 187–203.

Rice, W. R. 1983 b. Sexual reproduction: an adaptation reducing parent-offspring contagion. Evolution 37: 1317–1320.

Rice, W. R. 1986. On the instability of polygenic sex determination: the effect of specific selection. Evolution 40: 633–639.

Richards, A. J. 1973. The origin of *Taraxacum* agamospecies. Bot J. Linn. Soc. 66: 189–211.

Richards, C. M., and G. W. Nace. 1978. Gynogenetic and hormonal sex reversal used in the tests of the XX-XY hypothesis of sex determination in *Rana pipiens*. Growth 42: 319–331.

Robertson, D. R., and R. R. Warner. 1978. Sexual patterns in the labroid fishes of the western Carribean. II. The parrotfishes (Scaridae). Smithson. Contr. Zool. 255: 1–26.

Rockwood, L. L. 1973. The effect of defoliation on seed production in six Costa Rican tree species. Ecology 54: 1363–1369.

Rollins, R. C. 1949. Sources of genetic variation in *Parthenium argentatum* Gray (Compositae). Evolution 3: 358–368.

Rollinson, D., and F. Gebert. 1982. Some observations of genetic variation of *Schistosoma haematobium* and *Bulinus cernicus* in Mauritius. Parasitology 85, XXII.

Romero, G. A., and C. S. Nelson. 1986. Sexual dimorphism in *Catasetum* orchids: forcible pollen emplacement and male flower competition. Science 232: 1538–1540.

Roose, M. L., and L. D. Gottlieb. 1976. Genetic and biochemical consequences of polyploidy in *Tragopogon*. Evolution 30: 818–830.

Rosen, D., and P. DeBach. 1979. Species of *Aphytis* of the World (Hymenoptera: Aphelinidae). W. Junk, The Hague.

Ross, M. D. 1969. Digenic inheritance of male-sterility in *Plantago lanceolata*. Can. J. Genet. Cytol. 11: 739–744.

Ross, M. D. 1978. The evolution of gynodioecy and subdioecy. Evolution 32: 174–188.

Rossler, Y., and P. DeBach. 1972. The biosystematic relations between a thelytokous and a arrhenotokous form of *Aphytis mytilaspidis* (LeBaron) (Hymenoptera: Aphelinidae). 1. The reproductive relations. Entomophaga 17: 391–423.

Roth, S. 1973. A molecular model for cell interactions. Q. Rev. Biol. 48: 541–563.

Roughgarden, J. 1979. Theory of Population Genetics and Evolutionary Ecology: An Introduction. Macmillan, New York.

Roy, S. K. 1960. Interaction between rice varieties. J. Genet. 57: 137–152.

Rubenstein, D. I. 1982. Risk, uncertainty and evolutionary strategies. In: Current Problems in Sociobiology (ed. King's College Sociobiology Group), pp. 91–112. Cambridge University Press, Cambridge.

Rubin, D. A. 1985. Effect of pH on sex ratio in cichlids and poeciliids (Teleostei) Copeia: 233–235.

Rubini, P. G., M. G. Franco, and S. Vanossi Este. 1972. Polymorphisms for heterochromosomes and autosomal sex-determinants in *Musca domestica*. Atti. IX. Congr. ital. Ent: 341–352.

Ruttner-Kolisko, A. 1946. Ueber das Auftreten unbefruchteter Dauereier bei *Keratella quadrata*. Oest. zool. Z. 1: 179–181.

Sager, R., and Z. Ramanis. 1973. The mechanisms of maternal inheritance in *Chlamydomonas*: biochemical and genetic studies. Theor. appl. Genet. 43: 101–108.

Sakaguchi, Y. 1980. Karyotype and gametogenesis of the common liver fluke, *Fasciola sp.* Jap. J. Parasit. 29: 507–513.

Salt, G. 1961. Competition among insect parasitoids. Symp. Soc. exp. Biol. 15: 96–119.

Sanchez-Pescador, R., M. D. Power, P. J. Barr, K. S. Steimer, M. M. Stempien, S. L. Brownshimer, W. G. Gee, A. Renard, A. Randolph, J. A. Levy, D. Dina, and P. A. Luciw. 1985. Nucleotide sequence and expression of an AIDS-associated retrovirus (ARV-2). Science 227: 484–492.

Sandfaer, J. 1954. Virkningen af det naturlige udvalg ved foreaedlingsarbejtet med selvbefrugtende kornater. Tidsskr. PlAvl 58: 333–354.

Saunders, J. F., and W. M. Lewis Jr. Ms. Structure, dynamics, and control mechanisms in a tropical zooplankton community (Lake Valencia, Venezuela).

Sax, H. J. 1954. Polyploidy and apomixis in *Cotoneaster*. J. Arnold Arbor. 35: 334–365.

Schaffer, W. M. 1974. Optimal reproductive effort in fluctuating environments. Am. Nat. 108: 783–790.

Schaudinn, F. 1905. Neuere Forschungen über die Befruchtung bei Protozoen. Verh. d. zool. Ges.

Scheffer, R. P., and R. S. Livingston 1984. Host-selective toxins and their role in plant diseases. Science 223: 17–21.

Schemske, D. W. 1980a. Evolution of floral display in the orchid *Brassavola nodosa*. Evolution 34: 489–493.

Schemske, D. W. 1980b. Floral ecology and hummingbird pollination of *Combretum farinosum* in Costa Rica. Biotropica 12: 169–181.

Schemske, D. W. 1983. Breeding system and habitat effects on fitness components in three neotropical *Costus* (Zingiberaceae). Evolution 37: 529–539.

Schemske, D. W., and C. Fenster. 1983. Pollen grain interactions in a neotropical *Costus:* effects of clump size and competitors. In: Pollen: Biology and Implications for Plant Breeding (ed. D. Mulcahy), pp. 405–410. Elsevier, New York.

Schemske, D. W., and R. Lande. 1985. Fragrance, collection and territorial display by male orchid bees. Anim. Behav. 32: 1185–1193.

Schimke, R. T., (ed.) 1982. Gene Amplification. Cold Spring Harbor Laboratory, Cold Spring Harbor, New York.

Schlichting, C. D., A. G. Stephenson, L. E. Davis, and J. A. Winsor. MS. Pollen competition and offspring variance.

Schmid, M. 1983. Evolution of sex chromosomes and heterogametic systems in Amphibia. Differentiation 235: 513–522

Schmidt, G. D., and L. S. Roberts. 1985. Foundations of Parasitology. Times Mirror/Mosby, Toronto.

Schmitt, J., and J. Antonovics. 1986a. Experimental studies of the evolutionary significance of sexual reproduction. III. Maternal and paternal effects during seedling establishment. Evolution 40: 817–829.

Schmitt, J., and J. Antonovics. 1986b. Experimental studies of the evolutionary significance of sexual reproduction. IV. Effect of neighbor relatedness and aphid infestation on seedling performance. Evolution 40: 830–836.

Schoen, D. J., and D. G. Lloyd. 1984. The selection of cleistogamy and heteromorphic diaspores. Biol. J. Linn. Soc. 23: 303–322.

Schoen, D. J., and M. T. Clegg. 1985. The influence of flower color om outcrossing rate on male reproductive succes in *Ipomoea purpurea*. Evolution 39: 1242–1249.

Schoen, D. J., and S. C. Stewart. 1986. Variation in male reproductive success in white spruce. Evolution 40: 1109–1121.

Schrader, F. 1926. The cytology of pseudosexual eggs in a species of *Daphnia*. Z. indukt. Abstamm.- Vererb.-Lehre 40: 1–27.

Schrader, F., and S. Hughes-Schrader. 1931. Haploidy in metazoa. Q. Rev. Biol. 6: 411–438.

Schulman, S. R., and D. I. Rubenstein. 1983. Kinship, need and the distribution of altruism. Am. Nat. 121: 776–788.

Schultz, R. J. 1980. Role of polyploidy in the evolution of fishes. In: Polyploidy, Biological Relevance (ed. W. H. Lewis), pp. 313–340. Plenum Press, New York.

Schuster, P., and K. Sigmund. 1982. A note on the evolution of sexual dimorphism. J. theor. Biol. 94: 107–110.

Schwartz, S. S., and P. D. N. Hebert. 1986. Reproductive biology of *Daphniopsis ephemeralis:* adaptations to a transient environment. Hydrobiologia: in press.

Schwarzkopf, L., and R. J. Brooks. 1985. Sex determination in northern painted turtles: effect of incubation at constant and fluctuating temperatures. Can. J. Zool. 63: 2543–2547.

Scott, A. C. 1936. Haploidy and aberrant spermatogenesis in a coleopteran, *Micromalthus debilis*, Leconte. J. Morph. 59: 485–515.

Scott, A. C. 1938. Paedogenesis in the Coleoptera. Z. Morph. Oekol. Tiere 33: 633–653.

Scudo, F. M. 1964. Sex population genetics. Ric. scient. 34 II-B: 93–146.

Scudo, F. M. 1967 a. Criteria for the analysis of multifactorial sex determination. Monitore zool. ital. 1: 1–21.

Scudo, F. M. 1967 b. The adaptive value of sexual dimorphism: I. Anisogamy. Evolution 21: 285–291.

Seaton, A. P. C., and J. Antonovics. 1967. Population interrelationships. I. Evolution in mixtures of *Drosophila* mutants. Heredity 22: 19–33.

Seavey, S. R., and K. S. Bawa. 1986. Late-acting selfincompatibility in angiosperms. Bot. Rev. 52: 195–219.

Seger, J. 1983. Partial bivoltinism may cause alternating sex ratio biases that favor eusociality. Nature 301: 59–62.

Seger, J. 1985. Unifying genetic models for the evolution of female choice. Evolution 39: 1185–1193.

Shaw, R. F., and J. D. Mohler. 1953. The selective advantage of the sex ratio. Am. Nat. 87: 337–342.

Shields, W. M. 1982. Philopatry, Inbreeding, and the Evolution of Sex. State University of New York Press, Albany, New York.

Shimazu, T. 1981. Experimental completion of the life-cycle of the lung fluke, *Paragonimus westermani,* in the laboratory. Jap. J. Parasit. 30: 173–177.

Shine, R., and J. J. Bull. 1977. Skewed sex ratios in snakes. Copeia: 228–234.

Short, R. B. 1957. Chromosomes and sex in *Schistosomatium douthitti* (Trematoda: Schistosomatidae). J. Heredity 48: 2–6.

Short, R. B., and M. Y. Menzel. 1959. Chromosomes in parthenogenetic miracidia and embryonic cercariae of *Schistosomatium douthitti.* Exp. Parasit. 8: 249–264.

Sibley, C. G., and J. E. Ahlquist. 1984. The phylogeny of the hominoid primates, as indicated by DNA-DNA hybidizidation. J. mol. Evol. 20: 2–15.

von Siebold, C. T. E. 1857. On true parthenogenesis in moths and bees: a contribution to the history of reproduction in animals. Van Voorst, London.

Simmonds, N. W. 1959. Bananas, London.

Simmonds, N. W. 1962: Variability in crop plants, its use and conservation. Biol. Rev. 37: 442–465.

Singh, R. S., and L. Rhomberg. 1984. Allozyme variation, population structure and sibling species in *Aphis pomi.* Can. J. Genet. Cytol. 26: 364–373.

Skinner, S. W. 1985. Son-Killer: a third extra-chromosomal factor affecting the sex ratio of the parasitoid wasp *Nasonia (Mormoniella) vitripennes.* Genetics 109: 745–759.

Skinner, S. W. 1982. Maternally inherited sex ratio in the parasitoid wasp *Nasonia vitripennis.* Science 2156: 1133–1134.

Slatkin, M. 1978. Spatial patterns in the distribution of polygenic characters. J. theor. Biol. 70: 213–228.

Slatkin, M. 1984. Ecological causes of sexual dimorphism. Evolution 38: 622–630.

Smart, J. 1983. Undernutrition, maternal behavior and pup development. In: Parental Behavior of Rodents (ed. R. W. Elwood), pp. 205–234. Wiley, New York.

Smith, F. H., and F. Spencer. 1984. The Origins of Modern Humans. Liss, New York.

Smith, H. E. 1946. *Sedum pulchellum:* a physiological and morphological comparison of diploid, tetraploid, and hexaploid races. Bull. Torrey bot. Club 73: 495–541.

Smith, M. Y., and A. Frazer. 1976. Polymorphism in a cyclic parthenogenetic species: *Simocephalus serrulatus.* Genetics 84. 631–637.

Smith, R. L. (ed). 1984. Sperm Competition and the Evolution of Animal Mating Systems. 687 p. Academic Press, New York.

Snaydon, R. W. 1970. Rapid population differentiation in a mosaic environment. I. The response of *Anthoxanthum odoratum* populations to soils. Evolution 24: 257–269.

Snaydon, R. W., and M. S. Davies. 1972. Rapid population differentiation in a mosaic environment. II. Morphological variation in *Anthoxanthum odoratum.* Evolution 26: 390–405.

Snaydon, R. W., and M. S. Davies. 1976. Ibid. IV. Populations of *Anthoxanthum odoratum* at sharp boundaries. Heredity 37: 9–25.

Snell, T. W. 1979. Intraspecific competition and population structure in rotifers. Ecology 60: 494–502.

Snell, T. W. 1980. Blue-green algae and selection in rotifer populations. Oecologica 46: 343–346.

Snell, T. W., and B. C. Winkler. 1984. Isozyme analysis of rotifer proteins. Biochem. Syst. Ecol. 12: 199–202.

Snow, A. A. 1986. Pollination dynamics in *Epilobium canum* (Onagraceae): consequences for gametophytic selection. Am. J. Bot. 73: 139–151.

Snow, D. W. 1985. The Web of Adaptation. Cornell University Press, Ithaca. NY.

Sokal, R. R., and R. L. Sullivan. 1963. Competition between mutant and wildtype house-fly strains at varying densities. Ecology 63: 314–322.

Solbrig, O. T., S. Jain, G. B. Johnson, and P. H. Raven (eds). 1979. Topics in Plant Population Biology. Columbia University Press, New York.

Solbrig, O. T., and B. B. Simpson, 1974. Components of regulation of a population of dandelions in Michigan. J. Ecol. 62: 473–486.

Soreng, R. J. 1984. Dioecy and apomixis in the *Poa fendleriana* complex (Paceae). Am. J. Bot. 71: 189.

Sorensen, T., and G. Gudjonsson. 1946. Spontaneous chromosome-aberrants in apomictic *Taraxaca*. K. danske Vidensk. Sels. biol. Skr. 4: 1–48.

Sorenson, F. C. 1970. Self-fertility of a central Oregon source of ponderosa pine. USDA Forest Serv. Res. Publ. 1–109.

Sprague, G. F., and W. T. Federer. 1951. A comparison of variance components in corn yield. II. Agronom. J. 43: 535–541.

Springer, S. 1948. Oviphagous embryos of the sand shark, *Carcharias taurus*. Copeia: 430–436.

Staddon, J. E. R. 1980. Optimality analyses of operant behavior and their relation to optimal foraging. In: Limits to Action (ed. J. E. R. Staddon), pp. 101–142. Academic Press, New York.

Staddon, J. E. R. 1983. Adaptive Behavior and Learning. Cambridge University Press, Cambridge.

Staiger, H. 1956. Genetical and morphological variation in *Purpura lapillus* with respect to local and regional differentiation of population groups. Ann. Biol. T 33: 251–257.

Stanton, D. J. 1987. Evolution of asexuality in *Daphnia pulex:* implications of mitochondrial DNA analysis. PhD Thesis. University of Windsor.

Stanton, M. A., A. A. Snow, and S. N. Handel. 1986. Floral evolution: attractiveness to pollinators increases male fitness. Science 232: 1625–1627.

Starr, R. C. 1969. Structure, reproduction and differentiation in *Volvox carteri* f. *nagariensis Iyengar,* strains HK9 and 10. Arch. Protistenk. 111: 204–222.

Stearns, S. C. 1976. Life-history tactics: A review of the ideas. Q. Rev. Biol. 51: 3–47.

Stearns, S. C. 1985. The evolution of sex and the role of sex in evolution. Experientia 41: 1231–1234.

Stearns, S. C. 1986. Natural selection and fitness, adaptation and constraint. In: Patterns and Processes in the History of Life. (eds D. M. Raup and D. Jablonski), pp. 23-44. Dahlem Konferenzen. Springer-Verlag Berlin, Heidelberg, New York.

Stebbins, G. L. 1950. Variation and Evolution in Plants. Columbia University Press, New York.

Stebbins, G. L. 1958. Longevity, habitat, and release of genetic variability in the higher plants. Cold Spring Harb. Symp. quant. Biol. 23: 365–378.

Stebbins, G. L. 1971. Chromosomal Evolution in Higher Plants. Addison Wesley, New York.

Stebbins, G. L. 1980. Polyploidy in plants: unsolved problems and prospects. In: Polyploidy: Biological Relevance (ed. W. H. Lewis), pp. 495–520. Plenum, New York.

Stent, G. S., and R. Calendar. 1978. Molecular Genetics: An Introductory Narrative. W. H. Freeman, San Francisco. CA.

Stephenson, A. G. 1981. Flower and fruit abortion: proximate causes and ultimate functions. A. Rev. Ecol. Syst. 12: 253–279.

Stephenson, A. G. 1982. When does outcrossing occur in a mass flowering plant? Evolution 36: 762–767.

Stephenson, A. G. 1984. The regulation of maternal investment in an indeterminate flowering plant *(Lotus corniculatus)*. Ecology 65: 113–121.

Stephenson, A. G., J. A. Winsor, and L. E. Davis. 1986. Effects of pollen load size on fruit maturation and sporophyte quality in zucchini. In: Biotechnology and Ecology of Pollen (eds D. L. Mulcahy and G. B. Mulcahy), pp. 429–434. Springer Verlag, Berlin.

Stephenson, A. G., and J. A. Winsor. 1986. *Lotus corniculatus* regulates offspring quality through selective fruit abortion. Evolution 40: 453–458.

Stephenson, A. G., and R. I. Bertin. 1983. Male competition, female choice and sexual selection in plants. In: Pollination Biology (ed. L. Real), pp. 109–149. Academic Press, Orlando, FL.

Sterk, A. A., J. M. C. den Nijs, and W. Kreune. 1982. Sexual and agamospermous *Taraxacum spp.* in the Netherlands. Acta bot. neerl. 31: 227–238.

Stewart, F. M., and B. R. Levin. 1977. The population biology of bacterial plasmids: *A priori* conditions for the existence of conjugationally transmitted factors. Genetics 87: 209–228.

Stille, B. 1980. Meiosis and reproductive strategy in the parthenogenetic gall wasp *Diplolepis rosae* (Hymenoptera, Cynipidae). Hereditas 92: 353–362.

Stille, B. 1985a. Host plant specifity and allozyme variation in the parthenogenetic gall wasp. *Diplolepis mayri* and its relatedness to *D. rosae* (Hymenoptera: Cynipidae) Ent. Genet. 10: 87–96.

Stille, B. 1985b. Population genetics in the parthenogenetic gall wasp, *Diplolepis rosae* (Hymenoptera, Cynipidae). Genetica 67: 141–151.

von Stosch, H. A. 1951. Entwicklungsgeschichtliche Untersuchungen an zentrischen Diatomeen. I. Auxosporenbildung von *Melosira varians*. Arch. Mikrobiol. 16: 101–135.

von Stosch, H. A. 1956. Entwicklungsgeschichtliche Untersuchungen an Diatomeen. II: Geschlechtszellenreifung, Befruchtung und Auxosporenbildung einiger grundbewohnender Biddulphiaceen der Nordsee. Arch. Mikrobiol. 23: 327–365.

Strathern, J. N., and I. Herskowietz. 1979. Asymmetry and directionality in production of new cell types during clonal growth: the switching pattern of homothallic yeast. Cell 17: 371–381.

Strathmann, R. R., M. Strathmann, and R. H. Emson, 1984. Does limited brood capacity link adult size, brooding, and simultaneous hermaphroditism? A test with the starfish *Asterina phylactica*. Am. Nat. 123: 796–818.

Stribling, M. C., W. C. Hamlett, and J. P. Wourms. 1980. Developmental evidence of oophagy, a method of viviparous embryonic nutrition displayed by the Sand Tiger Shark *(Eugomphodus taurus)*. Proc. S. C. Acad. Sci. 42: 111.

Strobel, G. A. 1973. Basis of the resistance of sugarcane to eyespot disease. Proc. natl Acad. Sci. 70: 1693–1696.

Strobel, G. A. 1975. A mechanism of disease resistance in plants. Sci. Am 232: 80–88.

Stubblefield, J. W. 1980. Theoretical elements of sex ratio evolution. PhD Diss. Harvard University, Cambridge, MA.

Sullivan, J. A., and R. J. Schultz. 1986. Genetic and environmental basis of variable sex ratios in laboratory strains of *Poeciliopsis lucida*. Evolution 40: 152–158.

Sullivan, M., and R. I. Bertini. 1986. Effects of self-pollination on the success of cross-pollination in *Campsis radicans*. Am. J. Bot. 73: 663.

Sullivan, V. I. 1976. Diploidy, polyploidy, and agamospermy among species of *Eupatorium* (Compositae). Can. J. Bot. 54: 2907–2917.

Sumner, D. R., and R. H. Littrell. 1974. Influence of tillage, planting date, inoculum survival, and mixed populations on epidemiology of southern corn leaf blight. Phytopathology 64: 168–173.

Sutherland, M., and L. F. Delph. 1984. On the importance of male fitness in plants: patterns of fruit-set. Ecology 65: 1093–1104.

Sutherland, S. 1986. Patterns of fruit-set: What controls fruit-flower ratios in plants? Evolution 40: 117–128.

T.-W.-Fiennes, R. N. 1982. Infectious Cancers of Animals and Man. Academic Press, London.

Taborsky, M. 1985. On optimal parental care. Z. Tierpsychol. 70P: 331–336.

Tait, D. E. N. 1980. Abandonment as a reproductive tactic in grizzly bears. Am. Nat. 115: 800–808.

Tal, M. 1980. Physiology of polyploids. In: Polyploidy: Biochemical Relevance (ed. W. H. Lewis), pp. 61–75. Plenum, New York.

Taylor, M. G., M. B. A. Amin, and G. S. Nelson. 1969. Parthenogenesis in *Schistosoma mattheei*. J. Helminth. XLIII: 197–206.

Taylor, P. D. 1979. An analytic model of a short-term advantage for sex. J. theor. Biol. 97: 557–576.

Taylor, P. D. 1985a. A general mathematical model for sex allocation. J. theor. Biol. 112: 799–818.

Taylor, P. D. 1985b. Sex ratio equilibrium under partial sib mating. Heredity 54: 179–186.

Taylor, P. D., and A. Sauer. 1980. The selective advantage of sex ratio homeostasis. Am. Nat. 116: 305–310.

Taylor, P. D., and M. G. Bulmer. 1980. Local mate competition and the sex ratio. J. theor. Biol. 86: 409–419.

Templeton, A. R. 1982. The prophecies of parthenogenesis. In: Evolution and Genetics of Life Histories (eds H. Dingle and J. P. Hegmann), pp. 75–101. Springer Verlag, New York.

Tepedino, V. J., and P. F. Torchino. 1982. Temporal variability in the sex ratio of a non-social bee, *Osmia lignaria propinqua* Cresson: Extrinsic determination or the tracking of an optimum? Oikos 38: 177–182.

Thompson, P. E., and J. S. Bowen. 1972. Interactions of differentiated primary sex factors in *Chironomus tentans*. Genetics 70: 491–493.

Thomson, J. D., and S. C. H. Barrett. 1981. Selection for outcrossing, sexual selection and the evolution of dioecy in plants. Am. Nat. 118: 442–449.

Thornhill, R. 1981. *Panorpa* (Mecoptera: Panorpidae) scorpionflies: systems for understanding resource-defense polygyny and alternative male reproductive efforts. A. Rev. Ecol. Sys. 12: 335–386.

Thornhill, R., and J. Alcock. 1983. The Evolution of Insect Mating Systems. Harvard University Press, Cambridge, MA.

Thorpe, H. C. 1959. Wheat breeding in Kenya. Proc. 1st int. Wheat Genetics Symp., pp. 55–63.

Tomiuk, J., and K. Wohrmann. 1980. Enzyme variability in populations of aphids. Theor. appl. Genet. 57: 125–127.

Tomiuk, J., and K. Wohrmann. 1981. Changes of the genotype frequencies at the MDH locus in populations of *Macrosiphum rosae* L. Hemiptera, Aphididae. Biol. Zbl. 100: 631–640.

Tomiuk, J., and K. Wohrmann. 1982. Comments on the genetic stability of aphid clones. Experientia 38: 320–321.

Tomiuk, J., and K. Wohrmann. 1984. Genotype variability in natural populations of *Macrosiphum rosae* L. in Europe. Biol. Zbl. 103: 113–122.

Tomkins, D. J., and W. F. Grant. 1978. Morphological and genetic factors influencing the response of weed species to herbicides. Can. J. Bot. 56: 1466–1471.

Tonegawa, S. 1983. Somatic generation of antibody diversity. Nature 302: 575–581.

Triantophyllou, A. C., and D. J. Moncol. 1977. Cytology, reproduction, and sex determination of *Strongyloides ransomi* and *S. papillosus*. J. Parasit. 63: 961–973.

Trivers, R. L. 1985. Social Evolution. Benjamin/Cummings, Menlo Park, CA.

Trivers, R. L., and D. E. Willard. 1973. Natural selection of parental ability to vary the sex ratio of offspring. Science 179: 90–92.

Trivers, R. L., and H. Hare. 1976. Haploidy and the evolution of the social insects. Science 191: 249–263.

Tsubo, Y. 1961. Chemotaxis in *Chlamydomonas*. J. Protozool. 8: 114–121.

Turelli, M. 1984. Heritable genetic variation via mutation-selection balance. Lerch's zeta meets the abdominal bristle. Theor. Pop. Biol. 25: 138–193.

Turesson, G. 1943. Variation in the apomictic microspecies of *Alchemilla vulgaris* L. Bot. Notiser: 413–427.

Turesson, G., and B. Turesson. 1960. Experimental studies in *Hieracium pilosella* L. I. Reproduction, chromosome number, and distribution. Hereditas 46: 717–736.

Turkington, R., and J. L. Harper. 1979. The growth, distribution and neighbour relationships of *Trifolium repens* in a permanent pasture. IV. Fine-scale biotic differentiation. J. Ecol. 67: 245–254.

Twitty, V. C. 1961. Second generation hybrids of the species of *Taricha*. Proc. natl Acad. Sci. USA 47: 1461–1486.

Usberti, J. A. Jr, and A. K. Jain. 1978. Variation in *Panicum maximum:* a comparison of sexual and asexual populations. Bot. Gaz. 139: 112–116.

Uyenoyama, M. K. 1984. On the evolution of parthenogenesis: a genetic representation of the 'cost of meiosis'. Evolution 38: 87–102.

Uyenoyama, M. K., and B. O. Bengtsson. 1979. Towards a genetic theory for the evolution of the sex ratio. Genetics 93: 721–736.

Uyenoyama, M. K., and B. O. Bengtsson. 1981. Towards a genetic theory for the evolution of the sex ratio. II. Haplodiploid and diploid models with sibling and parental control of the brood sex ratio and brood size. Theor. Pop. Bio. 20: 57–79.

Uyenoyama, M. K., and M. W. Feldman. 1980. Theories of kin and group selection: a population genetics perspective. Theor. Pop. Biol. 17: 380–414.

Valdeyron, G., and D. G. Lloyd. 1979. Sex differences and flowering phenology in the common fig, *Ficus carica* L. Evolution 33: 673–685.

Van Alfen, N. K., R. A. Jaynes, S. L. Anagnostakis, and P. R. Day. 1975. Chestnut blight: Biological control by transmissible hypovirulence in *Endothia parasitica*. Science 189: 890–891.

Van Damme, J. M. M. 1982. Gynodioecy in *Plantago lanceolata* L. I. Polymorphism for plasmon type. Heredity 49: 303–318.

Van Damme, J. M. M. 1983. Gynodioecy in *Plantago lanceolata* L. II. Inheritance of three male-sterility types. Heredity 50: 253–273.

Van Damme, J. M. M. 1984. Gynodioecy in *Plantago lanceolata* L. III. Sexual reproduction and the maintenance of male steriles. Heredity 52: 77–93.

Van Damme, J. M. M., and W. V. Van Delden. 1984. Gynodioecy in *Plantago lanceolata* L. IV. Fitness components of sex types in different life cycle stages. Heredity 56: 355–364.

Van Damme, J. M. M., and R. van Damme. 1986. On the maintenance of gynodioecy: Lewis' results extended. J. theor. Biol. 121: 339–350.

Van Loenhoud, P. J., and H. Duyts. 1981. A comparative study of the germination ecology of some microspecies of *Taraxacum* Wigg. Acta bot. neerl. 30: 161–182.

Van Valen, L. 1973. A new evolutionary law. Evol. Theory 1: 1–30.

Van Winkle-Swift, K. P., and B. Aubert. 1983. Uniparental inheritance in a homothallic alga. Nature 303: 167–169.

Van den Ende, H. 1976. Sexual Interactions in Plants. Academic Press, London.

Van der Hoeven, N. 1984. A mathematical model for the coexistence of plasmids in a bacterial population. J. theor. Biol. 110: 411–423.

Van der Plank, J. E. 1975. Principles of Plant Infection. Academic Press, New York.

Vandel, A. 1928. La parthénogenèse géographique. Contribution à l'étude biologique et cytologique de la parthénogenèse naturelle. Bull. biol. Fr. Belg. 62: 164–281.

Vandel, A. 1940. La parthénogenèse géographique. IV. Polyploidie et distribution géographique. Bull. biol. Fr. Belg. 74: 94–100.

Vawter, L., and W. M. Brown. 1986. Nuclear and mitochondrial DNA comparisons reveal extreme rate variation in the molecular clock. Science 234: 194–196.

Venable, D. L., and L. Lawlor. 1980. Delayed germination and dispersal in desert annuals: escape in space and time. Oecologia 46: 272–282.

Verma, D. P. S., and T. Hohn (eds) 1984. Genes involved in Microbe-Plant Interactions. Springer-Verlag, Berlin, New York.

Verner, J. 1965. Selection for sex ratio. Am. Nat. 99: 419–421.

Victor, B. C. 1986. Larval settlement and juvenile mortality in a recruitment-limited coral reef fish population. Ecol. Monogr. 56: 145–160.

Viggiani, G. 1984. Bionomics of the Aphelinidae A. Rev. Ent. 29: 257–276.

Vogel, H. 1942. Ueber die Nachkommenschaft aus Kreuzpaarungen zwischen *Bilharzia mansoni* und *B. japonica*. Zentr. Bact. Parasit. 149: 319–333.

Vogt, R. C., and J. J. Bull. 1984. Ecology of hatchling sex ratio in map turtles. Ecology 65: 582–587.

Waage, J. K. 1982. Sib-mating and sex ratio strategies in scelionid wasps. Ecol. Ent. 7: 103–112.

Waage, J. K., and J. A. Lane. 1984. The reproductive strategy of a parasitic wasp: II. Sex allocation and local mate competition in *Trichogramma evanescens*. J. anim. Ecol. 53: 417–426.

Wachtel, S. S. 1983. H-Y Antigen and the Biology of Sex Determination. Grune and Stratton, New York.

Wade, M. J. 1978. A critical review of the models of group selection. Q. Rev. Biol. 53: 101–114.

Wade, M. J., and S. J. Arnold. 1980. The intensity of sexual selection in relation to male sexual behaviour, female choice and sperm precedence. Anim. Behav. 28: 446–461.

Wainscoat, J. S., A. V. S. Hill, A. L. Boyce, J. Flint, M. Hernandez, S. L. Thein, J. M. Old, J. R. Lynch, A. G. Falusi, D. J. Weatherall, and J. B. Clegg. 1986. Evolutionary relationships of human populations from an analysis of nuclear DNA polymorphisms. Nature 319: 491–493.

Wake, M. 1980. Diversity within a framework of constraints. Amphibian reproductive modes. In: Environmental Adaptation and Evolution (eds D. Mossakowskii and G. Roth), pp. 87–106. Gustav Fischer, Stuttgart and New York.

Walker, W. F. 1980. Sperm utilization strategies in nonsocial insects. Am. Nat. 115: 780–799.

Warner, R. R. 1978. The evolution of hermaphroditism and unisexuality in aquatic and terrestrial vertebrates. In: Contrasts in Behavior (ed. E. Reese), pp. 78–101. Wiley Interscience, New York.

Warner, R. R. 1984a. Mating behavior in coral reef fishes. Am. Sci. 72: 128–136.

Warner, R. R. 1984b. Deferred reproduction as a response to sexual selection in a coral reef fish: a test of the life historical consequences. Evolution 38: 148–162.

Warner, R. R., and S. G. Hoffman. 1980. Local population size as a determinant of mating system and sexual composition in two tropical marine fishes (*Thalassoma* spp.). Evolution 34: 508–518.

Warner, R. R., D. R. Robertson, and E. G. Leigh Jr. 1975. Sex change and sexual selection. Science 190: 633–638.

Warner, R. R., and R. D. Robertson. 1978. Sexual patterns in the labroid fishes of the western Caribbean I. The wrasses (Labridae). Smithson. Contr. Zool. 254: 1–27.

Watanabe, K., T. Fukuhara, and Y. Huziwara. 1982. Studies on the Asian Eupatorias. I. *Eupatorium chinese* var. *simplicifolium* from the Rokko Mountains. Bot. Mag. Tokyo 95: 261–280.

Watson, J. D. 1976. Molecular Biology of the Gene, 3rd. Ed. W. A. Benjamin, Menlo Park, CA.

Weatherhead, P. J., and R. H. Robertson. 1979. Offspring quality and the polygyny threshold: the 'sexy-son' hypothesis. Am. Nat. 113: 201–208.

Weatherhead, P. J., and R. H. Robertson. 1981. In defense of the 'sexy-son' hypothesis. Am. Nat. 117: 349–356.

Webb, C. J. 1981. Test of a model predicting equilibrium frequencies of females in populations of gynodioecious angiosperms. Heredity 46: 397–405.

Webb, G. J. W., R. Buckworth, and S. C. Manolis. 1983. *Crocodylus johnstoni* in the McKinlay River, N. T. VI. Nesting Biology. Aust. Wildl. Res. 10: 407–421.

Weber, G. 1985. Population genetics of insecticide resistance in the green peach aphid, *Myzus persicae* (Sulz) (Homoptera, Aphididae). Z. angew. Ent. 99: 408–421.

Weeks, S. C. 1986. Competition in phenotypically variable and uniform populations of the tadpole shrimp, *Triops longicaudatus*: a test of the Tangled Bank and Best-Man hypotheses for a short-term advantage to sex. Ms Thesis, University of California, Riverside.

Weider, L. J., and P. D. N. Hebert. 1987. Ecological and physiological differentiation among low-arctic clones of *Daphnia pulex*. Ecology: in press.

Weismann, A. 1886. Die Bedeutung der sexuellen Fortpflanzung. Fischer Verlag, Jena.

Weismann, A. 1893. The Germ-Plasm. A Theory of Heredity. W. Scott, London.

Weismann, A. 1902. Vorträge über Deszendenztheorie. Fischer Verlag, Jena.

Weld, L. H. 1952. Cynipoidae of the World. Privately printed, Ann Arbor.

Werren, J. H. 1980. Sex ratio adaptation to local mate competition in a parasitic wasp. Science 108: 1157–1159.

Werren, J. H., S. W. Skinner, and E. L. Charnov. 1981. Paternal inheritance of a daughterless sex ratio factor. Nature 293: 467–468.

Werren, J. H., and E. L. Charnov. 1975. Facultative sex ratios and population dynamics. Nature 272: 633–638.

Werren, J. H., and P. D. Taylor. 1984. The effects of the population recruitment on sex ratio selection. Am. Nat. 124: 143–148.

Werren, J. H. 1983. Sex ratio evolution under local mate competition in a parasitic wasp. Evolution 37: 116–124.

West-Eberhard, M. J. 1983. Sexual selection, social competition, and speciation. Q. Rev. Biol. 58: 155–183.

Westergaard, M. 1958. The mechanism of sex determination in dioecious flowering plants. Adv. Genet. 9: 217–281.

Westerman, J. M. 1971. Genotype-environment interaction and developmental regulation in *Arabidopsis thaliana*. Heredity 26: 93–206; 383–395.

Westerman, J. M., and H. Lawrence. 1970. Genotype-environment interaction and developmental regulation in *Arabidopsis thaliana*. Heredity 25: 609–627.

deWet, J. M. J. 1980. Origins of polyploids. In: Polyploidy: Biological Relevance (ed. W. H. Lewis), pp. 3–15. Plenum, New York.

deWet, J. M. J., and J. R. Harlan. 1970. Apomixis, polyploidy, and speciation in *Dicanthium*. Evolution 24: 270–277.

White, E. G., and C. B. Huffaker. 1969. Regulatory processes and population cyclicity in laboratory populations of *Anagasta kuehniella* (Zeller) (Lepidoptera: Phycitidae). I. Competition for food and predation. Res. Pop. Ecol. Kyoto 11: 57–83.

White, M. J. D. 1973. Animal Cytology and Evolution. 3rd edn. Cambridge University Press, Cambridge.

Whitfield, P. J., and N. A. Evans. 1983. Parthenogenesis and asexual multiplication among parasitic platyhelminths. Parasitology 86: 121–160.

Whiting, P. W. 1945. The evolution of male haploidy. Q. Rev. Biol. 20: 231–260.

Wohrmann, K. 1984. Population biology of the rose aphid *Macrosiphum rosae*. In: Population Biology and Evolution (eds K. Wohrmann and V. Loeschke), pp. 208–216. Springer Verlag, Berlin.

Wickler, W., and U. Seibt. 1983. Optimal maternal care. Z. Tierpsychol. 63: 201–205.

Wiese, L. 1976. Genetic aspects of sexuality in Volvocales. In: The Genetics of Algae (ed. R. A. Lewin). Blackwell, Oxford.

Wiese, L. 1981. On the evolution of anisogamy from isogamous monoecy and on the origin of sex. J. theor. Biol. 89: 573–580.

Wiese, L., W. Wiese, and D. A. Edwards. 1979. Inducible anisogamy and the evolution of oogamy from isogamy. Ann. Bot. 44: 131–139.

Wiese, L., L. A. Williams, and D. L. Baker. 1983. A general and fundamental molecular bipolarity of the sex cell contact mechanism as revealed by tunamycin and bacytracin in *Chlamydomonas*. Am. Nat. 122. 806-816.

Wiley, D. C., I. A. Wilson, and J. J. Skehel. 1981. Structural identification of the antibody-binding sites of Hongkong Influenza haemagglutinin and their involvement in antigenic variation. Nature 289: 373.

Wilkinson, L. E., and J. R. Pringle. 1974. Transient G1 arrest of *S. cerevisiae* cells of mating type alpha by a factor produced by cells of mating type a. Exp. Cell Res. 89: 175–187.

Williams, G. C. 1966. Adaptation and Natural Selection. Princeton University Press, Princeton, NJ.

Williams, G. C. 1975. Sex and Evolution. Princeton University Press, Princeton, NJ.

Williams, G. C. 1979. The question of adaptive sex ratio in outcrossed vertebrates. Proc. R. Soc. Lond. B 205: 567–580.

Williams, G. C., and J. B. Mitton. 1973. Why reproduce sexually? J. theor. Biol. 39: 545–554.

Willson, M. F. 1979. Sexual selection in plants. Am. Nat. 113: 770–790.

Willson, M. F., and P. W. Price. 1977. The evolution of inflorescence size in *Asclepias* (Asclepiadaceae). Evolution 31: 495–511.

Willson, M. F., and B. J. Rathke. 1974. Adaptive design of the floral display in *Asclepias syriaca* L. Am. Midl. Nat. 92: 47–57.

Willson, M. F., and D. W. Schemske. 1980. Pollinator-limitation, fruit production, and floral display in paw-paw. Bull. Torrey Bot. Club 107: 401–408.

Willson, M. F., and N. Burley. 1983. Mate Choice in Plants: Tactics, Mechanisms, and Consequences. Princeton University Press, Princeton, NJ.

Wilson, D. S. 1983. The group selection controversy: history and current status. A. Rev. Ecol. Syst. 14: 159–187.

Wilson, D. S., and R. K. Colwell. 1981. Evolution of sex ratio in structured demes. Evolution 35: 882–897.

Wilson, E. O. 1971. The Insect Societies. Belknap Press, Cambridge, MA.

Wilson, E. O. 1975. Sociobiology: The New Synthesis. Belknap, Cambridge, MA.

Wilson, E. O. 1978. On Human Nature. Harvard University Press, Cambridge, MA.

Winsor, J. A., Davis, L. E., and A. G. Stephenson. 1987. The relationship between pollen load and fruit maturation and its effect on offspring vigor in *Cucurbita pepo*. Am. Nat.: in press.

Witcombe, J. R., and W. J. Whittington. 1971. A study of the genotype by environment interaction shown by germinating seeds of *Brassica napus*. Heredity 26: 397–411.

Wohrmann, K. 1984. Population biology of the rose aphid Macrosiphum rosae. In: Population Biology and Evolution (eds K. Wohrmann and V. Loeschke). Springer-Verlag, Berlin.

Wolf, S. J. 1980. Cytogeographical studies in genus *Arnica* (Compositae: Senecioneae). I. Am. J. Bot. 67: 300–308.

Wolfe, M. S., J. A. Barrett, and J. E. E. Jenkins. 1981. The use of cultivar mixtures for disease control. In: Strategies for the Control of Cereal Diseases (eds J. F. Jenkins and R. T. Plumb), pp. 73–80. Blackwells, Oxford.

Wolfe, M. S., P. N. Minchin, and S. E. Wright. 1975. Effects of variety mixtures. A. Rep. Plant Breeding Inst.

Wolfe, M. S., and J. A. Barrett. 1980. Can we lead the pathogen astray? Plant Dis. 64: 148–155.

Wool, D., G. S. Bunting, and H. F. van Emden. 1978. Electrophoretic study of genetic variation in British *Myzus persicae* (Sulz.) (Hemiptera, Aphididae). Biochem. Genet. 16: 987–1006.

Wourms, J. P. 1977. Reproduction and development of chondrichthyan fishes. Am. Zool. 17: 379–410.

Wourms, J. P. 1981. Viviparity: the maternal-fetal relationship in fishes. Am. Zool. 21: 467–509.

Wright, S. 1946. Isolation by distance under diverse systems of mating. Genetics 31: 39–59.

Wright, S. 1968. Evolution and the Genetics of Populations. Vol. 1. Genetic and Biometric Foundations. University Chicago Press, Chicago, IL.

Wright, S. 1969. Evolution and the Genetics of Populations. Vol. 2. The Theory of Gene Frequencies. University of Chicago Press, Chicago, IL.

Wyatt, I. J. 1961. Pupal paedogenesis in the *Cecidomyiidae* (Diptera). Proc. R. ent. Soc. Lond. (A) 36: 133–143.

Wyatt, I. J. 1967. Pupal paedogenesis in the *Cecidomyiidae* (Diptera). A reclassification of the Heteropezini. Trans. R. ent. Soc. Lond. 119: 71–98.

Wynne-Edwards, V. C. 1962. Animal Dispersion in Relation to Social Behavior. Oliver and Boyd, Edinburgh and London.

Wynne-Edwards, V. C. 1986. Evolution Through Group Selection. Blackwell, Oxford.

Yamada, Y. and A. E. Bell. 1969. Selection for larval growth in *Tribolium* under two levels of nutrition. Genet. Res. 13: 175–195.

Yamaguchi, Y. 1985. Sex ratios of an aphid subject to local mate competition with variable maternal condition. Nature 318: 460–462.

Yamazaki, K., E. A. Boyse, V. Mike, H. T. Thaler, B. J. Mathieson, J. Abbott, J. Boyse, Z. H. Zayas, and L. Thomas. 1976. Control of mating preferences in mice by genes in the major histocompatibility complex. J. exp. Med. 144: 1324–1335.

Yamazaki, K., M. Yamaguchi, L. Baranoski, J. Bard, E. A. Boyse, and L. Thomas. 1979. Recognition among mice: evidence from the use of a Y-maze differentially scented by congenic mice of different major histocompatibility types. J. exp. Med. 150: 755–760.

Yampolsky, C., and H. Yampolsky. 1922. Distribution of sex forms in the phanerogamic flora. Biblio. Genet. 3: 1–62.

Yntema, C. L. 1976. Effects of incubation temperatures on sexual differentiation in the turtle, *Chelydra serpentina*. J. Morph. 150: 453–462.

Young, Y. P. W. 1981. Sib competition can favor sex in two ways. J. theor. Biol. 88: 755–756.

Young, G. B. 1953. A study of genotype-environment interaction in mice. J. agric. Sci. 43: 218–222.

Young, J. W. P. 1979a. Enzyme polymorphism and cyclical parthenogenesis in *Daphnia magna*. I. Selection and clonal diversity. Genetics 92: 953–970.

Young, J. W. P. 1979b. Enzyme polymorphism and cyclical parthenogenesis in *Daphnia magna*. II. Evidence of heterosis. Genetics 92: 971–982.

Young, J. W. P. 1983. The population structure of cyclic parthenogens. In: Protein Polymorphism: Adaptive and Taxonomic Significance (eds G. S. Oxford, and D. Rollinson). Academic Press, London.

Yu, P. 1972. Some host-parasite genetic interaction models. Theor. Pop. Biol. 3: 347–357.

Zaffagnini, F., and B. Sabelli. 1972. Karyological observations on the maturation of the summer and winter eggs of *Daphnia pulex* and *D. middendorffiana*. Chromosoma 36: 193–203.

Zahavi, A. 1975. Mate selection – a selection for a handicap. J. theor. Biol. 53: 205–214.

Zahavi, A. 1977. The cost of honesty (further remarks on the handicap principle). J. theor. Biol. 67: 603–605.

Zimmering, S., L. Sandler, and B. Nicoletti. 1970. Mechanism of meiotic drive. A. Rev. Genet. 4: 409–436.

Zuberi, M. I., and J. S. Gale. 1976. Variation in wild populations of *Papaver dubium*. X. Genotype-environment interaction associated with differences in soil. Heredity 36: 359–368.

Zuckerkandl, E., and L. Pauling. 1962. Horizons in biochemistry. In: Horizons in Biochemistry. (eds M. Kasha and B. Pullmann), pp. 189–225. Academic Press, New York.

# Index

## Subject

# Author

## Species
*Common names*

402

# New

# JOURNAL OF EVOLUTIONARY BIOLOGY

**The Journal of the European Society for Evolutionary Biology**

Editor: Stephen C. Stearns, Zoological Institute, University of Basel, CH–4051 Basel/Switzerland

**Aims**
- to provide an international forum for the integration of evolutionary research;
- to combine in one journal the perspectives of ecology, genetics, development paleontology, molecular evolution and behavioural ecology;
- to support the growth of evolutionary biology in Europe through its association with the European Society for Evolutionary Biology (ESEB). This Society will be founded at a meeting in Basel, 26–30 August 1987.

**Scope**
- original research on evolutionary aspects of ecology, genetics, development paleontology, molecular evolution and behavioural ecology;
- drawn from botany, zoology and theory;
- and including evolutionary ecology, genetical ecology, ecological genetics, population genetics, the role of development in evolution, the evolution of developmental mechanisms, the interplay of microevolution and macroevolution;
- but not excluding any interesting contribution to evolutionary biology.

**Emphasis**
- on papers integrating two or more fields, for example, ecology and genetics, genetics and development, development and paleontology.

**Print quality**
- first class reproduction of illustrations, including electron micrographs, and printing of equations;
- **Quarterly publication commencing early 1988.**

**Please write for the detailed leaflet and a specimen copy:**

**Birkhäuser Verlag AG**
P. O. Box 133
CH–4010 Basel/Switzerland

**Birkhäuser
Verlag**
Basel · Boston